SIMILARITY AND DIMENSIONAL METHODS IN MECHANICS

Similarity and Dimensional Methods in Mechanics

L. I. SEDOV
Member of the Academy of Sciences, U.S.S.R.

Translation Edited by
MAURICE HOLT
Associate Professor of Applied Mathematics
Brown University

Translation by
MORRIS FRIEDMAN
From the 4th Russian Edition

1959  **ACADEMIC PRESS**
New York and London

ACADEMIC PRESS INC.
111 FIFTH AVENUE
NEW YORK 3, N.Y.

Outside Western Hemisphere

INFOSEARCH LTD.
242 WILLESDEN LANE
LONDON, N.W.2.

Distributed outside the Western Hemisphere by

CLEAVER-HUME PRESS LTD.

Library of Congress Catalog Card Number: 59–11528

First Printing, 1959
Second Printing, 1961

Editor's Foreword

This book contains a complete development of the fundamental concepts of Dimensional Analysis and Similarity Methods, illustrated by applications to a wide variety of problems in mechanics, and particularly in fluid dynamics. The subject is developed from first principles and can be understood with the aid of an elementary knowledge of mathematical analysis and fluid dynamics. More advanced physical concepts are explained in the book itself. The first three chapters describe the basic ideas of the subject with illustrations from familiar problems in mechanics. The last two chapters show the power of Dimensional and Similarity Methods in solving new problems in the theory of explosions and astrophysics.

The book should be of interest to students who wish to learn dimensional analysis and similarity methods for the first time and to students of fluid dynamics who should gain further insight into the subject by following the presentation given here. The book as a whole and particularly the application to recent problems should appeal to all those connected with the many present-day aspects of gas dynamics including astrophysics, space technology and atomic energy.

The basic ideas behind Similarity and Dimensional Methods are given in the first chapter, which is general and descriptive in character. The second chapter consists of a series of examples of application of the methods, to familiar problems such as the motion of a simple pendulum, modelling in ship design, and the many scaling effects which arise in wind tunnel or water tank testing. Chapter III shows the use of Similarity and Dimensional Analysis in developing fundamental contributions to viscous fluid theory, such as the Blasius flat plate solution, and the various theories of isotropic turbulence.

Most of the material in Chapters IV and V is of recent origin and a great deal of it is the work of L. I. Sedov himself and his research group in Moscow University. Chapter IV is mainly concerned with self-similar solutions in the theory of spherical explosions, which proved to be a particularly fertile field for application. Self-similar solutions to a variety of explosion problems were developed independently during the war in Russia, England, and Germany and provided a very realistic description of the behaviour of atomic bombs, implosive waves and so on. Since the war these solutions have been extended to the cosmological field and have given very useful models of such phenomena as stellar flare-ups and pulsations. These very interesting and novel applications are described in Chapter V. The material in Chapters IV and V has not previously been presented in such a connected and unified form and should be of wide interest.

I wish to acknowledge the support of the Office of Naval Research and Air Force Office of Scientific Research in meeting the cost of literal translation by Morris D. Friedman. I am indebted to the U.S. National Aeronautics and Space Administration for providing the plate in Fig. 11 and to the U.S. Atomic Energy Commission for the plates in Figs. 59, 60 and 61. I am grateful to Garrett Birkhoff for his interest in the translation and, in particular for his comments on Chapter II. I should like to thank Mrs. Marion Porritt and other Secretaries in the Division who typed the manuscript, a number of students and colleagues, particularly Stanley A. Berger and William H. Reid, for advice at various stages of editing and the staff of Academic Press in preparing the book for the printer.

Finally I wish to record my deep appreciation of the excellent assistance given throughout by L. I. Sedov. Without his cooperation, in loaning original drawings and replying promptly to many queries, the task of completing the translation would have been much more difficult.

MAURICE HOLT

Division of Applied Mathematics,
Brown University,
   16 *February*, 1959

Author's Foreword to the English Translation

In the development of scientific knowledge the problem of discovering outstanding physical properties and of systematizing basic methods of investigation is of fundamental importance. This must be combined with the most useful formulation of new problems, as well as the construction of ideal models, or those which can be constructed with the means at our disposal.

Various methods of utilizing similarity and dimensional properties in mechanics and in physics are widely established in scientific practice; they are useful and even necessary in the analysis of specific theoretical or experimental problems.

In this book we have not only formulated certain problems which are interesting from the point of view of mechanics, but we have at the same time attempted to demonstrate certain techniques and to bring out the reciprocal relationship between the proposed ideal systems and the simple mathematical or experimental means which we have at our disposal for solving the problems under study.

There is reason to believe, that the methods developed and clearly described in this book are already being applied or will be applied in many other areas and to new problems.

I hope that this translation of the work into English will lead to a closer acquaintance with the results of research and experiment now being published in Moscow.

15 *December*, 1958   L. I. SEDOV

Foreword to First Edition

Physical similarity and dimensional properties play a very important part in experiments and calculations in physics and engineering. The construction of airplanes, ships, dams and other complicated engineering structures is based on preliminary, broad investigations, including the testing of models. Dimensional analysis and similarity theory determine the conditions under which the model experiments are to be carried out and the key parameters representing fundamental effects and processes. In addition, dimensional analysis and similarity theory when combined with the usual qualitative analysis of a physical phenomenon, can be a fruitful means of investigation in a number of cases.

Similarity theory and the use of models are encountered in the earliest study of physics in schools and in the initial stages of formulating new problems in research work. Moreover, these theories are of an extremely simple and elementary character. In spite of this it is only in recent years that the reasonings of similarity theory have been widely and consciously used; in hydrodynamics, for example, in the past thirty to forty years.

It is generally acknowledged that the explanation of these theories in textbooks and in actual teaching practice in colleges and universities is usually very inadequate; as a rule, these questions are only treated superficially or in passing. The fundamental concepts, such as those of dimensional and nondimensional quantities, the question of the number of basic units of measurement, etc., are not clearly explained. However, such confused and intuitive representations of the substance of the dimensions concept are often the origin of mystical or inherent physical meanings attributed to dimensional formulas. In some cases, this vagueness has led to paradoxes which were a source of confusion. We shall examine in detail one example of such a misunderstanding in connection with Rayleigh's conclusions on heat emission from a body in fluid flow. Often, relations and mathematical techniques not related to the substance of the theory are used to explain similarity theory. As in every general theory, it is desirable to construct the theories of dimensions and similarity by using methods and basic hypotheses which are adequate to the substance of the theory. Such a construction permits the limitations and possibilities of the theory to be clearly traced. This is necessary especially in dimensional analysis and similarity theory since they are regarded from widely different points of view: at one extreme they are considered to be all powerful while at the other they are only expected to give trivial results. Both of these extreme opinions are incorrect.

However, it should be noted that similarity theory gives the most useful results when used in combination with general physical assumptions which do not in themselves yield interesting conclusions. Consequently, to show the range of application more completely, we consider a whole series of mechanical problems and examples in which we combine dimensions methods with other reasonings of a mechanical and qualitative nature.

With this in mind the problems of turbulent fluid motion are considered in some detail. Similarity methods are the basic techniques used in turbulence theory, since we still do not have a closed system of equations in this field which would permit the mechanical problem to be reduced to a mathematical one. New results are contained in the section on turbulent fluid motion which supplement and explain certain questions of turbulence theory.

In addition to examples illustrating the use of methods of similarity and dimensional analysis, we discuss the formulation of a number of important mechanical problems, some of which are new and hardly worked out.

We dwell in some detail on the analysis of the fundamental equation of mechanics derived from Newton's second law. This is of interest on its own account and also helps to illuminate the usual reasoning about basic mechanical properties. Our viewpoint on this matter is not new; however, it differs radically from the treatment given in certain widely used textbooks on theoretical mechanics.

The number of familiar applications of the theories of dimensions and similarity in mechanics is very large; many of them are not touched upon here. The author hopes that the present book will give the reader an idea of typical methods and of their possibilities, which will be of assistance in the selection of new problems and in the formulation and treatment of new experiments.

A large part of the book does not require any special preparation by the reader. But in order to understand the material in the second half of the book, some general knowledge of hydrodynamics is necessary.

Moscow, 1943   L. I. SEDOV

Foreword to Third Edition

In recent years, scientific investigations of physical phenomena have relied more and more on the invariant character of the governing mathematical and physical laws, relative to choice of units for measuring the physical variables and scales.

The practical and theoretical power of these methods has been recognized more and more by scientific workers, contrary to the recently held opinion that the methods of similarity and dimensional analysis are of secondary value.

A certain analogy exists between the theories of dimensional analysis and similarity and the geometric theory of invariants relative to coordinate transformation, a fundamental theory in modern mathematics and physics.

Since the first edition of this book appeared, many new applications of the theories of dimensions and similarity have been made, to widely different problems in physics, continuum mechanics, to certain mathematical problems related to the use of group theory in solving differential equations (Birkhoff, 1950) and to statistical problems of selection and inspection of goods and finished products (Drobot and Warmus, 1954).

Some corrections and additions to emphasize better the basic ideas of the theories of similarity and dimensional analysis are introduced in this edition. One example of this is the discussion of the proof of the π-theorem. Furthermore, the definition of dynamic or physical similarity of phenomena has been given in more detail. This new definition is still not in general use to explain questions of similarity; however, from the practical viewpoint, it includes the essential peculiarities of physically similar processes; moreover, it is convenient for direct use and, apparently satisfies all the needs of different applications completely.

Beyond this, §§ 8–12 in Chapter IV and an entirely new Chapter V have been added. The additions in Chapter IV are devoted to certain problems of explosions and the attenuation of shock waves, besides a discussion of the general theory of one-dimensional gas motion. Applications of the theory of one-dimensional gas motion and the methods of dimensional analysis to certain astrophysical problems are considered in the new Chapter V.

It has become clear, at the present time, that to solve the fundamental problems of the internal structure of stars and to explain the spectacular and amazing phenomena observed in variable stars, we must turn to gas dynamics. New rational formulations of the problem and exact solutions of the equations of adiabatic gas motion and of the equations of gas equilibrium taking effects of radiation into account are given in the theory proposed. The appropriate problem can be idealized in certain cases, as a model simulating the actual gas dynamical effects in stars. This approach can be used to give us a picture of the mechanisms of stellar flare ups, stellar pulsations, internal structure of stars, the influence of various physical factors related to the liberation and absorption of energy within stars, the effect of variable density, the influence of gravitation, and possible motions dependent on the absence of an initial equilibrium pressure distribution, etc.

The theory developed in the additions to Chapter IV and in Chapter V is completely new in its basic approach. The proposed formulation and solution of the gas dynamics problems illustrate the applications of dimensional analysis methods to astronomy and provide a stock of simple, simulating, ideal motions which can be used to investigate problems of cosmology. Many of these results I obtained in collaboration with my young pupils in the course of the work of the hydromechanics seminar in Moscow University during the 1952–3 school year.

N. S. Mel'nikova and S. I. Sidorkin contributed to the preparation of § 14 of Chapter IV, V. A. Vasil'ev and M. L. Lidov to § 16, para. 1° of Chapter IV, and I. M. Iavorskaia to § 6 of Chapter V.

I express my deepest gratitude to them all.

Moscow, March 1954   L. I. SEDOV

Foreword to Fourth Edition

In this new edition some additions are made to Chapter II. Chapter IV and § 6 of Chapter V are extended, revised and improved in presentation.

Some calculations and diagrams were added under the supervision of N. S. Mel'nikova, to whom I wish to express my deepest gratitude.

Moscow, December 1956   L. I. SEDOV

ERRATA

Page	Line	Instead of	Read
6	10*	converging	conversion
54	19	Euchen	Eucken
54	24	Touloucian	Touloukian
58	13, 17, 5*	Strouhaille	Strouhal
59	1, 4	Strouhaille	Strouhal
61	14*	Strouhaille	Strouhal
163	4*	(22.2)	(2.22)
169	Eq. (3.8), 20	χ	κ
169	Eq. (3.8)	exponent: $[|\ \mu - \omega(\nu - 3)\delta\ |]$	$[\mu - \omega(\nu - 3)]\delta$
170	16	$\varepsilon = ab \cdots$	$\varepsilon = ab_1 \cdots$
170	17	arbitrary	dimensionless
170	10* (twice)	$\gamma p/(\gamma - 1)\rho$	$p/(\gamma - 1)\rho$
170	6*	$\lambda^{\nu+2}$	$\lambda^{\nu-k-1}$
170	1*	$\lambda^{\nu+2}[PV + (V - \delta)\left(\dfrac{PV^2}{2}\right.\cdots$	$\lambda^{\nu-k-1}[PV + (V - \delta)\left(\dfrac{RV^2}{2}\right.\cdots$
171	4	b^{1-1-k}	$b^{\nu-1-k}$
171	11	$\gamma p/[(\gamma - 1)\rho]$	$p/[(\gamma - 1)\rho]$
173	5*	$\gamma p/(\gamma - 1)\rho$	$p/(\gamma - 1)\rho$
173	Eq. (3.15)	$\dfrac{\gamma P}{\gamma - 1}$	$\dfrac{P}{\gamma - 1}$
173	4*	$\dfrac{\gamma P}{(\gamma - 1)R}$	$\dfrac{P}{(\gamma - 1)R}$
174	Eq. (3.16)	$P - (V - \delta)RV$	$P + (V - \delta)RV$

*counted from bottom

ERRATA (*Continued*)

Page	Line	Instead of	Read
174	Eq. (4.1), line 1	$\sim$	;
185	Eq. (5.13)	$\lambda^{\nu+2}$,	$\lambda^{\nu-k-1}$
185	3*	v	V
186	7	$2/[2 + \nu - \omega$	$2/\gamma[2 + \nu - \omega]$
214	5*	(1.2)	(11.2)
216	9	V	V_2
216	10	be	be on
217	7–8	symmetry on all the . . .	symmetry. On all the adjacent integral curves this is impossible.
228	2	are	is
228	Fig. 73	ρ/ρ_0 (in ordinate)	ρ/ρ_1
241	Eq. (12.7)	$q(\lambda, q)$	$g(\lambda, q)$
242	Eq. (12.11), last line	$\dfrac{q}{\gamma^\alpha \gamma}$	$\dfrac{q}{\gamma \alpha(\gamma)}$
243	1	unknown	known
249	Fig. 86, legend	at	as
355	15	Touloucian	Touloukian
360		Touloucian	Touloukian
363		Strouhaille number	Strouhal number

CONTENTS

Chapter I

General Dimensional Theory

1. INTRODUCTION

Every phenomenon in mechanics is determined by a series of variables, such as energy, velocity and stress, which take definite numerical values in given cases.

Problems in Dynamics or Statics reduce to the determination of certain functions and characteristic parameters. The relevant laws of nature and geometrical relations are represented as functional equations, usually differential equations.

In purely theoretical investigations, we use these equations to establish the general qualitative properties of the motion and to calculate the unknown physical variables by means of mathematical analysis. However, it is not always possible to solve a mechanical problem solely by the processes of analysis and calculation; sometimes the mathematical difficulties are too great. Very often the problem can not be formulated mathematically because the mechanical phenomenon to be investigated is too complex to be described by a satisfactory model. This situation arises in many important problems in aeromechanics, hydromechanics, and the theory of structures; in these cases we have to rely mainly on experimental methods of investigation, to establish the essential physical features of the problem. In general, we begin every investigation of a natural phenomenon by finding out which physical properties are important and looking for mathematical relations between them which govern the phenomenon.

Many phenomena can not be investigated directly and to determine the laws governing them we must perform experiments on similar phenomena which are easier to handle. To set up the most suitable experiments we must make a general qualitative analysis and bring out the essentials of the phenomenon in question. Moreover, theoretical analysis is needed when formulating experiments to determine values of particular parameters in the phenomenon. In general, and particularly in setting up experiments, it is very important to select the nondimensional parameters correctly; there should be as few parameters as possible and they must reflect the fundamental effects in the most convenient way.

This preliminary analysis of a phenomenon and the choice of a system of definite nondimensional parameters is made possible by dimensional

1

analysis and similarity theory: it can be used to analyse very complex phenomena and is of considerable help in setting up experiments. In fact it is out of the question to formulate and carry out experiments nowadays without making use of similarity and dimensionality concepts. Sometimes dimensional analysis is the only theoretical means available at the beginning of an investigation of some phenomenon. However, the potentialities of the method should not be overestimated. In many cases only trivial results are obtained from dimensional analysis. On the other hand, the widely held opinion that dimensional analysis rarely yields results of any importance is completely unjustified: quite significant results can be obtained by combining similarity theory with data obtained from experiment or from the mathematical equations of motion. In general, dimensional analysis and similarity theory are very useful in both theory and practice. All results derived from this theory are obtained in a simple and elementary manner. Nevertheless, in spite of their simple and elementary character, the methods of dimensional analysis and similarity theory require considerable experience and ingenuity on the part of the investigator, when probing into the properties of some new phenomenon.

In the study of phenomena which depend on a large number of parameters, dimensional analysis is especially valuable in determining which parameters are irrelevant and which are significant. We shall illustrate this point later by examples.

The methods of dimensional analysis and similarity theory play an especially large part in simulating phenomena.

2. DIMENSIONAL AND NONDIMENSIONAL QUANTITIES

We call quantities dimensional or concrete if their numerical values depend on the scale used; that is, on the system of measurement units. Quantities are called nondimensional or abstract, when their values are independent of the system of measurement units.

Typical dimensional quantities are length, time, force, energy, and moment. Angles, the ratio of two lengths, the ratio of the square of a length to an area, the ratio of energy to moment etc., are examples of nondimensional quantities.

However, the subdivision of quantities into dimensional and nondimensional is to a certain extent a matter of convention. For example, we have just called an angle nondimensional. It is known that angles can be measured in various units, such as, radians, degrees, or parts of a right angle; therefore, the number defining the angle depends on the choice of the measurement units: consequently, an angle can be considered as a dimensional quantity. On the other hand, suppose we define an angle as the ratio of the subtended arc of a circle to its radius; the radian—the angular unit of measurement—will then be defined uniquely. Now, if the angle is measured only in radians in all systems of units, then it can be considered as a nondimensional quantity. Exactly the same argument applies if a single fixed unit of measurement is introduced for the length in all systems of units. In these circumstances length can be considered nondimensional. But it is convenient to fix the unit of measurement for the angle and inconvenient for the length: this is explained by the fact that corresponding angles of geometrically similar figures are identical while corresponding lengths are not and, consequently, it is convenient to use different basic lengths in different problems.

Acceleration is usually considered as a dimensional quantity with dimensions of length divided by time squared. In many problems, the acceleration due to gravity g, which equals the acceleration of a body falling in a vacuum, can be considered as a constant ($9 \cdot 81$ m./sec^2). This constant acceleration g can be selected as a fixed unit of measurement for the acceleration in all systems of units. Then any acceleration will be measured by the ratio of its magnitude to the magnitude of the acceleration due to gravity. This ratio is called the weight factor, a numerical value which will not vary when a transformation is made from one unit of measurement to another; therefore, the weight factor is a nondimensional quantity. But the weight factor can be considered as a dimensional quantity at the same time, namely, as an acceleration, when the acceleration due to gravity is taken as the unit of measurement. In this latter case, we assume that the weight factor—the acceleration—can be taken as a unit of measurement which is not equal to the acceleration due to gravity.

On the other hand, abstract (nondimensional) quantities can be expressed in various numerical forms. In fact, the ratio of two lengths can be expressed as an arithmetic ratio, as a percentage or by other means.

The concepts of dimensional and nondimensional quantities are therefore relative. Each quantity is measured in certain units. When these units of measurement are identical in all systems the corresponding quantities are called nondimensional. Dimensional quantities are defined as those for which the units of measurement can vary in experimental or in theoretical investigations. Here it is irrelevant whether or not the investigations are actually carried out. It follows from this definition that certain quantities can be considered dimensional in some cases and nondimensional in others. We gave examples of these above and later we shall encounter a number of others.

3. FUNDAMENTAL AND DERIVED UNITS OF MEASUREMENT

If certain physical quantities are taken as basic with assigned units of measurement, then the units of measurement of all the remaining quantities can be expressed in a definite manner in terms of those of the fundamental quantities. The units of measurement taken for the fundamental quantities will be called fundamental or primary and all the rest will be derived or secondary.

In practice, it is sufficient to establish the units of measurement for three quantities; precisely which three depends on the particular conditions of the problem. Thus, in physical investigations it is convenient to take the units of length, time and mass as the fundamental units and in engineering investigations, to take the units of length, time and force. But the units of velocity, viscosity and density, etc., could also be taken as the fundamental units of measurement.

At the present time, the physical and engineering systems of units of measurement have become most widespread. The centimetre, gramme and second have been adopted as the fundamental units of measurement in physical systems (hence the abbreviation—CGS system of units) and the metre, kilogram and second in engineering systems (called the MKS system of units).

The units of length, one metre ($=100$ cm.), of mass, one kilogram ($=1000$ g.), and of time, one second, have been established experimentally by definite agreement. The length of a bar of platinum-iridium alloy, stored in the French Bureau of Weights and Measures, is taken as one metre; the mass of another bar of platinum-iridium alloy, stored in the same Bureau is taken as one kilogram. One second is assumed to be $1/(24 \times 3600)$ part of a mean solar day.

Once the fundamental units of measurement are established, the units of measurement for the other mechanical quantities, such as force, energy, velocity and acceleration, are obtained automatically from their definitions.

The expression of the derived unit of measurement in terms of the fundamental units is called its dimensions. The dimensions are written as a formula in which the symbol for the units of length, mass and time are denoted by L, M and T respectively (in the engineering system, the unit of force is denoted by K). When discussing measurements we must use a fixed system of units. For example, the dimensions of area are L^2; the dimensions of velocity are L/T or LT^{-1}, the dimensions of force in the physical system are ML/T^2 and in the engineering system, K.

We shall use the symbol $[a]$ to denote the dimensions of any quantity a.

For example, we shall write for the dimensions of force F in the physical system:

$$[F] = \frac{ML}{T^2} \quad \text{or} \quad \frac{ML}{T^2} = K.$$

Dimensional formulas are very convenient for converting the numerical values of a dimensional quantity when the units are transformed from one system to another. For example, in measuring the acceleration due to gravity in centimetres and seconds we have: $g = 981$ cm./sec². If we need to transform from these units to kilometres and hours, then the following relations should be used to convert the numerical value of the acceleration due to gravity:

$$1 \text{ cm.} = \frac{1}{10^5} \text{km.} \qquad 1 \text{ sec.} = \frac{1}{3600} \text{hour}$$

hence

$$g = 981 \text{ cm./sec}^2 = 981 \frac{1/10^5 \text{ km.}}{(1/3600)^2 \text{ hour}^2} = 98 \cdot 1 \times 36^2 \text{ km./hour}^2.$$

In general, if the units of length, mass and time in the new system of units are reduced by factors of α, β and γ respectively in terms of corresponding units in the old system, then the numerical value of the physical quantity a, with dimensions $[a] = L^l M^m T^n$, is increased by a factor of $\alpha^l \beta^m \gamma^n$ in the new system.

The number of fundamental units of measurement need not necessarily equal three: a greater number of units can be taken. For example, units of measurement for four quantities: length, time, mass and force, can be independently established by experiment. The Newton equation becomes in this case

$$F = cma$$

where F is the force; m is the mass; a is the acceleration and c is a constant with dimensions

$$[c] = \frac{KT^2}{ML}.$$

In the general case, four arguments will enter in the dimensional formulas of the mechanical quantities when the fundamental units are chosen in this way. The coefficient c in the above equation is a physical constant similar to the acceleration due to gravity g or to the gravitational constant γ in the universal gravitation law

$$F = \gamma \frac{m_1 m_2}{r^2}$$

where m_1 and m_2 are the masses of two particles and r is the distance between them. The numerical value of the coefficient c will depend on the choice of the fundamental units of measurement.

If the constant c is considered to be an abstract number (so that c will have the same numerical value in all systems of units) not necessarily equal to unity, then the dimensions of force are defined in terms of mass, length and time and the unit of measurement of force will be defined uniquely in terms of the units of measurement of mass, length and time.

In general, we can tentatively select independent units of measurement for n quantities ($n \geqslant 3$) provided that we introduce $n-3$ dimensional physical constants at the same time. In this case, the formulas of the derived quantities will generally contain n arguments.

When studying mechanical phenomena it is sufficient to introduce only three independent fundamental units, for length, mass (or force) and time. These units can also be used in studying thermal and even electrical phenomena. It is known from physics that the dimensions of thermal and of electrical quantities can be expressed in terms of L, M and T. For example, the quantity of heat and the temperature have the dimensions of mechanical energy. However, in many questions of thermodynamics and of gas dynamics, it is customary, in practice, to select the units of measurement for the quantity of heat and for the temperature independently of the units of measurement of the mechanical energy. The unit used to measure temperature is the degree Celsius and to measure heat we use the calory. These units of measurement have been established experimentally, independently of the units of measurement for mechanical quantities.

When studying phenomena in which a conversion of mechanical energy into heat occurs, it is necessary to introduce two additional physical dimensional constants; one of these is the mechanical equivalent of heat

$$J = 427 \text{ kg. m./cal.,}$$

and the other is either the specific heat c cal./m³deg., the gas constant R m²/sec²deg. or the Boltzmann constant $k = 1 \cdot 37 \times 10^{16}$ erg/deg. If we wish to measure the quantity of heat and the temperature in mechanical units, then the mechanical equivalent of heat and the Boltzmann constant will enter into the formulas as absolute nondimensional constants and they will be similar to converging factors in changing, for example, metres into feet, ergs into kilogram-metres, etc.

It is not difficult to see that fewer than three fundamental units of measurement can be taken. In fact, we can compare all forces with gravity, although this is inconvenient and unnatural when gravitation plays no part. A force in the physical system of units is generally defined by the equality

$$F = ma$$

and gravity by

$$F' = \gamma \frac{m_1 m_2}{r^2}$$

where γ is the gravitational constant with dimensions

$$\gamma = M^{-1} L^3 T^{-2}.$$

When measuring heat in mechanical units the dimensional constant in the mechanical equivalent of heat can be replaced by a nondimensional constant. In the same way the gravitational constant can be considered as an absolute nondimensional quantity. The dimensions of mass can then be expressed in terms of L and T by the relation

$$[m] = M = L^3 T^{-2}.$$

Therefore, the variation of the unit of mass in this case is determined completely by the variation of the units of measurement for length and time. Hence, if we regard the gravitational constant as an absolute, non-dimensional constant, we shall have a total of two independent units of measurement.

The number of independent units of measurement can be reduced to one if we regard some physical quantity such as the coefficient of kinematic viscosity of water, ν, or the velocity of light in a vacuum, c, as a nondimensional constant.

Finally, we can consider all physical quantities to be nondimensional if we regard appropriate physical quantities as absolute nondimensional constants. In this case, the possibility of using different systems of measurement units is ruled out. The single system of measurement units obtained is based on the physical quantities selected (for example, on the gravitational constant, the velocity of light and the coefficient of viscosity of water) and the values of these are taken as absolute universal constants.

There is a tendency to introduce such a system in scientific investigations since it permits the establishment of units of measurement of a permanent character. In contrast, the standards for the metre and the kilogram are substantially random quantities unrelated to the fundamental phenomena of nature.†

† According to the original idea of the commission of the French Academy of Science that established the metric system, a metre was to be defined as $1/4 \times 10^7$ of the length of the meridian through Paris and the kilogram was to be the weight of a cubic decimetre of distilled water at 40° C. Naturally, the measurement of the length of the meridian and the preparation of the standards for the metre and the kilogram were made with a certain error which appeared larger than the error admitted in later exact measurements. It was to be expected that new deviations would be detected in increasing the accuracy of finding the length of the meridian or the weight of a litre of pure water. Accordingly, it was agreed to take the magnitude of prepared prototypes of the standards as the fundamental units of measurement in order to avoid any constant variation in the standards of the metre and the kilogram. The relation of the units to the length of the meridian and the weight of a litre of water was thus rejected.

The introduction of a single system of units of measurement excluding all other systems, is equivalent to abandoning the dimensions concept completely. The numerical values of all the variables in a single universal system of units of measurement determine their physical magnitude uniquely.

A single universal system of measurement units of this type, that is, the use of identical measures, methods of calculating time, etc., would have certain definite advantages in practice since it would be one of the links standardizing measurement methods.

However, in many phenomena, such special constants as the gravitational constant, the velocity of light in a vacuum or the coefficient of kinematic viscosity of water, are completely irrelevant. Consequently, a single, universal system of measurement units related to the laws of gravitation, light propagation, and viscous friction in water or to any other physical processes, would often be artificial and impractical. Since phenomena in different branches of physics are of such a widely varied character, it is desirable to be able to carry the system of measurement units used to suit the conditions peculiar to each investigation.

It is convenient to take force, length and time as fundamental units in mechanics, using different units of force in celestial and engineering mechanics; it is more suitable to take the current intensity, resistance, length and time as fundamental units in electrical engineering (ampere, ohm, centimetre and second) etc.

Moreover, the numerical values of the characteristic quantities arising in the study of a particular phenomenon are often expressed advantageously as ratios of the most significant parameters in that phenomenon. These fundamental characteristic quantities can differ from one case to another.

4. DIMENSIONAL FORMULAS

The relation between the units of measurement of the derived quantities and those of the fundamental quantities can be represented as a formula. This formula is called the dimensional formula and it can be considered as a condensed definition and description of the physical nature of the derived quantity.

Dimensions can only be understood in application to a definite system of units of measurement. The dimensional formula for the same quantities in different measurement units may contain a different number of arguments and have different forms. The dimensional formulas of all the physical quantities in the CGS system of measurement units have the form of a monomial power $L^l M^m T^t$. We deduce this result from the physical property that the ratio of two numerical values of any derived quantity must be independent of the choice of the scale for the fundamental units. For example, suppose that we measure area, first in square metres, and secondly in square centimetres: then the ratio of two different areas measured in square metres would be the same as the ratio of the same areas measured in square centimetres. This condition imposed on the fundamental quantities is implied by the definition of the units of measurement.

Consider any derived dimensional quantity y; for simplicity, let us first assume that y is geometric and, consequently, depends only on length. Then

$$y = f(x_1, x_2, ..., x_n),$$

where $x_1, ..., x_n$ are certain distances. Let us denote the value of y corresponding to values of the arguments $x_1', ..., x_n'$ by y'. The numerical value of y, as well as of y', depends on the units of measurement for the distances $x_1, ..., x_n$. Let us diminish this unit or length scale by a factor α. According to the condition formulated above, we should then have:

$$\frac{y'}{y} = \frac{f(x_1', x_2', ..., x_n')}{f(x_1, x_2, ..., x_n)} = \frac{f(x_1' \alpha, x_2' \alpha, ..., x_n' \alpha)}{f(x_1 \alpha, x_2 \alpha, ..., x_n \alpha)}, \tag{4.1}$$

i.e. the y'/y ratio must be identical for any value α of the length scale.

We obtain from Equation (4.1):

$$\frac{f(x_1 \alpha, x_2 \alpha, ..., x_n \alpha)}{f(x_1, x_2, ..., x_n)} = \frac{f(x_1' \alpha, x_2' \alpha, ..., x_n' \alpha)}{f(x_1', x_2', ..., x_n')}$$

or

$$\frac{y(\alpha)}{y(1)} = \frac{y'(\alpha)}{y'(1)} = \phi(\alpha). \tag{4.2}$$

Therefore, the ratio of the numerical values of the derived geometric quantities measured in different length scales depends only on the ratio of the length scales.

It is easy to find the form of the function $\phi(\alpha)$ from relation (4.2). In fact, we have:

$$\frac{y(\alpha_1)}{y(1)} = \phi(\alpha_1), \quad \frac{y(\alpha_2)}{y(1)} = \phi(\alpha_2).$$

Hence, we obtain:

$$\frac{\phi(\alpha_1)}{\phi(\alpha_2)} = \phi\left(\frac{\alpha_1}{\alpha_2}\right), \tag{4.3}$$

1*

since we have for $x_1' = x_1\alpha_2,\ x_2' = x_2\alpha_2,\ \ldots,\ x_n' = x_n\alpha_2$

$$\frac{y(\alpha_1)}{y(\alpha_2)} = \frac{y_1'(\alpha_1/\alpha_2)}{y_1'(1)} = \phi\left(\frac{\alpha_1}{\alpha_2}\right).$$

Differentiating Equation (4.3) with respect to α_1 and putting $\alpha_1 = \alpha_2 = \alpha$, we obtain

$$\frac{1}{\phi(\alpha)}\frac{d\phi}{d\alpha} = \frac{1}{\alpha}\left(\frac{d\phi}{d\alpha_1}\right)_{\alpha_1 = \alpha_2} = \frac{m}{\alpha}.$$

Integrating, we find

$$\phi = C\alpha^m.$$

Since $\phi = 1$ when $\alpha = 1$, then $C = 1$; therefore

$$\phi = \alpha^m. \tag{4.4}$$

This result holds for any dimensional quantity depending on several fundamental quantities, if we vary just one scale. It is easy to see that if the α, β, γ scales of the three fundamental quantities are altered, then the function ϕ will be

$$\phi = \alpha^m \beta^n \gamma^t.$$

This proves that the dimensions formulas of physical quantities must be monomial powers.

5. ON NEWTON'S SECOND LAW

When investigating mechanical or physical phenomena, we introduce, first, a system of quantities characterizing the various aspects of the processes being studied (let us call them simply characteristic quantities) and, secondly, a system of units of measurement which is used to determine the numerical values of these quantities.

A number of relations exist between the characteristic quantities of the phenomena. Some of these relations apply only to a specific system and to a particular part of the process: other relations may be valid for certain classes of systems and motions. Relations of the latter type have special value and to look for them is an important object of physical investigations.

The methods of similarity and dimensional analysis provide one way of determining relations between variables. In what follows we intend to show ways and means of applying these methods. But before explaining them directly we shall illustrate certain important aspects by means of examples. In this connection, we consider the fundamental relation of mechanics known as Newton's second law.

Certain relations between the variables are simple consequences of their definition. For example, the magnitude of the velocity v equals the ratio of the path travelled to the corresponding time interval: the magnitude of the kinetic energy of a material point E equals $mv^2/2$ where m is the mass of a particle, and so on.

In addition to these trivial relations, theoretical and experimental investigations establish functional relations between the numerical values of the variables. These reflect the peculiar properties of the phenomenon, or class of phenomena, under consideration. The Kepler laws on planetary motion and the law of universal gravitation are examples of such relations. Let us illustrate the relation between these laws briefly.

As the result of observing the motion of planets extensively over many years, Kepler formulated the following general laws in 1609 and 1619:

(1) The planets describe ellipses around the sun, and the sun is always at one of the foci.

(2) The radius vector connecting the sun with the planet traverses equal areas in equal time intervals.

(3) The square of the period of rotation of a planet around the sun is proportional to the cube of the corresponding mean distance of the planet from the sun.

If the magnitude of the force of interaction between the sun and the planet is defined as the mass multiplied by the acceleration, then the law of universal gravitation can be derived mathematically from the Kepler laws, namely

$$F = \gamma \frac{m_1 m_2}{r^2} \tag{5.1}$$

where F is the force of attraction, r is the distance between two particles of masses m_1 and m_2. This law was established by Newton in 1682 and was subsequently checked and verified by a comparison of the numerous results obtained from it with observations in nature and in specially formulated experiments.

Another example is Hooke's law which relates the spring tension F with its extension x.

This law is derived from static and dynamic observations of a load suspended on a spring, using the definition of the magnitude of force as the product of mass with acceleration and the rule of addition of forces.

In mathematical terms, this law is written

$$F = kx, \tag{5.2}$$

where k is the spring constant.

Using this law, the law of motion, (i.e. the expression of all the mechanical quantities as functions of time) the period of the oscillations, etc., can be determined theoretically in a number of different special cases (suspending the load from several springs, varying the mass or the spring constant, varying the initial conditions, etc.).

The solution of these and of other similar problems of mechanics is based on the investigation of the equation of motion of a particle

$$\overline{F} = m\overline{a}, \tag{5.3}$$

where $\overline{a}$ is the acceleration vector, m is the mass of the particle and $\overline{F}$ is the force vector. Very often the force $\overline{F}$ is the vector sum of several forces

$$\overline{F} = \overline{F}_1 + \overline{F}_2 + ..., \tag{5.4}$$

representing the total of a series of effects. The possibility of replacing several forces acting simultaneously by one force, defined by formula (5.4), is an experimental fact.

Now, let us consider quantities in (5.3) in more detail. The acceleration $\overline{a}$ is a kinematic quantity which can always be determined experimentally independently of (5.3). The mass m defines the inertial property of the body. The idea of mass of a particle can be introduced by means of Newton's third law (every action has an equal and opposite reaction). In fact, the value of a constant magnitude, its mass, can be attributed to each particle so that the following relations will hold for the motion of any two isolated, interacting particles M_1 and M_2 or M_1 and M_3:

$$m_1 \overline{a}_1 + m_2 \overline{a}_2 = 0, \quad m_1 \overline{a}_1' + m_3 \overline{a}_3' = 0. \tag{5.5}$$

Therefore, the ratio of the masses can always be determined experimentally, independently of (5.3), by measuring the ratio of the accelerations when the interacting bodies move.

The constancy of mass, defined by (5.5) for every kind of motion, is an experimental fact expressing a law of nature which, in general, can be made more precise.

If the motion is known, then (5.3) is a simple equation to determine the value of the total force. In practice, (5.3) is often used to calculate forces. Relation (5.3) can be used to determine the motion only when the relation between the force and the quantity characterizing the motion (time, position vector of the point, velocity, etc.) is known. This relation can either be obtained theoretically, using additional hypotheses which must certainly be verified by experiment, or directly by experimental means.

The relation between force and various physical quantities is obtained from (5.3) in both theoretical and experimental approaches. The relation between the product $m\overline{a}$ and the other parameters in the motion is established by the observation and study of simple motions. The relations obtained are then generalized to apply to the more complex classes of motion, and the validity of the generalization must be checked experimentally by comparing the conclusions obtained from the equations of motion with the results of experiment. Hence, the general procedure of deriving the universal gravitation law from the Kepler laws is typical for the determination of a force as a function of the parameters of a motion.

The Coulomb law on force between interacting electric charges, the Biot-Savart law on magnetic intensity, the Weber law on capillary force, the Coulomb friction law on frictional forces between solid bodies, Hooke's law on the relation between the stresses and deformations in an elastic body, Newton's law on viscous friction in a fluid, etc., are all determined in an analogous manner.

It is often stated that force can be determined by static means independently of (5.3). Actually, in a number of important cases, in particular when it can be assumed that force depends only on position, the dependence of force on the coordinates can be determined by comparing the required force with forces known from an analysis of the particular case of motion when $\bar{a} = 0$.†

However, in this connection, the following must be kept in mind. Firstly, we use Equation (5.3) at $\bar{a} = 0$ in comparing the desired force, given by the static definition, with the known forces; secondly, the statement that relations between forces holding in the static case are also valid in the dynamic case, is an additional hypothesis requiring experimental confirmation supplied by (5.3). The experimental check often does not confirm the assumption made. For example, this is the case with frictional forces; it appears that friction at rest (for a velocity $v = 0$) and in motion (for $v \neq 0$) can be different; such is the case with the force of a spring acting on a suspended load. The law relating force to the extension [formula (5.2)], valid for springs of any mass in static measurements, ceases to be correct for motion, in which the deflections increase as the mass of the spring is increased. If the mass of spring is small in comparison with the loaded mass, then (5.2) can be considered correct for the motion of the load.

Often it is impossible to define force as a function of time, position, velocity and acceleration. For example, consider the total force acting on the submerged part of a ship which is performing a complex loop-shaped motion relative to the water surface. The force acting on the submerged part of a ship depends on the state of motion of the water, which is determined by the whole law of motion of the ship. Let the

† In practice, force is often determined by use of the fact that a system of forces is unchanged by a uniform translation. In this case accelerations do not vary.

position, velocity and acceleration at the same instant, be identical in two different motions of the ship (the time can be measured from the start of the motion when the ship and water are at rest). It is clearly impossible to say that the forces acting on the submerged part of the ship at this moment will be identical; the forces can differ considerably. The ship could agitate the water intensely in the first motion while the motion of the water may be calmer at the place considered during the second motion of the ship. Evidently, the forces acting on the ship in this example will depend on the law of motion, i.e. on the whole history of the motion: in other words, the phenomenon will be hereditary in character.

We have seen that experimental laws of nature, such as the laws of universal gravitation, Hooke's law, etc., are obtained from an analysis of wide classes of motion in which the magnitude of force is defined as the product of mass and acceleration.

Therefore, in particular problems of kinematics, we may not introduce forces which violate the equation $\bar{F} = m\bar{a}$, which is backed by sound experimental evidence.

The investigation of mechanical phenomena can be made by similar means if another concept, for example, the kinetic energy of the system, is taken as the fundamental quantity instead of the force. The equation

$$E = \sum \frac{mv^2}{2} \tag{5.6}$$

can be considered as the definition of the kinetic energy of a mechanical system. Investigating certain classes of motion of this mechanical system experimentally, we can record the variation of the magnitude of the energy E with a number of other mechanical properties. For example, it is established in the motion of a conservative system that the kinetic energy can be represented as a certain function of position and an additive constant h which isolates a known sub-class among all the possible motions of the system:

$$E = -V + h. \tag{5.7}$$

The quantity V is called the potential energy of the system. Equation (5.6) and the condition (5.7) characterizing a conservative system lead to the equation

$$\sum \frac{mv^2}{2} + V = h \tag{5.8}$$

which expresses the law of the conservation of mechanical energy.

At the present time, there are still a number of mechanical phenomena which cannot be investigated by means of (5.3) or (5.7) owing to incomplete knowledge of the behaviour of the forces concerned and the kinetic energy.

In analytical mechanics, it is always understood that the laws of force or an expression for the potential energy are known. The fundamental problems of analytical mechanics are related to the mathematical techniques to be employed; to the methods of integrating the equations of motion and to the establishment of various equivalent or broader principles which can replace the initial experimental laws.†

The main problem of mechanical or, in general, of physical investigations is to establish the laws which express forces in terms of the basic variables of the state of motion. Further, the significance of these variables must be explained and the practical value of the laws governing them must be assessed.

One of Newton's principal achievements is to have shown that the product of mass by acceleration is a quantity taking the same value for various bodies and various motions occurring at different places with different velocities. Further, it is, in general, a quantity which can be determined experimentally as a function of the time, position and velocity of the points of a system in a number of cases.

However, as we saw, to determine force as a function of the simplest variables of the motion is not always possible in principle. In these cases, it may be more convenient to replace the product of mass by acceleration by other variables and to investigate their behaviour instead.

Let us consider briefly the question of inertial forces. Suppose we have a set of different coordinate systems moving relative to each other. The acceleration has a different magnitude and direction in the various coordinate systems, since their motions are different. The relation

---

† At the end of the eighteenth century, the main attention and effort of theoretical scientists were directed to the investigation and overcoming of mathematical difficulties (the problems of celestial mechanics, the development of a general theory of differential equations, variational principles etc.). The initial equations of motion were analysed in general form, giving special attention to the convergence of physical phenomena to mechanical motion and the completeness of mechanics as a science. The fundamental difficulty was seen in the integration of the differential equations of mechanics. As Laplace stated: "Give the initial conditions and this is sufficient to predict the whole future of a motion and to reproduce its whole past." However, it must be noted that it is impossible to consider the theoretical problem of formulating the differential equations of motion as simple, even within the scope of classical mechanics and, in principle, it is still unsolved. In fact, the problem of the formulation of the equations of motion, the problem of the effective forces, i.e. of determining the right-hand sides of the differential equations of motion, are the fundamental problems of physical investigations. In many cases, this problem has not been solved even in classical mechanics. In the simplest applications, existing solutions are approximate and are in continued need of improvement.

between the accelerations of a point in coordinate systems with different relative motions is established in kinematics.

We can establish the relation between force and the basic variables of a motion experimentally in a certain specific coordinate system, usually one fixed in the earth or at the centre of gravity of the solar system.

If we know the laws of force in one coordinate system, then we can easily find the product of mass and acceleration, that is, the force, in any other coordinate system, the motion of which relative to the original system is given. In this case, of course, we must introduce what is known as the inertial force. For an observer stationary relative to a fixed system, the effective forces are composed of forces determined in the coordinate system used for experiments (initial coordinate system) and of the inertial forces. To a fixed observer these have the same mechanical behaviour as any other forces.

6. NATURE OF THE FUNCTIONAL RELATIONS BETWEEN PHYSICAL QUANTITIES

Physical laws, established either theoretically or directly from experiments, are functional relations between the quantities characterizing the phenomenon under investigation. The numerical values of these dimensional physical quantities depend on the choice of a system of units of measurement, which has no connection with the substance of the phenomenon. Consequently, the functional relations, which express the physical facts themselves, and which are independent of the system of units, must have a certain special structure.

Consider the dimensional quantity a, defined as a function of the independent dimensional quantities $a_1, a_2, ..., a_n$ by

$$a = f(a_1, a_2, ..., a_k, a_{k+1}, ..., a_n). \qquad (6.1)$$

Some of these parameters vary, others are constant in the process being considered.

To be more specific we suppose that the function $f(a_1, a_2, ..., a_n)$ represents a certain physical law independent of the choice of the system of units.†

---

† Let us stress that the functional relation (6.1) is, by hypothesis, just one essential physical relation defining a as a function of the independent quantities $a_1, a_2, ..., a_n$. Hence, it is not the general form of a relation between dimensional quantities which is independent of the choice of the system of units.

For example, the relation

$$a = f(a_1, a_2, ..., a_n) + \Phi(a_1, a_2, ..., a_n) \ln \frac{P}{mg}, \qquad (6.1')$$

is also independent of the choice of the system of units where the function $\Phi(a_1; a_2, ..., a_n)$ is arbitrary. Both (6.1') and (6.1) are satisfied by the quantities $a_1, a_2, ..., a_n, g, P, m$ where

Let the first k $(k \leqslant n)$ of the dimensional quantities $a_1, ..., a_n$ have independent dimensions (the number of basic units should be larger than or equal to k).

Independence of dimensions means that the dimensions of one quantity cannot be represented as a combination, in the form of a monomial power, of the dimensions of the other quantities. For example, the dimensions of length L, velocity L/T, and energy ML^2/T^2 are independent; the dimensions of length L, velocity L/T and acceleration L/T^2 are dependent. With mechanical quantities, usually not more than three have independent dimensions. We assume that k is the largest number of parameters with independent dimensions, consequently, the dimensions of the quantities $a, a_{k+1}, ..., a_n$ can be expressed in terms of the dimensions of the parameters $a_1, a_2, ..., a_k$.

We take the k independent quantities $a_1, a_2, ..., a_k$ as the basic quantities and introduce the following notation for their dimensions:

$$[a_1] = A_1, \qquad [a_2] = A_2, ..., \qquad [a_k] = A_k.$$

The dimensions of the remaining quantities will be

$$[a] = A_1^{m_1} A_2^{m_2} ... A_k^{m_k},$$

$$[a_{k+1}] = A_1^{p_1} A_2^{p_2} ... A_k^{p_k},$$

$$\cdot \quad \cdot \quad \cdot \quad \cdot \quad \cdot \quad \cdot \quad \cdot \quad \cdot$$

$$\cdot \quad \cdot \quad \cdot \quad \cdot \quad \cdot \quad \cdot \quad \cdot \quad \cdot$$

$$[a_n] = A_1^{q_1} A_2^{q_2} ... A_k^{q_k}.$$

g is the acceleration due to gravity, P is the weight and m is the mass of a certain body. However, (6.1′) resolves into two different physical laws: namely (6.1) and the relation $P = mg$; the latter is superfluous in this problem. Other artificial examples of this kind could be given.

We shall take our physical law in the form (6.1). In what follows we shall only assume that such a relation, which may be many valued, exists.

Questions concerning the actual theoretical or experimental method of establishing this relation are irrelevant to the argument which follows. Consequently, the investigation of relations between dimensional quantities in implicit form

$$\Phi(a, a_1, a_2, ..., a_n) = 0$$

or in the form of several implicit functions for the quantities a, b, c

$$\Phi_1(a, b, c, ..., a_1, a_2, ..., a_n) = 0$$

$$\Phi_2(a, b, c, ..., a_1, a_2, ..., a_n) = 0$$

$$\Phi_3(a, b, c, ..., a_1, a_2, ..., a_n) = 0$$

is not an analysis of the question in any more general form. The role of the "parasitic" relations grows in such a treatment. The value of dimensional analysis in producing fundamental results in a simple manner is now obscured by complex irrelevant questions, such as the possibility of solving a system of implicit equations.

We now alter the units of measurement of the quantities a_1, a_2, ..., a_k by factors of α_1, α_2, ..., α_k respectively; the numerical values of these quantities and the quantities a, a_{k+1}, ..., a_n, in the new system of units, will be given by

$$a_1' = \alpha_1 a_1, \qquad a' = \alpha_1^{m_1} \alpha_2^{m_2} ... \alpha_k^{m_k} a,$$
$$a_2' = \alpha_2 a_2, \qquad a_{k+1}' = \alpha_1^{p_1} \alpha_2^{p_2} ... \alpha_k^{p_k} a_{k+1},$$
$$\cdots \cdots \qquad \cdots \cdots \cdots$$
$$\cdots \cdots \qquad \cdots \cdots \cdots$$
$$a_k' = \alpha_k a_k \qquad a_n' = \alpha_1^{q_1} \alpha_2^{q_2} ... \alpha_k^{q_k} a_n.$$

respectively. Relation (6.1) becomes, in the new system of units,

$$\begin{aligned} a' &= \alpha_1^{m_1} \alpha_2^{m_2} ... \alpha_k^{m_k} a \\ &= \alpha_1^{m_1} \alpha_2^{m_2} ... \alpha_k^{m_k} f(a_1, a_2, ..., a_n) \\ &= f(\alpha_1 a_1, ..., \alpha_k a_k, \alpha_1^{p_1} \alpha_2^{p_2} ... \alpha_k^{p_k} a_{k+1}, ..., \alpha_1^{q_1} \alpha_2^{q_2} ... \alpha_k^{q_k} a_n). \end{aligned} \qquad (6.2)$$

This relation shows that the function f is homogeneous in the scales α_1, α_2, ..., α_k. The scales α_1, α_2, ..., α_k are arbitrary; we choose them so as to cut down the number of arguments in the function f. Let us assume:

$$\alpha_1 = \frac{1}{a_1}; \qquad \alpha_2 = \frac{1}{a_2}, ..., \qquad \alpha_k = \frac{1}{a_k},$$

i.e. select the system of units of measurement in such a way that the values of the first k arguments on the right-hand side of (6.2) would equal unity.† In other words, using the fact that (6.1) (by hypothesis) is independent of the system of units, we establish a system such that k arguments of the function f would have the fixed constant value, unity.

The numerical values of the parameters a, a_{k+1}, ..., a_n are determined in this relative system of units by the formulas

$$\Pi = \frac{a}{a_1^{m_1} a_2^{m_2} ... a_k^{m_k}},$$

$$\Pi_1 = \frac{a_{k+1}}{a_1^{p_1} a_2^{p_2} ... a_k^{p_k}}$$

$$\cdots \cdots \cdots \cdots$$
$$\cdots \cdots \cdots \cdots$$

$$\Pi_{n-k} = \frac{a_n}{a_1^{q_1} a_2^{q_2} ... a_k^{q_k}}$$

† For simplicity, we assume that the parameters a_1, a_2, ..., a_k are finite and different from zero. The final conclusions can be extended to the case when a_1, ..., a_k become zero or infinite if the function f is continuous for these values of the arguments.

where $a, a_1, a_2, ..., a_n$ are the numerical values of the quantities considered in the original system of units. It is not difficult to see that the values $\Pi, \Pi_1, ..., \Pi_{n-k}$ are independent of the choice of the original system of units of measurement since they have zero dimensions relative to the units $A_1, A_2, ..., A_k$. It is also evident that the values $\Pi, \Pi_1, ..., \Pi_{n-k}$ are generally independent of the choice of the system of units adopted for the k quantities $a_1, a_2, ..., a_k$; therefore, these quantities can be considered nondimensional.

Using the relative system of units, (6.1) can be written:

$$\Pi = f(1, 1, ..., \Pi_1, ..., \Pi_{n-k}). \tag{6.3}$$

Hence, the relation between the $n + 1$ dimensional quantities $a, a_1, ..., a_n$, reduces to one between the $n + 1 - k$ quantities $\Pi, \Pi_1, ..., \Pi_{n-k}$, which are a nondimensional combination of $n + 1$ dimensional quantities.† This relation is independent of choice of units. This general conclusion of dimensions theory is known as the Π-theorem.

If it is known that a certain nondimensional quantity is a function of a number of dimensional quantities, then this function can depend only on nondimensional combinations of these dimensional quantities.

By changing the function f, the system of nondimensional parameters in (6.3) can be replaced by another system which are functions of the $n-k$ parameters $\Pi_1, ..., \Pi_{n-k}$. Clearly, at most $n-k$ independent, nondimensional power combinations can be formed from the n parameters $a_1, ..., a_n$ of which not more than k parameters have independent dimensions. This follows directly from (6.3) if we take any nondimensional combination of the quantities $a_1, a_2, ..., a_n$ as the quantity a.

Every physical relation between the dimensional quantities can be formulated as a relation between nondimensional quantities. This fact is the basic reason why dimensions theory is useful in the investigation of mechanical problems.

Reducing the number of parameters defining the quantity to be studied, restricts the functional relation and simplifies the investigation. In particular, if the number of basic units of measurement equals the number of characteristic parameters with independent dimensions, then the relation will be determined completely, to within a constant factor, by using dimensional analysis.

In fact, if $n = k$, i.e. all the dimensions are independent, then it is

---

† This conclusion must be corrected if the function $f(a_1, a_2, ..., a_n)$ is discontinuous for zero or infinite values of the first k arguments. The first k arguments in (6.3) can be replaced by unity only at those points at which $a_1, a_2, ... a_k$ are not zero or infinite; consequently, the number of essential arguments in the right-hand side of (6.3) can exceed $n-k$ near, such points.

impossible to form a nondimensional combination of the parameters $a_1, ..., a_n$ and, consequently, the functional relation (6.3) can be written

$$a = ca_1^{m_1} a_2^{m_2} ... a_n^{m_n}$$

where c is a nondimensional constant and the exponents $m_1, m_2, ..., m_n$ are easily determined by using the dimensions formula for a. The nondimensional constant c, can be determined either by experiment or theoretically by solving the appropriate mathematical problem.

Evidently, dimensional analysis will be of greater use, the more freedom we have in selecting the basic units of measurement.

We saw above that the number of basic units of measurement can be selected arbitrarily. However, increasing this number means introducing additional physical constants, which must also be included among the characteristic parameters. We increase the number of dimensional constants by increasing the number of basic units of measurement; in the general case, the difference $n + 1 - k$, equal to the number of nondimensional parameters in which the physical relation is formulated, remains constant.

An increase in the number of basic units is advantageous only if the entire physical constants introduced are not essential. For example, if we consider a phenomenon in which mechanical and thermal processes occur, then we can introduce two different units, the calory and the joule, to measure the quantity of heat and mechanical energy, but the nondimensional constant J, the mechanical equivalent of heat, must enter the analysis. Suppose, now, that we are analysing the phenomenon of heat transfer in a moving, incompressible, ideal fluid: neither the transformation of thermal energy into mechanical energy nor the reverse process occurs in this case and, consequently, the thermal and mechanical processes will proceed independently of the value of the mechanical equivalent of heat. If we admit that the value of the mechanical equivalent of heat can be changed, then this in no way affects the values of the characteristic quantities. Therefore, the constant J does not enter the physical relation and the increase in the number of basic units permits additional important information to be obtained by using dimensional analysis.

Later, we shall illustrate these conclusions by examples.

7. PARAMETERS DEFINING A CLASS OF PHENOMENA

We start every study of mechanical phenomena with a survey, picking out the basic factors which define the quantities of interest to us and, in the broadest sense, drawing on already familiar examples of phenomena to construct a model of the processes under investigation.

A sound survey is very often a difficult problem which requires, on the part of the investigator, a great deal of experience, intuition and a preliminary qualitative explanation of the mechanism of the process being studied. The essence of some problems consists in checking the validity of a plausible hypothesis.

Singling out distinguishing features and developing a real understanding of connections and laws is the basis of the conscious use and control of natural phenomena to solve successfully the multitude of problems presented in the life of mankind.

The properties of matter and the elementary physical laws which play a substantial role in controlling phenomena are characterized by a number of quantities which can be dimensional or nondimensional, variable or constant.

A mechanical system and the state of its motion are determined by a number of dimensional and nondimensional parameters and functions.

Suppose that several different mechanical systems perform a certain motion; then we can always restrict the number of admissible systems and motions to those which can be defined by a finite number of dimensional and nondimensional parameters. The limitation of the class of admissible systems and motions can always be attained by imposing additional conditions on the abstract parameters and the type of function associated with the problem in nondimensional form.

Dimensional analysis enables us to draw conclusions by using arbitrary or special systems of units of measurement to describe physical laws. Consequently, when listing the parameters defining a class of motions, it is necessary to include all the dimensional parameters related to the substance of the phenomena independently of whether these parameters are constant (in particular, they can be physical constants) or can vary for different motions of the class isolated. It is important that the dimensional parameters should be able to assume various numerical values in different systems of units, although they are possibly identical for all the motions being considered. For example, when considering the motion in which the weight of a body is of importance, we must certainly take into account the acceleration due to gravity g as a physical dimensional constant although the value of g is constant under all actual conditions. After the acceleration due to gravity g has been introduced as a characteristic parameter, we can, without introducing complications, extend the class of motions artificially by introducing those in which the acceleration due to gravity g assumes different values. Such a method permits qualitative conclusions of practical value to be obtained in a number of cases.

We now consider how to find the system of parameters defining a class of phenomena.

If the problem is formulated mathematically, a table of the parameters defining the phenomenon is always easily extracted. To do this, we need to note all the dimensional and nondimensional quantities which are required to determine the numerical values of all unknowns from the equations of the problem. A table of the characteristic parameters can be formed without writing down these equations in a number of cases. It is possible to establish simply those factors needed to find all the required quantities; sometimes the numerical values of these can only be found experimentally.

It is necessary, when making up the system of characteristic parameters, to form a clear picture of the phenomenon, just as when formulating the equations of the problem.

However, less need be known when using dimensional analysis than when formulating the equations of motion of a mechanical system. Several equations of motion can exist for the same system of characteristic parameters. The equations of motion show not only on which parameters the required quantities depend, but also contain all the functional relations determined by the mathematical form of the problem.

These arguments show that dimensional analysis is limited in its scope. We cannot determine functional relations between nondimensional quantities by use of dimensional analysis alone.

The conclusions of dimensional analysis cannot be altered if we multiply the various terms in the equations of motion by positive or negative nondimensional numbers or functions depending on the system of characteristic parameters; yet modifications to the equations of this sort can influence the character of the physical laws substantially.†

Every system of equations which includes a mathematical description of the controlling phenomenon can be formulated as a relation between nondimensional quantities. None of the conclusions of dimensional analysis are changed by variation of the physical laws, when they are in the form of relations between identical nondimensional quantities.

The system of characteristic parameters must be complete. Some of the characteristic parameters, including dimensional physical constants, must have dimensions, in terms of which, the dimensions of all the dependent parameters can be expressed.‡ As an example of this

---

† For example, changing the sign of certain forces in the equations of motion can have an important effect on the laws of motion; all the conclusions of dimensional analysis remain invariant under this operation.

‡ If the system of characteristic parameters is incomplete and it is impracticable to extend it, then the defining quantity equals either zero or infinity. We often encounter such cases when assigning initial conditions of the "source type" by using δ functions.

requirement, let us consider the parameters which can determine the static state of a gas. It is wrong to assert that the state of a gas is determined only by the two dimensional quantities, absolute temperature, Θ ($[\Theta] = C°$) and density ρ, ($[\rho] = M/L^3$) because the pressure p is finite, not zero, and has dimensions independent of the dimensions of the temperature and the density.

Now let us assume that the state of the gas is determined by the values of the temperature, the density and a physical constant, say the coefficient of specific heat c_v', measured in mechanical units ($[c_v'] = L^2/T^2\, C°$). Denoting the mechanical equivalent of heat by J kg./cal., we shall have:

$$c_v' = Jc_v$$

where c_v is the specific heat in thermal units ($[c_v] = $ cal./mass $\times C°$). The dimensions of the pressure can be expressed in terms of the dimensions of $\Theta°$, ρ and c_v'; consequently, the assumption made is admissible from the point of view of dimensional analysis. Since the dimensions of $\Theta°$, ρ and c_v' are independent, then from the assumption that

$$p = f(\Theta, \rho, c_v'),$$

we at once deduce the Clapeyron equation

$$\frac{p}{c_v' \rho \Theta} = c \quad \text{or} \quad p = \rho R \Theta$$

where c is a nondimensional constant and R denotes the dimensional constant $cc_v' = cJc_v$.

Hence, the Clapeyron equation can be considered as a consequence of the single hypothesis that the pressure, density, temperature and specific heat are connected, independently of the values of the other characteristics, by a relation which has physical meaning. Examples discussed in later chapters illustrate methods of combining dimensional analysis with reasoning resulting from symmetry, from linearity of the problem, from the mathematical properties of the functions for small or large values of the defining parameters, and so on.

CHAPTER II

Similarity, Modelling and Various Examples of the Application of Dimensional Analysis

1. MOTION OF A SIMPLE PENDULUM

As the first example, we shall consider the classical problem of motion of a simple pendulum.

A simple pendulum (Fig. 1) is a heavy particle suspended by a weight-less, inextensible string which is fixed at its other end. We shall assume that the motion of the pendulum is two-dimensional.

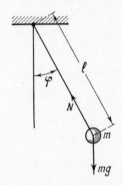

FIG. 1. Simple pendulum.

We introduce the following notation: l is the length of the pendulum; ϕ is the angle between the string and the vertical; t is time, m is the mass of the particle; and N is the tension in the string. If resistance forces are neglected then the problem reduces to the solution of the equations

$$\frac{d^2\phi}{dt^2} = -\frac{g}{l}\sin\phi \tag{1.1}$$

$$m\left(\frac{d\phi}{dt}\right)^2 l = N - mg\cos\phi \tag{1.2}$$

with the initial conditions

$$\phi = \phi_0 \quad \text{and} \quad \frac{d\phi}{dt} = 0 \quad \text{when } t = 0,$$

24

i.e. the initial instant is taken as that at which the pendulum is deflected by the angle ϕ_0 and the velocity is zero.

It is evident from Equations (1.1), (1.2) and the initial conditions that we may choose as our system of characteristic parameters, the quantities

$$t, \, l, \, g, \, m, \, \phi_0.$$

The numerical values of all the remaining quantities are determined completely by the values of these parameters. Therefore, we can write

$$\begin{aligned} \phi &= \phi(t, \, \phi_0, \, l, \, g, \, m), \\ N &= mgf(t, \, \phi_0, \, l, \, g, \, m), \end{aligned} \tag{1.3}$$

where ϕ and f are nondimensional functions.

The numerical values of the functions ϕ and f should not depend on the system of units of measurement. The form of these functions can be determined either by solving Equations (1.1) and (1.2) or by experiment.

From these general considerations, it follows that the five dimensional arguments of the functions ϕ and f can be reduced to only two. These are nondimensional combinations of t, l, g, m and ϕ_0 since there are three independent units of measurement.

Two independent nondimensional combinations

$$\phi_0 \quad \text{and} \quad t\sqrt{(g/l)} \tag{1.4}$$

can be formed from the quantities t, l, g, m and ϕ_0. All the other non-dimensional combinations formed from t, l, g, m and ϕ_0 or, generally, from any quantities determined by these parameters will be functions of the combinations (1.4). Therefore, we can write

$$\phi = \phi(\phi_0, t\sqrt{[g/l]}), \tag{1.5'}$$

$$N = mgf(\phi_0, t\sqrt{[g/l]}). \tag{1.5''}$$

The formulas (1.5), obtained by dimensional methods, show that the law of motion is independent of the mass of the particle but the tension in the string is directly proportional to the mass. These conclusions also follow from Equations (1.1) and (1.2). The quantity $t\sqrt{(g/l)}$ can be considered as the time expressed in a special system of units, in which the length of pendulum and the acceleration due to gravity are taken equal to unity.

Let us denote any characteristic time interval, for example, the time taken by the pendulum to move between the extreme and vertical positions or between two identical phases, i.e. the period of oscillation, by T (the existence of a periodic motion can be taken as a hypothesis or as a result known from additional data). We shall have

$$T = f_1(\phi_0, \, l, \, g, \, m) = \sqrt{(l/g)} f_2(\phi_0, \, l, \, g, \, m).$$

The function f_2 is a nondimensional quantity and since a nondimensional combination cannot possibly be formed from l, g and m, it is clear that the function f_2 is independent of l, g and m. Therefore

$$T = \sqrt{(l/g)} \cdot f_2(\phi_0). \qquad (1.6)$$

Formula (1.6) establishes the relation between the time T and the length of the pendulum. It is impossible to determine the form of the function $f_2(\phi_0)$ by dimensional analysis: this must either be found theoretically from Equation (1.1) or experimentally.

Formula (1.6) can be obtained directly from relation (1.5'). In fact, (1.5') gives, for the period of oscillation,

$$\phi_0 = \phi(\phi_0, \; T\sqrt{[g/l]}).$$

Solving this equation, we obtain formula (1.6).

If T is the period of oscillation, then it is evident, from symmetry considerations, that the value of the period T is independent of the sign of ϕ_0, that is

$$f_2(\phi_0) = f_2(-\phi_0).$$

Therefore, the function f_2 is an even function of the argument ϕ_0. Assuming that the function $f_2(\phi_0)$ is regular for small ϕ_0, we can write

$$f_2(\phi_0) = c_1 + c_2\phi_0^2 + c_3\phi_0^4 + \dots \qquad (1.7)$$

The terms in ϕ_0^2 and higher order terms can be discarded for small oscillations and we obtain the following for the period T

$$T = c_1\sqrt{(l/g)}. \qquad (1.8)$$

The solution of Equation (1.1) shows that $c_1 = 2\pi$. Hence, we see that the formula for the period of the oscillations can be obtained by dimensional analysis to within the accuracy of a constant factor, when the amplitude is small.

Formulas (1.5) and (1.6) still remain valid if we take

$$\frac{d^2\phi}{dt^2} = -\frac{g}{l}f(\phi),$$

instead of (1.1), where $f(\phi)$ is any function of ϕ. In general, the validity of formulas (1.5) and (1.6) results from the single condition that the motion is determined by the parameters

$$t, \, l, \, g, \, m, \, \phi_0.$$

In order to establish this system of parameters, we had to work with the equations of motion but it can be derived independently of these. We must choose l and m for the pendulum characteristics; furthermore, g must be involved since the nature of the phenomenon is determined by gravity. Finally ϕ_0 and t must appear since the actual motion is determined by the maximum angle of deflection ϕ_0 and by the time t.

2. FLOW OF A HEAVY FLUID THROUGH A SPILLWAY

We next consider the problem of flow of a heavy fluid through a spillway (Fig. 2) consisting of a vertical wall with a triangular orifice which is symmetrical about the vertical and includes an angle α equal to 90°. The fluid flows under a head h which equals the height of the fluid level above the vertex of the triangle at a large distance from the spillway exit.

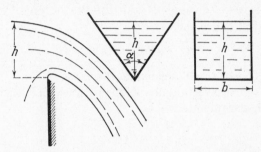

Fig. 2. Flow of a heavy fluid through a spillway.

We assume, for simplicity, that the vessel holding the fluid is very large and, consequently, the fluid motion can be considered steady.

The properties of inertia and weight, characterized by the density ρ and by the acceleration of gravity g, have special significance in the streaming motion of a fluid.

Our flow problem is defined completely by the parameters

$$\rho, g, h.$$

The weight of fluid Q which flows through the spillway opening per unit time can be a function only of these parameters, so that

$$Q = f(\rho, g, h).$$

Using dimensional analysis, it is not difficult to find the form of this function. Actually, the dimensions of Q are kg./sec. The combination $\rho g h^3 \sqrt{(g/h)}$ also has the dimensions of kg./sec. Consequently, the ratio

$$\frac{Q}{\rho g^{3/2} h^{5/2}}$$

is nondimensional. This ratio is a function of the quantities ρ, g, h from which a nondimensional combination cannot possibly be formed. Consequently, we can write

$$\frac{Q}{\rho g^{3/2} h^{5/2}} = C$$

or

$$Q = C \rho g^{3/2} h^{5/2} \tag{2.1}$$

where C is an absolute constant which is determined, most readily, by experiment. This formula determines the relation between the mass flow, the head h and the density ρ completely.

We can extend this analysis to cover spillways with different included angles α. In this case, the angle is added to the system of characteristic parameters and (2.1) becomes

$$Q = C(\alpha) \rho g^{3/2} h^{5/2}, \tag{2.2}$$

that is, the coefficient C will depend on the angle α.

If the spillway has a rectangular shape with width b, then the system of characteristic parameters will be

$$\rho, g, h, b.$$

All the nondimensional quantities are determined by the parameter h/b. In this case, we replace (2.1) by

$$Q = f(h/b) \rho g^{3/2} h^{5/2}. \tag{2.3}$$

The function $f(h/b)$ can be determined experimentally by observing the flow through spillways of various widths b but with constant h. Formula (2.3) can be applied to cases in which the width b is constant but the head h, varies, i.e. to cases for which no experimental results are available.

This example shows the usefulness of dimensional analysis in cutting down the number of experiments with a consequent saving in time as well as cost of apparatus. In experiments the particular quantities to be investigated can be replaced by another quantity. Complete information about flow of oil, mercury, etc., can be derived from experiments on water.

3. FLUID MOTION IN PIPES

The value of dimensional analysis and similarity theory was first demonstrated with special clarity in hydraulics during the study of fluid motion in pipes. Despite the practical importance and the simplicity of the reasoning of dimensional analysis, its use in hydraulics problems, leading to a considerable advance in the subject only occurred at the end of the nineteenth century, following the work of Osborne Reynolds (1883, 1886).

Empirical formulas, proposed by various authors, have long been used in hydraulics. These formulas contained a number of dimensional constants the values of which were determined by special experiments and by fluid properties.

The reasoning of dimensional analysis, besides giving a clearer and more general formulation of the problem, led to empirical laws governing the motion of fluids in pipes with different temperatures, diameters and speeds.

We now describe and formulate our problem. We consider a cylindrical pipe of uniform cross-section (Fig. 3). The geometry of the pipe is then defined completely by the cross-sectional area or by a characteristic length a. The radius or diameter is usually taken as the characteristic

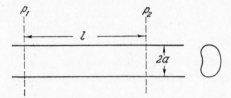

FIG. 3. Motion of an incompressible fluid in a cylindrical pipe.

dimension for circular pipes. The pipe length is assumed to be sufficiently large to justify neglect of end effects, and we can therefore assume that the pipes are infinitely long. We assume that the motion is steady. We neglect compressibility but take account of inertia and viscosity represented by the density ρ and the coefficient of viscosity ν. Since the coefficient of viscosity depends on the temperature, we can allow for temperature effects by varying the viscosity.†

In order to determine the motion of the fluid, we need to know the pressure drop along the pipe, the fluid discharge per unit time through the pipe cross-section, the average fluid velocity $\bar{u}$ over the pipe cross-section, or other equivalent data.

Therefore, the complete motion in the pipe is determined by the system of parameters

$$\rho, \mu, a, \bar{u}.$$

All the physical variables of the motion are functions of these parameters.

† Here, however, we assume that the temperature is constant throughout the whole fluid.

For example, consider the pressure drop along the pipe. The pressure drop per unit length of pipe is given by the quantity

$$\frac{p_1 - p_2}{l}$$

where p_1 and p_2 are the pressures in sections of the pipe a distance l apart. The combination

$$\frac{p_1 - p_2}{l \cdot \dfrac{\rho \overline{u}^2}{2a}} = \psi$$

is a nondimensional quantity and is called the resistance (drag) coefficient of the pipe.

The resistance of a section of pipe of length l is

$$P = (p_1 - p_2)S = \psi \frac{l}{a} S \frac{\rho \overline{u}^2}{2} \tag{3.1}$$

where S is the area of the pipe cross-section.

Only one independent nondimensional combination

$$\frac{\overline{u} a \rho}{\mu} = \mathbf{R},$$

called the Reynolds number, can be formed from the four characteristic parameters ρ, μ, a and u. All nondimensional quantities depending on these four parameters are functions of the Reynolds number.

In particular

$$\psi = \psi(\mathbf{R}). \tag{3.2}$$

The problems of determining the pipe resistance or the fluid discharge as a function of the pressure drop reduce to finding the function $\psi(\mathbf{R})$: this function can be found experimentally by measuring the variation of resistance with the velocity (or with the mass flow of the stream) when water moves along the pipe. The results obtained can be used to analyse the motion of other fluids in pipes with different diameters. For example, the properties of motion of air in pipes, etc. (see Fig. 4) can be deduced from experimental data on water under certain conditions (when compressibility can be neglected, i.e. at speeds considerably less than the speed of sound).

Experiment shows that fluid motion in pipes is of two very different types; namely laminar motion and turbulent motion. The fluid particles in laminar motion in a cylindrical pipe move in lines parallel to the generators of the pipe; in turbulent motion, there is disordered mixing of the fluid in a direction perpendicular to the generators. Turbulent flow can be considered as steady motion only on the average.

In some cases laminar motion in a pipe is slightly stable but, more generally, it is unstable and degenerates into turbulent motion.

The stability property is characteristic of fluid motion as a whole and, consequently, must be defined by the Reynolds number **R** for smooth pipes. Experiment is in good agreement with this conclusion. Laminar motion is stable for low values of the Reynolds number and unstable for large values. The Reynolds number determines the regime of the motion. The boundary of stability of laminar motion is defined by a certain value

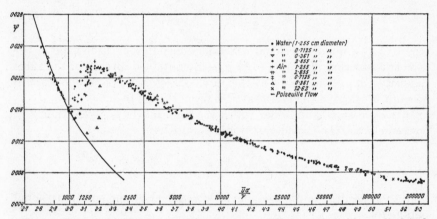

FIG. 4. Experimental data on the resistance coefficient of circular cylindrical pipes. (Measurements of Stanton and Pannel.)

of the Reynolds number known as the critical value. The critical Reynolds number for circular pipes is of the order of $\mathbf{R}_{cr} = 1000\text{–}1300$.

The laminar regime is found in slow motions of a fluid of high viscosity in pipes of small diameter (for example, in capillary tubes). The turbulent regime is found in fluids with low viscosity moving at high speed in pipes of large diameter.

Experimental data show that the function $\psi(\mathbf{R})$ has two branches, one of which corresponds to the laminar and the other to the turbulent regime of the motion. A transition region exists close to the critical value of Reynolds number.

All the fluid particles in laminar motion in a cylindrical pipe move in straight lines parallel to the pipe axis at a constant speed, i.e. with zero acceleration: this fluid motion in pipes is called Hagen-Poiseuille flow. The inertia property of the fluid represented by the parameter ρ, can only be felt if the accelerations are not zero,† consequently, the resistance must be independent of ρ in laminar motion. Therefore, the right side of (3.1) must be independent of ρ in laminar motion, and the function $\psi(\mathbf{R})$ must be

$$\psi = \frac{C}{\mathbf{R}} = \frac{C\mu}{\rho a \overline{u}} \tag{3.3}$$

where C is a nondimensional constant determined by the geometric shape of the pipe cross-section; a is the pipe radius. For a circular pipe, C is easily calculated theoretically: $C = 16$.

The pipe resistance in the case of laminar motion is then given by

$$P = \frac{1}{2}\frac{S}{a^2}C\mu\overline{u} = C_1\mu l\overline{u}, \tag{3.4}$$

where C_1 is a nondimensional constant depending on the shape of the pipe cross-section. Formula (3.4) is easily obtained directly if the three quantities a, μ, $\overline{u}$ are taken as the only characteristic parameters and if account is taken of the fact that P is proportional to l.

If the pressure drop under which the fluid moves is given, then it is convenient to take the characteristic parameters:

$$\rho, \mu, a \quad \text{and} \quad i = \frac{p_1 - p_2}{l}.$$

In this case, the motion regime is determined by the nondimensional parameter

$$\frac{\rho i a^3}{\mu^2} = J.$$

It is easy to see, from (3.1), that

$$J = \tfrac{1}{2}\mathbf{R}^2\psi(\mathbf{R}). \tag{3.5}$$

This relation gives the variation of J with $\mathbf{R}$ in terms of the function $\psi(\mathbf{R})$. Let us denote the volume of fluid flowing through the pipe cross-section per unit time (the volume discharge of the pipe) by

$$Q = \overline{u}S.$$

The nondimensional combination

$$\frac{Q\rho}{\mu a} = \mathbf{R}\frac{S}{a^2}$$

---

† As is known, the density and acceleration only enter into the derivatives in the equations of motion.

is a function of J, i.e.

$$Q = \frac{\mu a}{\rho} f(J). \tag{3.6}$$

The form of the function $f(J)$ is easily determined for laminar motion. From (3.3) and (3.5), we find

$$Q = \frac{2}{C} \frac{S}{a^2} \frac{ia^4}{\mu} = C_2 \frac{ia^4}{\mu}, \tag{3.7}$$

where the nondimensional constant C_2 depends on the shape of the pipe cross-section. For a circular pipe

$$C_2 = \frac{\pi}{8}.$$

Formula (3.7) is Poiseuille's law which was established experimentally by Hagen in 1839 and by Poiseuille in 1840. The very good agreement of this law with experiment is one of the main confirmations of the validity of the law of viscous friction in fluids and of the initial survey of the phenomenon.

4. MOTION OF A BODY IN A FLUID

A survey of the problem of the motion of an airplane, submarine, etc., leads to the problem of forward uniform motion of a solid body in an infinite fluid.

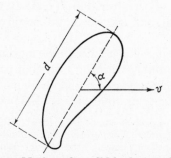

FIG. 5. Motion of a solid body in a fluid.

If we regard the geometric shape of the boundary of the body as fixed, then it will be completely specified by a characteristic length d in order to give the surface of the body completely.

Consider the forward motion of the body parallel to a fixed plane. Let us denote the velocity of the motion and the angle which defines the velocity direction (Fig. 5) by v and α, respectively; the quantities v and α can be variable.

2

We assume that the fluid is incompressible, but take account of the inertia and viscosity of the fluid. For simplicity, we assume that body forces are absent. The pressure distribution on the body surface and the total forces exerted by the fluid on the body depend on the state of the disturbed fluid motion.

The steady state motion of a fluid is defined for a body of given shape by a system of five parameters.†

$$d, v, \alpha, \rho, \mu.$$

All the nondimensional, mechanical quantities related to the motion can be considered as a function of $5 - 3 = 2$ nondimensional parameters: the angle of attack α and the Reynolds number

$$\frac{v \, d\rho}{\mu} = \mathbf{R}.$$

Let us denote the force exerted by the fluid on the body by W (in the subsequent discussion, it does not matter whether we understand W to be the total resistance (drag) or one of its components, the profile drag, directed opposite to the fluid motion or the lift, which is normal to the velocity direction). It follows from the general theorems on dimensional analysis that the nondimensional quantity $M/\rho d^2 v^2$ is a function of the angle of attack α and the Reynolds number $\mathbf{R}$. Consequently

$$W = \rho \, d^2 v^2 f(\alpha, \mathbf{R}). \tag{4.1}$$

The determination of the function $f(\alpha, \mathbf{R})$ is the most important problem of theoretical and experimental aerodynamics and hydrodynamics.

Evidently, the influence of viscosity on the motion is felt only through that of the Reynolds number with the system of parameters chosen.

Certain general conclusions can be made about the role of the fluid viscosity when the velocity or dimensions of the body increase, from the

---

† The pressure at infinity, p_0, which can be assigned arbitrarily, is not introduced into this system of parameters for the following reasons. The fluid is incompressible, consequently,.the variation of p_0 cannot influence the velocity field. The pressure difference $p - p_0$ can always be considered instead of the value of the total pressure p. Hence, it is evident that the quantity p_0 is not essential and, consequently, it is not necessary to introduce it as a characteristic parameter. However, when the fluid motion involves cavitation phenomena resulting from vaporization in the lower pressure regions, it is necessary to include the quantity $p_0 - p'$ among the characteristic parameters, where p' is the vapour pressure at a given temperature. The quantity p_0 or an equivalent parameter must be included among the number of characteristic parameters for a compressible fluid. The nondimensional parameter $\kappa = 2\dfrac{p_0 - p'}{\rho v^2}$ is also important in motions accompanied by cavitation. When studying the influence of the cavitation number κ in experiments, its value can be varied either by varying p_0, v or, artificially, by varying p'. Different fluids can also be used and the density ρ can be varied.

form of the formula $\mathbf{R} = (vd\rho/\mu)$. For example, the Reynolds number increases as the velocity or the linear dimensions of the body increase. But the Reynolds number must remain constant in order to preserve the role of viscosity since the whole variation of the Reynolds number can be referred to the variation of the coefficient of viscosity; if the product $vd\rho$ increases, then the coefficient of viscosity μ must be increased in order to keep the Reynolds number constant. Therefore, the motion of honey (large μ/ρ), caused by the motion of a large body, is similar to the motion of water (small μ/ρ) caused by the motion of a small body at identical velocities. Or the motion of a body in honey at a high velocity is the same as the motion of the same body in water at low velocity. The similarity of the motions is expressed by the fact that all the non-dimensional quantities are identical in these motions.

Furthermore, these considerations show that the effect of viscosity on a body moving in the same fluid decreases as the velocity and size of the body increase.† Theoretical investigations and experimental results show that the influence of fluid viscosity decreases for high values of the Reynolds number and becomes unimportant in certain cases. Neglecting the viscosity, i.e. putting $\mu = 0$, we arrive at the concept of an ideal fluid.

The number of characteristic parameters in problems concerning the motion of a body in an ideal fluid is cut down to four:

$$d, \alpha, \rho, v.$$

All the nondimensional characteristics in an ideal fluid are determined by the angle of attack α, consequently, formula (4.1) is replaced by

$$W = \rho d^2 v^2 f_1(\alpha). \tag{4.2}$$

Therefore, the forces acting on a body moving in an ideal incompressible fluid are proportional to the velocity squared. This law is approximately correct for a viscous fluid at high enough values of the Reynolds number.

The functions $f(\alpha, \mathbf{R})$ and $f_1(\alpha)$ in (4.1) and (4.2) depend, for bodies of different shapes, on abstract parameters determined by the body shape as well as on the angle of attack. Figures 6 and 7 show experimental results on the variation of the drag coefficient with the Reynolds number. Figure 8 shows the nature of the influence of the angle of attack on the drag and lift of a wing.

Now let us consider the case of very slow motion which corresponds to low values of the Reynolds number.

As the Reynolds number decreases, the role of viscosity increases. If we neglect the inertial forces in comparison with the viscous forces, then

---

† We assume in this case that, other things being equal, the viscous effect is reduced as the coefficient of viscosity decreases.

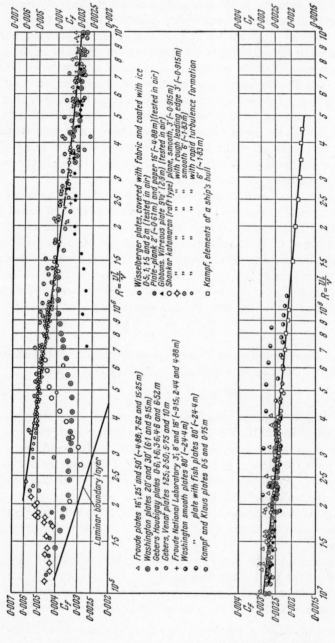

FIG. 6. Drag coefficient $c_f = \dfrac{W}{l_b \rho v^2/2}$ of plane square plates towed parallel to their planes. (l is a dimension parallel to the velocity direction. b is the plate width.)

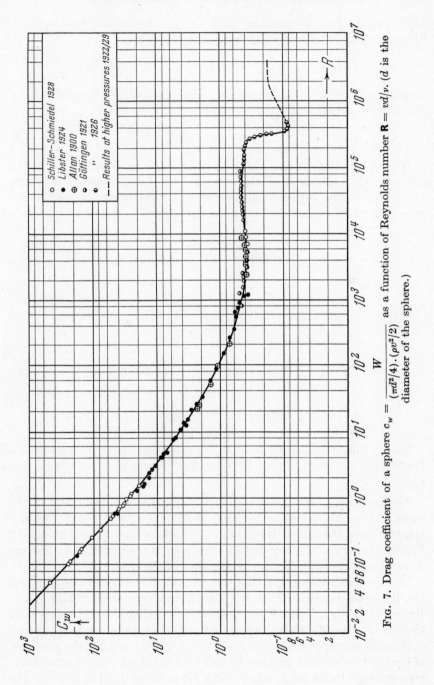

FIG. 7. Drag coefficient of a sphere $c_w = \dfrac{W}{(\pi d^2/4) \cdot (\rho v^2/2)}$ as a function of Reynolds number $\mathbf{R} = vd/\nu$. (d is the diameter of the sphere.)

this is equivalent to assuming that the parameter ρ is unimportant. There are four characteristic parameters in this case:

$$d, \alpha, v, \mu,$$

consequently all the nondimensional variables will also depend only on the angle of attack α. Therefore,

$$W = \mu \, dv \, f_2(\alpha). \tag{4.3}$$

Hence, it is evident that the drag and the lift are proportional to the velocity, the coefficient of viscosity and the linear scale d. This law, called Stokes' law, is in good agreement with experiment on small bodies at low speeds; for example, in the precipitation of fine particles in a fluid.

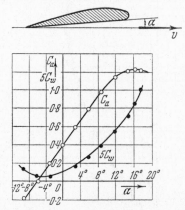

FIG. 8. Typical curves of the lift $C_a = 2A/\rho Sv^2$ and drag $C_w = 2w/\rho Sv^2$ as a function of the angle of attack for a wing. (S is the wing planform area.)

The function $f_2(\alpha) = \text{const} = c$ for a sphere, i.e. is independent of the angle α. The theoretical value of the coefficient c for slow motions of a sphere under the above assumptions (which reduce to neglecting the inertia terms in the Navier-Stokes equations) was calculated by Stokes; it appears that $c = 3\pi$ (if d is the diameter of the sphere).

We see that dimensional analysis permits the form of the function $f(\alpha, \mathbf{R})$ to be determined for very small and very large values of the Reynolds number. When $\mathbf{R} \to \infty$, we arrive at an ideal fluid; in this case, the function $f(\alpha, \mathbf{R})$ tends to a certain function $f_1(\alpha)$ which is independent of Reynolds number. At low values of the Reynolds number, the influence of the fluid viscosity is paramount and the inertia property is secondary, giving a high value for μ, a low value for ρ. Formula (4.3) holds in the limit as $\rho = 0$; assuming $\rho \neq 0$, we obtain from this formula

$$W = \rho\, d^2\, v^2 \cdot \frac{f_2(\alpha)}{v\, d\rho/\mu}.$$

Hence, it follows that

$$f(\alpha, \mathbf{R}) = \frac{f_2(\alpha)}{\mathbf{R}}$$

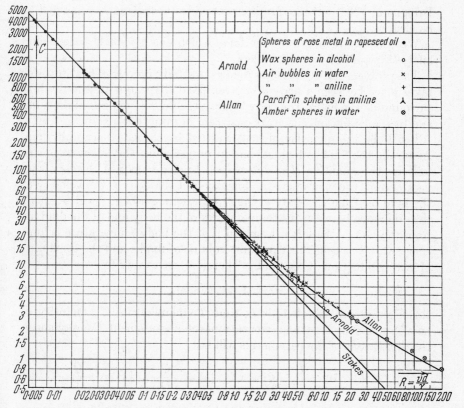

FIG. 9. Drag coefficient of a sphere at low Reynolds numbers.

at low values of Reynolds number. This relation is a consequence of Stokes' law. A comparison of experimental results with Stokes' law for a sphere is shown on Fig. 9.

Using dimensional analysis, we have shown that Stokes' law (4.3) is correct for bodies of any shape if the inertia terms in the Navier-Stokes equations are neglected.

The function $f_2(\alpha)$ can be determined experimentally or theoretically by solving the simplified Navier-Stokes equations.

5. HEAT TRANSFER FROM A BODY IN A FLUID FLOW FIELD

In 1915, Rayleigh (1915a) applied dimensional analysis to the Boussinesq problem of heat transfer from a body to a fluid flowing around the body. Subsequently, the Rayleigh reasoning was the subject of remarks by a number of authors (Riabouchinsky, 1915; Rayleigh, 1915b; Bridgman, 1934) but the questions raised in these remarks remained to be cleared up.

We consider in detail all the facts which emerge when dimensional analysis is applied to this problem.

The problem can be stated as follows: a steady process of heat transfer takes place between a body of a given fixed shape and an infinite fluid surrounding the body. The body is fixed, the fluid flows around the body and has constant velocity v far upstream in front of the body.

Let H be the quantity of heat emitted by the body per unit time. Assuming the fluid to be ideal and incompressible, Rayleigh reasons as follows: the quantity H is determined by the values of the following parameters: the characteristic dimension of the body l, the velocity of the fluid far from the body v, the temperature gradient θ which equals the difference between the temperatures of the body and of the fluid far from the body, where it is assumed that the body temperature is kept constant, the specific heat per unit volume of fluid c and the coefficient of heat conduction of the fluid λ. Therefore, we can write

$$H = f(l, v, \theta, c, \lambda).$$

Rayleigh chooses the length L, time T, temperature $C°$, quantity of heat Q, and mass M as fundamental units of measurement. Consequently, the dimensions of the parameters will be

$$[l] = L, \quad [v] = \frac{L}{T}, \quad [\theta] = C°, \quad [c] = \frac{Q}{L^3 C°}, \quad [\lambda] = \frac{Q}{L\, C°\, T}.$$

We note that all these dimensions are independent of the mass.

Only one independent nondimensional combination $(lvc)/\lambda$ can be formed from the five characteristic dimensional parameters. The dimensions of H will be Q/T.

It is easy to see that the combination $H/(\lambda l\theta)$ is a nondimensional quantity, consequently

$$H = \lambda l\theta f\left(\frac{lvc}{\lambda}\right). \tag{5.1}$$

Rayleigh obtained this formula. It follows that the rate of heat transfer is proportional to the temperature gradient θ and has identical values for different values of v and c provided that the product vc is constant.

Riabouchinsky made the following remark: since the quantity of heat and the temperature have the dimensions of energy (the temperature is defined in the kinetic theory of gases as the average kinetic energy of the molecules in random motion), then only the units of measurement for length, time and mass can be taken as the fundamental units. The dimensions of the characteristic parameters will then be

$$[l] = L, \quad [v] = \frac{L}{T}, \quad [\theta] = \frac{ML^2}{T^2}, \quad [c] = \frac{1}{L^3}, \quad [\lambda] = \frac{1}{LT}.$$

Now, two independent nondimensional combinations

$$\frac{lvc}{\lambda} \quad \text{and} \quad cl^3$$

can be formed from the characteristic parameters. Therefore, in this case dimensional analysis leads to the formula

$$H = \lambda l \theta f\left(\frac{lvc}{\lambda}, cl^3\right), \tag{5.2}$$

which clearly yields less information than formula (5.1).

In his answer to Riabouchinsky, Rayleigh writes (Rayleigh, 1915b):

[Question raised by Dr. Riabouchinsky belongs rather to the logic than to the use of the principle of similitude with which I was mainly concerned (Rayleigh, 1915a). It would be well worthy of discussion. The conclusion I gave follows on the basis of the usual Fourier Equations for conduction of heat; in which heat and temperature are regarded as *sui generis*. It would indeed be a paradox if further knowledge of the nature of heat afforded by the molecular theory put us in a worse position than before in dealing with a particular problem. The solution would seem to be that the Fourier Equations embody something as to the nature of heat and temperature which is ignored in the alternative argument of Dr. Riabouchinsky.]

Bridgman (1934) correctly noted that Rayleigh's answer is hardly satisfactory and does not clear up the question at all.

The misunderstanding is explained as follows: there are three different units of measurement for the energy in the system used by Rayleigh to derive (5.1): the erg $= ML^2/T^2$, the degree C° and calory Q. The definition of heat and temperature as mechanical energy is given in the kinetic theory of gases. The conversion of the quantity of heat and the temperature into mechanical units is related to the values of the mechanical

2*

equivalent of heat $J = 427$ ($[J] = ML^2/T^2Q$) kgm./cal. and the Boltzmann constant $k = 1\cdot38 \times 10^{-16}$ erg/deg. ($[k] = ML^2/T^2C°$). These must be considered as physical constants in the independent units of measurement for the mechanical energy, the quantity of heat and the temperature.

Since the fluid is ideal and incompressible it follows that the velocity field is determined by kinematic conditions and the phenomenon is not accompanied by a conversion from thermal into mechanical energy. The mechanical processes occur independently of the thermal. Hence, it follows that the value of the density of the fluid does not affect all the thermal quantities and the value of the mechanical equivalent of heat is generally not essential because of the absence of a conversion of thermal energy into mechanical. Furthermore, if it is assumed that the density ρ and the quantity J do not influence the process of heat transfer, then it can be shown by dimensional analysis that the value of the Boltzmann constant k is not at all essential since the dimensions of k include the unit of mass which does not appear in the dimensions of H and the characteristic parameters. The insignificance of the quantities ρ, J and k under the above assumptions can also be easily established from the mathematical formulation of the heat transfer problem. These facts justify the omission of ρ, J and k from the group of characteristic parameters considered by Rayleigh.† However, if we maintain the assumption that ρ is insignificant‡ but make no such assumption about J and k, then the quantities k and J must be added to Rayleigh's table of characteristic parameters, so that the complete system is

$$l,\ v,\ \theta,\ c,\ \lambda,\ J,\ k.$$

From these seven dimensional quantities we can only form the two independent nondimensional combinations

$$\frac{lvc}{\lambda} \quad \text{and} \quad \frac{Jcl^3}{k}.$$

In this case, formula (5.1) is replaced by

$$H = \lambda l \theta f \left(\frac{lvc}{\lambda}, \frac{Jcl^3}{k} \right). \tag{5.3}$$

Formula (5.3) reduces to (5.1) if we take account of the fact that the

---

† If we analyse the same problem in the case of a viscous, compressible fluid, then the quantities ρ, J and k become significant and these parameters or their equivalents must be included in the table of characteristic parameters.

‡ Riabouchinsky retains this assumption in his reasoning.

mechanical equivalent of heat is insignificant and, therefore, that the same is true of the parameter

$$\frac{Jcl^3}{k}.$$

Now, if thermal quantities are defined in mechanical terms, following Riabouchinsky, then k and J will be nondimensional universal constants and formula (5.3) transforms into (5.2). The derivation is weaker because the methods of analysis here do not take into account the additional reasoning on the mechanism of the phenomena.

6. DYNAMIC SIMILARITY AND MODELLING OF PHENOMENA

Dimensional and similarity theory has great value in making models of various phenomena. This modelling is to replace the study of the natural phenomenon which interests us by the study of an analogous phenomenon in a model of smaller or greater scale, usually under special laboratory conditions. The basic idea of modelling is that the information required about the character of the effects and the various quantities related to the phenomenon under natural conditions can be derived from the results of experiments with models.

Modelling is based on an analysis of physically similar phenomena in the majority of cases. We replace the study of the natural phenomenon which interests us by the study of a physically similar phenomenon which is more convenient and easier to reproduce. Mechanical or, generally, physical similarity can be considered as a generalization of geometric similarity. Two geometric figures are similar if the ratios of all the corresponding lengths are identical. If the similarity ratio, the scale, is known, then simple multiplication of the dimensions of one geometric figure by the scale factor yields the dimensions of the other, its similar, geometric figure.

There are various ways of defining dynamical or physical similarity. Below, we shall give a definition of physical similarity in a form required in practical application and which is ready for direct use.

Two phenomena are similar, if the characteristics of one can be obtained from the assigned characteristics of the other by a simple conversion, which is analogous to the transformation from one system of units of measurement to another.

The "scaling factor" must be known in order to accomplish the conversion.

The numerical properties of two different, but similar phenomena can be considered as the numerical properties of the same phenomenon expressed in two different systems of measurement units. All the nondimensional variables (nondimensional combinations of dimensional quantities) of a set of similar phenomena have the same numerical values. It is not difficult to see that the converse is also correct, i.e. if all the nondimensional characteristics of two motions are identical, then the motions are similar.

The similarity of two phenomena can sometimes be understood in a wider sense by assuming that the above definition refers only to a certain special system of parameters. These are to define the phenomena completely and enable any other characteristics to be found except those obtained by simple scaling, when transforming from one to the other "similar" phenomenon. For example, any two ellipses can be considered as similar in this sense when Cartesian coordinates directed along the principal axes of the ellipses are used. The Cartesian coordinates of points of any ellipse can be obtained in terms of the coordinates of points of some particular ellipse (affine similarity) if we use the above conversion.

In order to maintain similarity in modelling, it is necessary to comply with certain conditions. However, quite often in practice, these conditions, which guarantee the similarity of the phenomena as a whole, are not fulfilled, and we must then consider the question of the magnitude of the errors (scale effect) which arise in applying the model results to actual conditions.

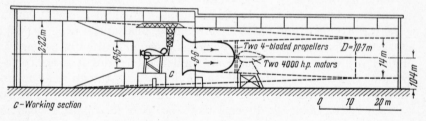

FIG. 10. Longitudinal section of the NACA wind-tunnel for testing under actual conditions.

After the system of parameters defining the particular class of phenomena has been established, it is not difficult to establish the similarity conditions of the two phenomena.

In fact, let the phenomenon be defined by n parameters, some of which can be nondimensional and others dimensional physical constants. Furthermore, let us assume that the dimensions of the variable parameters and of the physical constants are expressed by means of k fundamental units ($k \leqslant n$). In the general case, it is evident that not more than $n-k$ independent nondimensional combinations can be formed from the n quantities. All the nondimensional variables of the phenomenon

FIG. 11. Photograph of the actual wind-tunnel. Tunnel width 18·3 m. Return channel not shown. Two 4-bladed propellers. Two 4,000 h.p. motors.

can be considered as a function of these $n-k$ independent nondimensional combinations. Therefore, a certain basic system, which defines all the remaining quantities, can be selected from all the nondimensional quantities formed by the characteristics of the phenomenon.

The particular class of phenomena arising in the formulation of the problem contains phenomena which are not generally similar. The phenomena in a class are similar if the following condition is satisfied:

The necessary and sufficient conditions for two phenomena to be similar, are that the numerical values of the nondimensional coefficients forming the basic system are constant. These conditions are called similarity criteria.

When these similarity conditions are fulfilled, to calculate the characteristics of the full-scale phenomenon from model data, we need to know the scaling factors for corresponding quantities.

If, among n characteristic parameters, k quantities have independent dimensions, then the scaling factors for all k quantities must be given, or derived from experiments, in any particular problem. Corresponding factors for all remaining quantities are easily found from formulae connecting the dimensions of these quantities with those of the k quantities.

All the nondimensional quantities in the problem of steady, uniform motion of a body in an incompressible, viscous fluid are defined by two parameters: the angle of attack α and the Reynolds number **R**. The conditions of physical similarity—the similarity criteria—are represented by the relations

$$\alpha = \text{const} \quad \text{and} \quad \mathbf{R} = \frac{v\,d\rho}{\mu} = \text{const}.$$

When modelling this phenomenon, the experimental results for the model can be converted to actual conditions only for identical values of α and **R**. The first condition is always easy to realize in practice. It is more difficult to satisfy the second condition (**R** = const) especially in those cases when the streamlined body is of large size as, for example, the wing of an airplane. If the model is smaller than in reality, then either the velocity of the undisturbed stream, which is usually restricted in practice, or the density and viscosity of the fluid, must be altered substantially in order to maintain the magnitude of the Reynolds number.

In practice, these circumstances introduce great difficulties in the study of aerodynamic drag. The necessity of a constant Reynolds number led to the construction of gigantic aerodynamic wind-tunnels in which airplanes could be investigated under actual conditions (Figs. 10 and 11) as well as of closed-type tunnels in which compressed, i.e. more dense, air circulates at high speed (Fig. 12).

Special theoretical and experimental investigations show that on many streamlined bodies the Reynolds number noticeably affects only the nondimensional profile drag coefficient and affects the nondimensional lift coefficient and certain other quantities of practical importance very slightly. Therefore, the difference in the values of Reynolds number for the model and the actual phenomenon is not very important in certain cases.

We mentioned the similarity conditions for wing motion without taking into account the property of compressibility of air which is not essential for velocities which are low in comparison with the speed of sound. Later, we shall consider the similarity conditions taking compressibility into account.

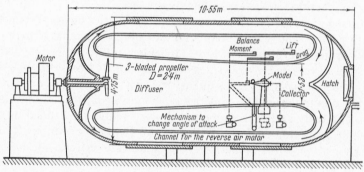

FIG. 12. Cross-section of the NACA variable density wind-tunnel. The pressure within the tunnel can reach 21 atmospheres and the velocity of the air flow 23 m./sec. Motor 3-bladed propeller diameter 2·4 m.

As another example, let us analyse the problem of modelling the equilibrium of elastic structures.

Consider a structure of homogeneous material, for example, a bridge girder. The elastic properties of an isotropic material are determined by two constants, Young's modulus $E(\text{kg.}/\text{m}^2)$ and the nondimensional Poisson's ratio σ. We consider geometrically similar structures and form a table of characteristic parameters.

In order to define all the model dimensions, it is sufficient to assign a certain characteristic dimension B. If the weight of the structure is essential in the equilibrium state, then the specific gravity $\gamma = \rho g$ $(\text{kg.}/\text{m}^3)$ must appear as a characteristic parameter. External loads distributed in a certain way over the components of the structure act upon it in addition to the weight of its parts. Let the magnitude of these loads be determined by the force P (kg.). Then the system of characteristic parameters will be

$$\sigma, E, B, P, \rho g.$$

In this case, we have $n = 5$, $k = 2$, therefore three nondimensional parameters will form the basis for mechanical similarity of the states of elastic equilibrium namely

$$\sigma, \frac{E}{\rho g B}, \frac{P}{E B^2}.$$

The similarity criteria require that these parameters are constant on the model and in the actual structure. All the deformations will be similar when these conditions are satisfied. If the model is n times smaller than the actual structure, then the deformations on the model will be n times smaller than in reality.

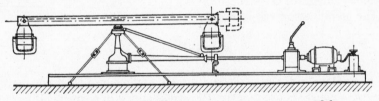

Fig. 13. Schematic diagram of a centrifuge to test models.

If the model and the actual structure are produced from the same material, then the values ρ, σ and E are identical on the model and in reality and, consequently, the following conditions must be satisfied for mechanical similarity

$$g B = \text{const.}$$

Under ordinary conditions, $g = \text{const}$; therefore, B must be a constant in order to conserve mechanical similarity, i.e. the model must coincide with the actual structure: in other words modelling is impossible for constant g.†

The variation of g can be realized artificially if the model is forced to rotate at a constant angular velocity by being placed in a so-called centrifuge (Fig. 13). The centrifugal forces of inertia of the model elements can be considered to be parallel for small enough model dimensions and large enough radius of rotation. Performing the rotation about the vertical axis, we find that constant mass forces, similar to gravitation but with another acceleration, will act on the model which is in a state of

† The increase in the specific gravity ρg required in practice when the model dimensions are diminished can be accomplished by applying an additional load to the model elements.

relative equilibrium (with respect to the centrifuge). Any large value can be obtained for the acceleration by choosing the right angular rotational velocity.

The idea of using centrifuges for modelling various processes was proposed by Bucky (1931) and independently by N. N. Davidenkov (1933) and G. I. Pokrovskii (1933). Prior to this in 1929, Davidenkov (Pokrovskii, 1940) proposed to use a box falling on a rigid spring, for this purpose, but this method proved inconvenient and was abandoned.

At the present time, centrifuges have been constructed to investigate various processes occurring in solids by means of models.†

Consider the stress τ kg./m² which develops in the deformation of an elastic structure under the action of a weight and of a given load distribution. We can interpret τ to be the maximum value of some stress component or, in general, to be a certain stress component acting on a specific element of the structure.

The combination τ/E is nondimensional, consequently, we can write

$$\frac{\tau}{E} = f\left(\sigma, \frac{E}{\rho g B}, \frac{P}{EB^2}\right).$$

If the model and the actual structure are made out of the same material, then $E = $ const; consequently, the stress states in corresponding points will be identical for mechanical similarity.

If we assume that the stressed states are dynamically similar and that the failure is determined by the values of the maximum stresses, then it is evident that failure occurs both on the model and in reality for similar states. If the magnitudes of the external loads are large but the intrinsic weight of the structure is small enough to be neglected, then the parameter $\gamma = \rho g$ and, therefore, the parameter $E/\rho g B$ is not essential. The preceding relation becomes, in this case

$$\frac{\tau}{E} = f\left(\sigma, \frac{P}{EB^2}\right),$$

and the similarity conditions will reduce to the two conditions

$$\sigma = \text{const} \quad \text{and} \quad \frac{P}{EB^2} = \text{const}.$$

Hence, it follows that the external loads must be proportional to the square of the linear dimension when modelling with the material properties conserved.

---

† The condition $E/\rho g B = $ const must be satisfied when modelling processes in which the parameters ρ, g, B and E as well as the other essential parameters are encountered. Consequently, modelling is possible in all such cases by using a centrifuge.

We denote the variation of the length during the deformation of a certain element of an elastic system by l. For a structure of the above type we have

$$\frac{l}{B} = \phi\left(\sigma, \frac{\rho g B}{E}, \frac{P}{EB^2}\right).$$

In a number of cases, it is seen at once from physical considerations that the quantity l/B will diminish as the specific gravity of the basic structure decreases, i.e. as the parameter $\rho g B/E$ decreases.

Now, consider two geometrically similar structures of certain dimensions which have been made from the same material (E and σ are identical). Let us assume that the magnitude of the external loads are proportional to the square of the dimensions, i.e.

$$\frac{P}{EB^2} = \text{const.}$$

Evidently, the parameter $\rho g B/E$ decreases in this case as the dimensions of the structure decrease, therefore, dynamical similarity will be violated. The relative deformations will be less in structures of smaller dimensions, consequently, a structure of small dimensions will have greater strength. However, this conclusion is valid only when the specific gravity of the material $\gamma = \rho g$ plays an essential part. If the intrinsic weight (γ) is not essential and $P/EB^2 = \text{const}$, then the relative deformations have the same values for bodies of different scales.

Consider the case when γ is not essential and it is known that the l/B ratio for a given structure decreases as the external loads P decrease. If the external loads are proportional to the cube of the linear dimension, the l/B ratio will be less than for structures of large dimensions. Therefore, the decrease in the dimensions also increases the strength in this case.

An interesting example in which the scale of the model is defined uniquely is found in hydrostatic models of dirigibles and balloons (Vorob'ev, 1928; Katanskii, 1936).

In practice, it is very important to know the shape and deformation of elements of the fabric of the balloon after it has been filled with gas. The geometric shape of the airship determines its hydrostatic and aerodynamic properties; information on the deformation of the material is necessary to guarantee the strength of the balloon.

We note the parameters which determine the static state of the airship.

It can be considered that the air and the gas have constant specific gravities over the height range between the upper and lower surface of the airship. The difference between the air pressure and the gas pressure within the airship, acting on its skin (only the differences in these pressures are essential in this problem) is determined by the magnitude of γ^*, which equals the difference in the specific gravities of the air and of the gas, $\gamma^* = \gamma_{air} - \gamma_{gas}$.

Experiments investigating the relation between stresses and deformations in materials show that the relative deformations are identical for a given material when the stresses are identical. The stresses in the fabric are defined as forces calculated per unit cross-sectional length.

We denote by τ kg./m. a certain stress of the properties of the material, and by l a characteristic linear dimension. Furthermore, let us introduce the weight per unit area of material q kg./m^2 and also given external forces Q kg., applied to various elements of the skin (the direction of these forces must correspond in different cases).

We obtain the following system of characteristic parameters for the geometric similarity of skins prepared from materials with the same stress strain relations,

$$\gamma^*, \tau, l, q, Q.$$

The similarity conditions are

$$\frac{\gamma_1^* l_1^2}{\tau_1} = \frac{\gamma_2^* l_2^2}{\tau_2} \tag{a}$$

$$\frac{\gamma_1^* l_1}{q_1} = \frac{\gamma_2^* l_2}{q_2} \tag{b}$$

$$\frac{\gamma_1^* l_1^3}{Q_1} = \frac{\gamma_2^* l_2^3}{Q_2}. \tag{c}$$

Condition (c) can easily be satisfied by an appropriate choice of the external force Q by using loads and blocks. If the model and the actual object $(l_2 \neq l_1)$ are prepared from identical material, then $q_1 = q_2$ and $\tau_1 = \tau_2$, and conditions (a) and (b) contradict each other. Consequently, we confine ourselves to the case when weight of the skin is ignored so that condition (b) drops out.

In order to satisfy condition (a), it is necessary to put

$$\frac{l_1}{l_2} = \sqrt{\frac{\gamma_2^*}{\gamma_1^*}};$$

hence, we must increase the specific lift $\gamma_2^* > \gamma_1^*$ as the scale $l_2 < l_1$ is decreased. This effect can be accomplished by using a heavy fluid and by riveting the model. The hydrostatic models of dirigibles and airships use this idea. Water, mercury, glycerine, etc., can be taken as the filling for the model. The usual shape of such a model is shown in Fig. 14.

If the flight altitude, the temperature, pressure and the gas used are known for the actual case, then by choosing the fluid for the model, we fix the ratio of the specific gravities, which defines the scale of the model.

For example, we have under normal conditions:

for hydrogen........ $\gamma_1^* = 1 \cdot 1 \text{ kg./m}^3$

for water.......... $\gamma_2^* = 1000 \text{ kg./m}^3$

for mercury........ $\gamma_2^* = 13{,}600 \text{ kg./m}^3.$

Hence, we find the following value for the model scale when using water

$$\frac{l_1}{l_2} = n = \sqrt{\left(\frac{1000}{1 \cdot 1}\right)} = 30 \cdot 1$$

when using mercury, we obtain

$$n = \sqrt{\left(\frac{13{,}600}{1 \cdot 1}\right)} = 111.$$

Modelling with mercury leads to smaller models, which generally is inconvenient.

The weight of the actual skin and the weight of the model skin act in opposite ways; consequently, the influence of the weight will violate similarity. The calculation of the influence of the weight of the skin can be accomplished with the aid of special instrumentation by applying external forces acting vertically upward and distributed over the skin in conformity with the similarity condition (b), taking into account the intrinsic weight of the skin directed downwards.

Sometimes, modelling can be carried out when certain similarity criteria, $\pi_1, \pi_2, \ldots$ have different values on the model and in reality, provided that the variation of the desired nondimensional quantities with these parameters is known beforehand. In such cases, it is only necessary to keep those similarity criteria constant, the influence of which is not known.

The above type of modelling can be used when the relations between the required quantities and the parameters $\pi_1, \pi_2, \ldots$ are assumed in advance. Their validity can be confirmed or refuted by means of the model investigations.

As we mentioned above, an example of this is modelling at different values of the Reynolds number when its influence on the unknown in question is insignificant. However, this same method can be used in those cases when the Reynolds number is essential but its effect is known beforehand.

An investigation with models is often the only possible method of experimental study and of solving important practical problems. Such is the case when studying natural phenomena which proceed through decades, centuries or millennia; a similar phenomenon can be produced under model experiment conditions which takes only hours or days to complete. This is the case in modelling the phenomenon of escaping gas which is capable of exploitation and of being pumped out through a well.

Conversely, instead of investigating a phenomenon which occurs very rapidly in nature, we can study a similar phenomenon which occurs much more slowly in a model.

Modelling is a recognized scientific technique which has general value, in principle and in practice, but it must be considered only as the initial approach to the main problem. This is the actual determination of the laws of nature, the search for general properties and characteristics of various classes of phenomena, the development of experimental and theoretical methods of investigation and of solutions of various problems and, finally, obtaining systematic techniques, methods, rules and recommendations to solve specific practical problems.

7. STEADY MOTION OF A SOLID BODY IN A COMPRESSIBLE FLUID

We now consider the general problem of the steady, forward motion of a solid body at a constant velocity in an infinite fluid. We shall take into account the inertia, viscosity, compressibility and heat conduction. For simplicity, we shall not consider the weight of the fluid† and heat transfer due to radiation.

Let us formulate the problem mathematically in order to determine the system of characteristic parameters. First, let us write down the equations of motion of a compressible viscous fluid, which we shall consider to be a perfect gas. We shall have

(1) The Navier-Stokes equation

$$\rho \frac{d\bar{\mathbf{v}}}{dt} = -\operatorname{grad} p - \tfrac{2}{3}\operatorname{grad}\mu\operatorname{div}\bar{\mathbf{v}} + 2\operatorname{div}\mu\operatorname{def}\bar{\mathbf{v}} \qquad (7.1)$$

where $\operatorname{def}\bar{\mathbf{v}}$ is the tensor of the deformation velocity;

(2) The continuity equation

$$\frac{\partial \rho}{\partial t} + \operatorname{div}\rho\bar{\mathbf{v}} = 0; \qquad (7.2)$$

† The weight of fluid is essential in a number of cases since it can cause intense convective motion which arises from the nonuniform heating of the fluid.

FIG. 14. Testing of a dirigible model.

(3) The equation of state of the gas (Clapeyron equation)

$$p = \rho R \Theta, \tag{7.3}$$

where R is the gas constant;

(4) The equation of heat transfer

$$Jc_v \rho \frac{d\Theta}{dt} + p \operatorname{div} \bar{\mathfrak{v}} = \operatorname{div} J\lambda \operatorname{grad} \Theta - \tfrac{2}{3}\mu \left| \operatorname{div} \bar{\mathfrak{v}} \right|^2$$

$$+ \mu \left\{ 2 \left[\left(\frac{\partial u}{\partial x}\right)^2 + \left(\frac{\partial v}{\partial y}\right)^2 + \left(\frac{\partial w}{\partial z}\right)^2 \right] \right.$$

$$\left. + \left(\frac{\partial w}{\partial y} + \frac{\partial v}{\partial z}\right)^2 + \left(\frac{\partial u}{\partial z} + \frac{\partial w}{\partial x}\right)^2 + \left(\frac{\partial v}{\partial x} + \frac{\partial u}{\partial y}\right)^2 \right\} \tag{7.4}$$

The left side of this equation gives the variation of the internal energy due to temperature variations and the work of the pressure forces. The right side gives the variation of the energy due to the influx of heat from heat conduction and from the work of the internal frictional forces.

Further, we shall assume that the coefficients of heat conduction and of viscosity vary with temperature according to the Sutherland formula

$$\frac{\lambda}{\lambda_0} = \frac{\mu}{\mu_0} = \frac{1 + (C/273 \cdot 1)}{1 + (C/\Theta)} \sqrt{\left(\frac{\Theta}{273 \cdot 1}\right)} \tag{7.5}$$

(Jeans, 1925) where λ_0 and μ_0 are the coefficients of heat conduction and viscosity at the temperature $\Theta = 273 \cdot 1°$ corresponding to zero in the Centigrade scale, and C is Sutherland's constant, which has the dimensions of a temperature. The constant C has a different value for different gases. For air $C \doteqdot 113°C$.

Besides these equations, we have the condition at infinity and the boundary conditions: at infinity in front of the body and in directions perpendicular to the body velocity, the fluid is at rest $|\bar{\mathfrak{v}}|_\infty = 0$ but the density and the absolute temperature have the assigned values ρ_0 and Θ_0.

The shape of the body surface is fixed, consequently, all the body dimensions are determined by the value of a certain characteristic dimension l.

The kinematic conditions of the flow are defined completely by the given value of the forward velocity of the body v and by the two angles α and β which the velocity vector forms with the axes of the coordinate system fixed in the body.

Finally, the boundary conditions for the temperature at the body surface must be given.† Let us assume that the body temperature is constant on the surface and equal to Θ_1.

---

† In many technical problems and in aerodynamics in particular, we encounter problems of fluid flow around a strongly heated or cooled body; for example, the flow around motor radiators, the investigation of the flow around cooled fuselages of aircraft, etc.

Equations (7.1)–(7.5), the condition at infinity and the boundary conditions, show that the system of characteristic parameters is

$$l, \alpha, \beta, v, \Theta_0, \Theta_1, \rho, \mu_0, R, c_v, J, \lambda_0, C.$$

The dimensions of the thirteen quantities listed are expressed in terms of five basic units of measurement L, T, M, Q and $C°$.

The number of parameters can be cut down by one if account is taken of the fact that the constants J, c_v and λ enter in (7.4) only as the products Jc_v and $J\lambda$. If the three parameters J, c_v and λ were to be replaced by the two parameters Jc_v and $J\lambda$, then the dimensions of all the quantities would be expressed in terms of only the four fundamental units L, T, M and $C°$; consequently, such a reduction in the number of parameters is not essential. The type of motion and the class of similar motions are determined by the values of eight independent nondimensional parameters

$$\frac{R}{Jc_v} = \gamma - 1, \quad \frac{\mu c_v}{\lambda} = P, \alpha, \beta, \quad \frac{\rho_0 l v}{\mu_0} = \mathbf{R}, \quad \frac{v}{\sqrt{(\gamma R \Theta_0)}} = \mathbf{M}, \quad \frac{\Theta_1}{\Theta_0}, \quad \frac{C}{\Theta_0}.$$

It is shown in the kinetic theory of gases that the constants γ and P depend only on the structure of the gas molecule. The following theoretical formulas are known for a perfect gas: $c_p/c_v = \gamma = 1 + (2/m)$; $\lambda/\mu c_v = (9\gamma - 5)/4$ (A. Euchen), where m is the number of degrees of freedom of the gas molecule. For a monatomic gas, $m = 3$; for a diatomic gas, $m = 5$. The experimental value for many gases (helium, hydrogen, and many others) is close to the theoretical values mentioned, but the influence of the temperature and pressure is noticeable when they vary over a wide range (Hilsenrath and Touloucian, 1954).† In practice, we can assume that for air at normal temperature and pressure

$$\gamma = 1 \cdot 4, \quad \frac{\lambda}{\mu c_v} = P^{-1} = 1 \cdot 9, \quad \frac{\mu c_p}{\lambda} = 0 \cdot 737.$$

The condition that motions of geometrically similar bodies be dynamically similar is that the above eight parameters be constant. Let us denote the total drag by W and the lift by A. For gases with identical constant values of the numbers γ and P we have

$$\frac{2W}{\rho l^2 v^2} = c_x = f_1\left(\alpha, \beta, \mathbf{R}, \mathbf{M}, \frac{\Theta_1}{\Theta_0}, \frac{C}{\Theta_0}\right),$$

$$\frac{2A}{\rho l^2 v^2} = c_y = f_2\left(\alpha, \beta, \mathbf{R}, \mathbf{M}, \frac{\Theta_1}{\Theta_0}, \frac{C}{\Theta_0}\right).$$

† Many references to investigations on this subject are contained here.

In addition to the above eight nondimensional parameters, it is necessary to introduce the nondimensional coordinates of the point in space x/l, v/l, z/l (to be definite, we assume that the coordinate system is fixed in the body) when studying the pressure distribution, temperature field, velocity field, etc. For example, for the pressure and temperature distribution we have

$$\frac{2p}{\rho_0 v^2} = f_3\left(\frac{x}{l}, \frac{y}{l}, \frac{z}{l}, \alpha, \beta, \mathbf{R}, \mathbf{M}, \frac{\Theta_1}{\Theta_0}, \frac{C}{\Theta_0}\right),$$

$$\frac{\Theta}{\Theta_0} = f_4\left(\frac{x}{l}, \frac{y}{l}, \frac{z}{l}, \alpha, \beta, \mathbf{R}, \mathbf{M}, \frac{\Theta_1}{\Theta_0}, \frac{C}{\Theta_0}\right).$$

The determination of the form of the functions f_1, f_2, f_3 and f_4 is the basic problem of experimental and theoretical aerodynamics.

The parameter C/Θ_0 appears to be essential only when heat conduction and the viscosity play a noticeable part, and then the phenomenon is characterized by noticeable temperature variations. But even when heat conduction and viscosity are important, the influence of the temperature on the coefficients of viscosity and heat conduction can usually be represented by the following formula (which does not contain the dimensional constant C) as a good approximation in place of formula (7.5)

$$\frac{\lambda}{\lambda_0} = \frac{\mu}{\mu_0} = \sqrt{\left(\frac{\Theta}{273 \cdot 1}\right)},$$

or by the more general formula

$$\frac{\lambda}{\lambda_0} = \frac{\mu}{\mu_0} = \left(\frac{\Theta}{273 \cdot 1}\right)^k,$$

where k is a certain arbitrary constant. These show that the parameter C/Θ_0 does not play an essential part in practice and, therefore, it cannot be taken into account in the simulation.

When the velocity of the body is small (small $\mathbf{M}$), the temperatures Θ_1 and Θ_0 only differ slightly and in such cases, the ratio Θ_1/Θ_0 is close to unity.

In the previous discussions, we assumed that the body temperature Θ_1 has a given value. If the transfer of heat from the body to the fluid can be neglected, then the boundary conditions on the body surface can be written

$$\frac{\partial \Theta}{\partial n} = 0.$$

This corresponds to the case when the body surface is non heat-conducting; under this condition, the parameter Θ_1/Θ_0 is eliminated. The same applies when heat conduction is neglected completely and adiabatic processes are considered.

Sometimes the parameter Θ_1/Θ_0 is eliminated because the temperature Θ_1 becomes a characteristic quantity. For example, in some cases, because of heat transfer from the fluid to the body, the body temperature builds up and attains a certain value different from the fluid temperature at infinity, independently of the thermal properties of the body. Such a case arises when measuring the temperature of a gas moving at very high speed with a liquid thermometer. The thermometer readings depend on **R** and **M** and on the method of installing the thermometer relative to the flow; the body (thermometer) temperature will differ from the undisturbed stream temperature far from the body.

When the above additional data are taken into account in the formulation of the problem, then it is evident that the constancy of the Reynolds and Mach numbers, the Mach number being essential only if compressibility is significant, is of special value for similarity of perturbed gas motions.

All the above conclusions can easily be extended to the case of propeller motion in a fluid. Besides the forward motion of the propeller, rotation about the axis also occurs in the propeller problem. Consequently, still one more parameter, the angular velocity, which can be defined by the number of revolutions of the propeller n per unit time, is added in the steady motion of a propeller with constant forward and angular velocities. The nondimensional parameter v/nl, which is called the advance ratio of the propeller, is added to the characteristics of the forward motion of the body in the propeller case. This parameter is fundamental when studying the aerodynamic or hydrodynamic properties of propellers.

If the fluid is ideal and incompressible, then the advance ratio is the single, nondimensional parameter defining the flow due to a propeller moving forward along its axis at a uniform rate.

8. UNSTEADY MOTION OF A FLUID

The conclusions of dimensional analysis which were obtained for steady motion in the previous paragraph, can be generalized to unsteady motion.

In the general case, when studying unsteady motion, we must introduce the time t, which is variable, into the series of characteristic parameters. In analysing dynamically similar motion we find that t varies due to scale variations and due to time variation during the motion. In this connection, we shall first make some general remarks about kinematically similar unsteady motions.

In such motions all the appropriate nondimensional combinations formed from the kinematic quantities are identical for all motions with kinematic similarity. The class of motions can be broadened to include kinematically dissimilar motions, if it is assumed that certain nondimensional kinematic parameters which characterize the whole motion, can assume different values for different motions.

Each particular motion and the type of kinematically similar motions as a whole, are defined by three parameters. Two parameters isolate the motion of the system as a whole and one parameter fixes the particular state of the motion.

A length characterizing the geometric dimensions and a characteristic time can be taken as the parameters defining the particular motion. These parameters determine the kinematic scale of the motion.

For example, it is natural to select the radius of the circle as the characteristic dimension in motion about a circle; and the amplitude of the oscillations of a certain point in oscillatory motion. It is natural to select the period as the characteristic time for periodic processes. The velocity or mean velocity can be taken instead of the characteristic time for a certain particular state.

The value of the time t determines the instantaneous state of a given unsteady motion.†

Let d, v and t be the characteristic length, the characteristic velocity and the moment of time under consideration. Similar, or in other words, corresponding states of the motion, for motions similar as a whole, are determined by the value of the nondimensional variable parameter vt/d which can be considered as a nondimensional time.

The relation $(v_1 t_1)/d_1 = (v_2 t_2)/d_2$, if satisfied for similar states of motion of two systems, can be considered as the time-conversion condition when changing from one system to the other.

As an example, let us consider the system of harmonic oscillations of a point on the arc of a circle of radius d. The law of motion is

$$s = a \cos kt$$

where s is the arc length. For kinematically similar motions, we must have

$$\frac{a}{d} = \text{const.}$$

The particular motion is determined by the amplitude a and by the frequency k. The state of the motion is determined by the time t.

---

† The time origin and the origin for linear coordinates are selected so that the position of the system and the state of the motion at $t = 0$ would be similar for different motions.

Similar states for different motions are characterized by identical values of the nondimensional parameter kt.

If ratios a_1/d_1 and a_2/d_2 have different values in two motions, then such motions are kinematically dissimilar. The ratio a_1/a_2 defines the length scale. For kinematically similar harmonic oscillations around a circle, we must have $a_1/a_2 = d_1/d_2$. It is impossible to choose a characteristic dimension for a straight line; consequently, any two harmonic motions in a straight line are kinematically similar.

Let us consider the rectilinear motion

$$s = vt + a \cos kt$$

representing uniform motion combined with a harmonic oscillation. In this case, the class of similar motions is characterized by a constant value of the nondimensional parameter $\mathbf{S} = (ak)/v$, called the Strouhaille number. Similar states of motion are characterized by the value of the parameter kt or by the value of the parameter $(vt)/a = (kt)/\mathbf{S}$.

The class of motions can be broadened and dissimilar motions can be considered by assuming that the Strouhaille number varies for different motions, while the basic law of the oscillations is retained.

Now let us consider the unsteady motion of a body in a fluid, assuming that the motion of the body itself is known. A certain length d and velocity v can be taken as the dimensional parameters defining a particular motion. By comparison with the steady motion case, the system of characteristic parameters in the unsteady case, with a given law of motion, is supplemented only by the value of the length d characterizing the law of motion and by a variable parameter, the time t. Consequently, the system of nondimensional parameters defining the motion as a whole and each state of the motion is supplemented only by the two parameters d/l and $(vt)/l$, where it is necessary to impose the condition $d/l = \text{const}$ for the two motions to be similar: the constancy of the parameter $(vt)/l$ only defines the appropriate value of the time (time scale) for similar states of motion.

If the unsteady motions are oscillations with a definite mode and frequency k which can be given arbitrarily, then the table of characteristic parameters is supplemented by the parameter k, with the consequent addition of the Strouhaille number

$$\mathbf{S} = \frac{kl}{v}$$

as a nondimensional parameter defining the motion.

Consider the combination of motion of a fluid with forward velocity and an oscillatory motion of fixed mode, but with variable frequency k.

The Strouhaille number must be kept constant to ensure the similarity of different motions, if k, l and v are given beforehand in terms of the data of the problem considered. If the frequency k is a specific quantity, then the condition of constant Strouhaille number is obtained as a consequence of the similarity conditions formed from the quantities assigned. In many cases we have to study unsteady motion of a body in a fluid when the body motion is not known beforehand. As an example, consider the problem of the oscillations of an elastic wing in an advancing stream of fluid (wing flutter).

For simplicity, let us assume that the wing, with a longitudinal plane of symmetry, is clamped rigidly along its centre section.

The elastic properties of wings composed of homogeneous material and of fixed structural design, are determined by two physical constants.

We omit the analysis of the practical methods of establishing the elastic properties of airplane wings. A similar and detailed formulation of the problem permits the elastic properties of various structures to be characterized by certain functions and parameters which are only essential from the viewpoint of the problem under consideration. Thus, classes of wings with equivalent elastic properties but different geometric properties and, in general, with different structures can be separated out.

In the general formulation of the problem, we can assume that all the elastic characteristics, within the limits of accuracy of Hooke's law, of each wing from the different systems of geometrically similar wings with equivalent elastic properties, are defined by the value of the characteristic dimension l, Young's modulus E and the shear modulus G.

The distribution of mass can influence the oscillations radically. The mass distribution law can be expressed so that the mass of each element would be proportional to the total mass and would be defined completely by the value of a certain mass m.

It is well known that the dynamic properties of an inelastic solid body are defined completely by the total mass of the body, by the position of the centre of gravity and by the inertia tensor at the centre of gravity of the body.

A detailed analysis in the practical approximate, formulation of the problem shows that elastic wings with different mass distributions can, just as in the case of an inelastic body, be dynamically equivalent.

In practice, a system of abstract parameters characterizing the distribution of mass is established, determining the oscillatory motions which arise (the position of the centres of gravity of the various wing sections, the section inertias, etc.). All subsequent conclusions can be extended to the case of different, but dynamically equivalent, mass distributions for an elastic wing.

Let us assume that the fluid is ideal, homogeneous and incompressible in the phenomenon under consideration. The mechanical properties of the fluid are determined completely by the density ρ in this case. The effect of gravity on the fluid cannot be taken into account. Furthermore, let us assume that the fluid is infinite in extent. The wing centre-section is fixed. The fluid at infinity has a forward velocity v which is constant in time and parallel to the longitudinal plane of symmetry of the wing, and the velocity v has a fixed direction relative to the fixed wing section.

Summarizing, we find that the unsteady motion of the wing-fluid system is determined, within the limits of our assumptions, by the parameters

$$l, E, G, m, \rho, v$$

and the quantities giving the initial disturbance. Moreover, each separate state of motion is determined by the value of the time t.

Those properties of disturbed unsteady motions which are unaffected by the nature of the initial disturbances, are evidently independent of the parameters defining these disturbances.

Therefore, these properties are determined, for dynamically similar motions, by the system of nondimensional parameters

$$\frac{\rho v^2}{E}, \frac{G}{E}, \frac{m}{\rho l^3}, \frac{vt}{l};$$

where it is understood that the mass distribution and the wing structure are identical in different cases.

The last property can be considered in a broadened sense and only the equivalence of the elastic and dynamic properties of the wings need be required while retaining the geometric similarity of the external wing surfaces adjacent to the fluid.

It is evident that the properties characterizing the motion as a whole are independent of the parameter $(vt)/l$, whatever the separate states of each motion.

Experiment shows that steady flow about an elastic wing can be stable or unstable.

The displacements due to an oscillatory disturbance of a wing in a moving stream decrease in stable motion. An increase in displacements occurs in unstable flow with possible wing failure as a result.

Stability and instability of the streamlines for sufficiently small disturbances can be considered as properties independent of the initial conditions and of the separate states of motion. Consequently the properties of stability of motion must be defined by the system of parameters

$$\frac{\rho v^2}{E}, \frac{G}{E}, \frac{m}{\rho l^3}.$$

In the usual formulation of the problem, it is impossible to divide the set of all motions into motions with increasing and decreasing amplitudes; cases are possible when maximum deflections are either constant or variable during the oscillations but, in practice, they retain small enough values for any t.

If the motions with increasing and decreasing amplitudes can be clearly separated, then the boundary between these regimes is defined by the relation

$$F\left(\frac{\rho v^2}{E}, \frac{G}{E}, \frac{m}{\rho l^3}\right) = 0,$$

which can be written

$$v_{\mathrm{cr}} = \sqrt{\left(\frac{E}{\rho}\right)} f\left(\frac{G}{E}, \frac{m}{\rho l^3}\right).$$

This formula defines the critical flutter speed. The critical value of the velocity separates the stable and unstable flow regions for variable undisturbed stream velocity, keeping the rest of the parameters constant.

The wing stiffnesses are proportional to E and G. An increase of E and G by a factor of n is equivalent to an increase in the stiffness coefficients by a factor of n. It is seen from the formula for v_{cr} that the critical velocity will be increased by a factor of $\sqrt{n}$, when a wing stiffness is increased by a factor of n, if the wing mass, shape and dimensions, remain the same.

The Strouhaille number $(kl)/v_{\mathrm{cr}}$ corresponding to the critical state is defined by the values of the abstract parameters

$$\frac{G}{E} \quad \text{and} \quad \frac{m}{\rho l^3}$$

which also define all the abstract quantities independently of the initial data characterizing the critical motion.

9. SHIP MOTION

Methods of dimensional analysis and similarity are of great practical value in numerical and experimental solutions of problems of motion on a water surface.

Consider the steady, rectilinear forward motion of a ship over the surface of a semi-infinite fluid which is at rest at large depth and at a large distance ahead of the ship. The motion of a floating body causes a disturbance to the free surface. The disturbed fluid motion is of wave type dependent on the effect of gravity.

We take into account the important effects of density ρ, the gravity g and the viscosity μ of the water. The compressibility of the water is of no practical significance, and is neglected. Capilliarity is also not essential to the motion of ordinary ships.

The dimensions and shape of the ship hull have a big effect on the principal mechanical characteristics. Let us first analyse the motion of a ship with a definite hull shape. All the geometric dimensions are determined by the value of the ship length L. Geometrically similar hulls correspond to different values of L. For ordinary, heavy ships, it can be assumed that the total weight completely determines the hull position relative to the water. Evidently, the position of the ship relative to the water influences the drag, etc. Consequently, we take the weight or the displacement of the ship A as a characteristic parameter.† Instead of the displacement A in tons, the volume displacement D, in cubic metres, can be taken since $A = \rho g D$, where ρ is the water density. For fresh water, we have

$$\rho g = 1 \text{ ton}/m^3.$$

Let us denote the speed of the ship by v. The system of characteristic parameters will be‡

$$L, \ D, \ \rho, \ g, \ \mu, \ v.$$

All the geometrical and mechanical quantities, for example, the wetted area§ S, the drag W, etc., can be considered as functions of these parameters. Dynamically similar motions and the state of each motion are defined by the three nondimensional parameters

$$\psi = \frac{L}{\sqrt[3]{D}} \qquad \text{(fineness coefficient)}\|$$

$$\mathbf{F} = \frac{v}{\sqrt{(gL)}} \quad \text{(Froude number)}$$

$$\mathbf{R} = \frac{\rho v L}{\mu} \quad \text{(Reynolds number)}$$

† If different motions occur for similar ship locations relative to the water level, then the weight is proportional to L. It is sufficient to take one of the parameters L and A in this case.

‡ The motion occurs at the interface between the water and the air, consequently, the air density and viscosity must enter as characteristic parameters (air compressibility is not significant at the usual speeds). However, these parameters exert slight influence on the phenomenon and taking them into account does not alter the subsequent conclusions since here the nondimensional quantities ρ'/ρ and μ'/μ are added, which can be considered as constants for every class of motion.

§ The quantity S differs slightly from the magnitude of the wetted area in the static state, which is determined by only two parameters L and D.

‖ The fineness coefficient can be considered as the length of a geometrically similar hull with a one ton displacement.

Hence, we can write for the drag

$$W = f(\psi, \mathbf{F}, \mathbf{R})\, \rho S v^2. \qquad (9.1)$$

The similarity criteria are

$$\frac{L_1}{\sqrt[3]{D_1}} = \frac{L_2}{\sqrt[3]{D_2}}, \quad \frac{v_1}{\sqrt{(gL_1)}} = \frac{v_2}{\sqrt{(gL_2)}}, \quad \frac{v_1 L_1 \rho_1}{\mu_1} = \frac{v_2 L_2 \rho_2}{\mu_2}.$$

It is not difficult to see that if an identical fluid is used both for the model motion and the actual motion, then $\mu_1/\rho_1 = \mu_2/\rho_2$ and simulation of the phenomenon is impossible. Actually, since the Froude number is constant the velocity must decrease as the linear dimensions of the ship decrease and since the Reynolds number is constant it follows that the velocity must increase as the linear dimensions of the ship decrease. Consequently, complete similarity is not maintained when simulating this phenomenon with a change of length scale and the magnitude of the drag coefficient of the model is not equal to the magnitude of this coefficient in the actual motion.

The determination of ship drag using model experiments is based on the practical possibility of separating the drag into two components: one is determined by the viscosity and the other is determined by gravity. It appears that formula (9.1) can be replaced approximately by the following:

$$W = W_1 + W_2 = c_f(\mathbf{R}) \frac{\rho S v^2}{2} + c_w(\psi, \mathbf{F})\, \rho g D. \qquad (9.2)$$

The form of (9.2) can be established by theoretical arguments which we shall not touch upon here.

The drag $W_1 = c_f \rho (S v^2/2)$ is called frictional drag. Frictional drag for the model motion and for the actual motion is determined by a calculation based on various semi-empirical formulas. The value of the coefficient of frictional drag is determined by the Reynolds number $\mathbf{R}$. Moreover, this coefficient depends on the roughness and, to a certain degree, on the shape of the ship's hull contour. The coefficient of frictional drag decreases as the Reynolds number increases. In practice, the coefficient c_f is taken equal to the corresponding coefficient on a flat plate. Experimental results on flat plates are given in Fig. 6.

The value of the coefficient c_f for smooth plates is determined according to the Prandtl-Schlichting formula

$$c_f = \frac{0 \cdot 455}{(\log \mathbf{R})^{2 \cdot 58}}. \qquad (9.3)$$

The value of the coefficient c_f for a rough plate will be considerably higher.

The drag W_2 is called the residual drag. The coefficient $c_w = W_2/(\rho g D)$ gives the residual drag per ton displacement. This coefficient can be determined experimentally, by testing geometrically similar models while complying with the following conditions

$$\frac{L_1}{\sqrt[3]{D_1}} = \frac{L_2}{\sqrt[3]{D_2}} \quad \text{and} \quad \frac{v_1}{\sqrt{(gL_1)}} = \frac{v_2}{\sqrt{(gL_2)}}.$$

These conditions are the Froude similarity laws.

The residual drag depends on the hull shape. When studying the influence of the hull shape, it is necessary to broaden the class of motions and to study the motion of a family of hulls the shapes of which depend on several geometric parameters.

In practice, it is very important to pick out the geometric parameters, which have a significant effect on the residual drag. Experiments show that the basic parameters for all possible geometrically dissimilar ship hull contours of the usual types which determine the coefficient c_w, are the Froude number and the fineness coefficient. The volume-displacement Froude number

$$\mathbf{F}_D = \frac{v}{\sqrt{(g \sqrt[3]{D})}} = \mathbf{F}\sqrt{\psi}$$

can be taken instead of the linear Froude number $\mathbf{F} = v/\sqrt{(gL)}$.

The graph of Doyère (Doyère, 1927) giving averaged experimental data on the dependence of the coefficient c_w on ψ and F_D for hulls without projecting parts (rudders, cantilevers for the propeller shafts, etc.) is shown on Fig. 15. Using the Doyère graph and the value of the coefficient of friction as a function of the Reynolds number, the ship hull drag can be calculated easily as a function of the velocity. This computation often gives very good results as a first approximation.

It appears that the following parameters are the most important in making a more exact determination of the residual drag†:

$$\chi = \frac{D}{L\kappa} \quad \text{and} \quad \frac{B}{T},$$

where κ is the midships area, B is the hull width and T is the draught. The coefficient χ is called the prismatic fullness coefficient.

Besides the fundamental parameters mentioned, parameters determining the influence of various quantities characterizing the midships shape, the bow, the stern, etc., on the drag can also be taken into account.

---

† Experimental results on the influence of $\mathbf{F}$, ψ, χ, and B/T on the residual drag for a certain series of hulls is contained in Taylor (1933).

When ship motion is studied, the tow drag of the ship without propellers W' and the drag W'' with rotating propellers, which cause a thrust producing motion, is considered. The drag W in the last case equals the

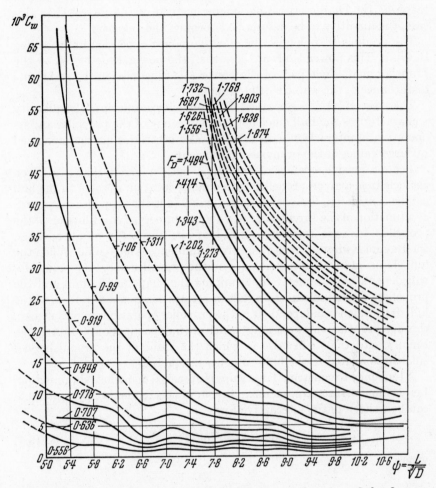

FIG. 15. Residual drag per ton displacement as a function of the fineness coefficient ψ and the Froude number $\mathbf{F_D}$ corresponding to the displacement.

horizontal component of the propeller pull in uniform steady motion. The drag W' is a hull characteristic independent of the motor properties. Because of the interaction between the hull and the propeller, the inequality $W' < W''$ usually holds.

3

If the hull, the location, velocity, and displacement of the propeller, and all dimensions are given, then the drag is determined and so are the required number, n, of propeller revolutions and the power E needed at the propeller shaft. The power E or the number of revolutions n can be taken as the characteristic parameter instead of the velocity; in this case, the quantity to be determined becomes the velocity.

The product $W''v$ defines the effective power expended in propelling the ship. This power is always less than the power E delivered to the propeller shaft since part of the power E is consumed in the additional disturbance of the water by the propeller.

The ratio $(W'v)/E = \eta$ is called the propulsive coefficient. The quantity η characterizes the hull efficiency, the efficiency of the propeller and its operation in interaction with the hull. The best ships are characterized by large values of the propulsive coefficient.

The propulsive coefficient for ships with a given shape, a given coefficient ψ, a given propeller with definite location relative to the hull can be considered, for a constant ratio d/L (d is the propeller diameter), as a function of the Froude and Reynolds numbers or as a function of the propeller advance ratio $v/(nd)$ and of the Reynolds number; when the Reynolds number varies slightly its influence is negligible. When the hull shape and the geometric data on the propeller vary, the value of η will depend on parameters determining the shape of the hull and the propeller. These influences are felt in certain cases in a variation of the hull drag which is independent of the propeller operation; in other cases, through propeller characteristics which are independent of the hull shape. Finally, situations can arise in which the value of the propulsive coefficient is related to the hull-propeller interaction.

A tendency to construct large ships is observed in modern shipbuilding. Let us demonstrate arguments in favour of this tendency. Formula (9.2) can be written in the alternative form

$$W = [c_f(\mathbf{R}) + c_w'(\psi, \mathbf{F})]\rho \frac{Sv^2}{2}, \qquad (9.4)$$

where

$$c_w' = 2c_w \frac{D}{LS} \cdot \frac{1}{\mathbf{F}^2}.$$

Take two geometrically similar ships with displacements proportional to the cube of the linear dimension:

$$\frac{D_1}{L_1^3} = \frac{D_2}{L_2^3},$$

which is a natural assumption about the similarity of the submerged parts. Evidently the wetted area S is proportional to the square of the linear dimension.

Let L_1 and L_2 be the corresponding characteristic lengths, where $L_2 > L_1$. If the velocities are identical, we have

$$\mathbf{R}_2 > \mathbf{R}_1 \quad \text{and} \quad \mathbf{F}_2 < \mathbf{F}_1.$$

It is known from experiment that the coefficient c_f decreases as the Reynolds number increases (see formula (9.3)) and that the coefficient c_w' also decreases as the Froude number decreases (at least, in the range of small Froude number values $\mathbf{F} < 0\cdot5$ common in practice). A typical variation of c_w' with Froude number is shown in Fig. 16.

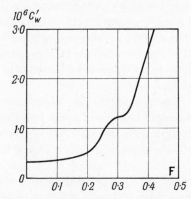

FIG. 16. Typical curve for the residual drag coefficient c_w' as a function of the Froude number $\mathbf{F}$.

If the motion occurs at the same velocity, then the drag ratio W_1/W_2 equals the ratio of the powers or the ratio of fuel consumption per unit time, if the efficiencies are identical. The value of the transportable load is proportional to the displacement, i.e. to the cube of the linear dimension. The cost of transporting one ton is defined by the ratio of the weight of the fuel consumed to the weight of the transported load. The ratio of the costs Q_2 and Q_1 of transporting one ton a distance of one kilometre is given by the following formula if the velocities are identical:

$$\frac{Q_2}{Q_1} = \frac{W_2 L_1^3}{W_1 L_2^3}.$$

The last ratio is one of the most important elements in the economics of ship transportation. Using formula (9.4), we can write

$$\frac{Q_2}{Q_1} = \frac{c_f(\mathbf{R}_2) + c_w'(\mathbf{F}_2)}{c_f(\mathbf{R}_1) + c_w'(\mathbf{F}_1)} \cdot \frac{L_1}{L_2} = \kappa \cdot \frac{L_1}{L_2}, \tag{9.5}$$

where κ is a certain quantity less than unity and decreasing as the ratio L_2/L_1 increases.

Formula (9.5) shows that the quantity Q_2 decreases more rapidly than the first inverse power in the ship dimensions. The ratio of the powers increases more slowly, for the same velocities, than the ratio of lengths squared.

By analogous reasoning, it can be shown that if the power increases in proportion to the cube of the linear dimension, then the velocity is increased and the sailing time and the cost of transporting a ton load one kilometre decrease.

The above considerations can be applied not only to ships moving over the surface of the water but also to airplanes, since the air drag increases in proportion to the square of the length at a fixed flight speed and the airplane weight and the useful load increase approximately in proportion to the length cubed. The fuel capacity and the range of airplanes therefore increase with their size: this explains the increase in the size and weight of airplanes intended for long-range flights.

In addition, the question of the suitable dimensions for aviation motors, hydraulic machines, etc., is of great practical interest.

The ratio of the weight of a reactive motor to the thrust it develops (weight-thrust ratio) is a most important characteristic of the motor for use in airplanes.

It is more advantageous to use several motors of small dimensions in order to obtain a given thrust, than to use one motor of large size since the weight-thrust ratio of the motor increases as its size increases. For the thrust increases proportionally to the length squared but the motor weight is approximately proportional to the length cubed.

Hence, if the weight-thrust ratio of the motor, the use of expensive material and actual operation are kept in mind, then it is more advantageous to construct several small motors than one large one. These arguments become less valid for motors of very small size since mechanical similarity is lost when the size decreases sharply, hence the thrust and useful power decrease very rapidly.

Besides these general qualitative considerations of similarity theory, the choice of the suitable size of a motor and of hydraulic machines is also limited, by the properties and dimensions of the auxiliary automatic mechanisms, by economic, technological, structural and certain other requirements which must be taken into account in making a final decision.

The questions of the logical dimensions of aviation engines, hydraulic turbines and many other machines, must be analysed and studied thoroughly. Similarity considerations are of great value in this analysis.

10. PLANING OVER THE WATER SURFACE

Planing can be visualized as sliding over the water surface. The supporting force during planing is specified almost entirely by the dynamic reaction of the water. When displacing ships move, the supporting force is visualized, just as at rest, as an Archimedean force specified by the increase of the hydrostatic pressure with depth.

The planing principle is used in high-speed ships such as modern high-speed torpedo cutters. The run of a seaplane in water at take-off and the run after landing are achieved by planing.

The planing of a ship of given shape can occur for different orientations of the ship relative to the water surface. The ship's orientation relative to the water is of key importance.

The number of parameters defining the motion of a hydrofoil or seaplane with given geometric shape, is larger than in the case analysed above of the motion of a ship which displaces water. In planing, besides the immersion or the wetted area, the angle of incidence θ (the angle a certain direction fixed in the ship makes with the horizontal) must also be given. The loading due to water pressure Δ, the position of the centre of gravity of the ship and the moment of the external forces about the centre of gravity can be given instead of the wetted area and the angle of incidence θ, but not the hydrodynamic and aerodynamic forces. In practice, it is convenient to select the loading in the water and the angle of incidence as the decisive quantities.

If we formulate the planing problem in the same way as the problem of motion of a ship with displacement, we find that steady planing of a ship of given geometric shape is determined by the following system of parameters

$$B, \Delta, \theta, v, \rho, g, \mu.$$

The set of dynamically similar motions and all the nondimensional combinations formed from the various mechanical quantities are defined by the values of the nondimensional parameters

$$\theta, \quad \frac{\Delta}{\rho g B^3} = C_\Delta, \quad \frac{vB\rho}{\mu} = \mathbf{R}, \quad \frac{v}{\sqrt{(gB)}} = \mathbf{F}. \tag{10.1}$$

By comparison with the motion of ships which displace water the angle of incidence θ, which can have different values in the motions being compared, must be introduced. The difference in the angles of incidence for such ships can be neglected in all cases of practical interest.

The effect of weight of water is expressed by means of parameters containing the acceleration due to gravity. The two parameters containing g in (10.1) are C_Δ and $\mathbf{F}$. This system can be replaced by the system

$$\theta, \quad C_B = \frac{2\Delta}{\rho B^2 v^2} = \frac{2C_\Delta}{\mathbf{F}^2}, \quad \mathbf{R}, \quad \mathbf{F}, \tag{10.2}$$

in which the acceleration due to gravity enters only in the Froude number $\mathbf{F} = v/\sqrt{(gB)}$.

The planing phenomenon has a definite impulsive character. In front of a planing boat, the water is practically at rest, then the water is set into motion by the approaching bottom of the boat in a short interval of time. This justifies the assumption that the inertial forces are the main forces compared to which the forces due to the weight of water particles are small and can be neglected.

The assumption that the weight of the water can be neglected is equivalent to the assumption that the parameter g and, therefore, the Froude number, can be neglected, in the system of parameters θ, C, R and $\mathbf{F}$.

The unimportance of the influence of the weight of the water on a number of fundamental characteristics of motion in the planing regime has been established theoretically (Sedov, 1937; Kochin, 1938; Chaplygin, 1940). This is confirmed by a great deal of experimental material (Epshtein, 1940; Sedov and Vladimirov, 1941a, b). Consequently, the Froude similarity law is inadequate for simulation of purely planing motions.

It is incorrect to say that both parameters involving Froude number $\mathbf{F}$ and the coefficient C_Δ can be neglected in the absence of the influence of the fluid weight. The parameters C_Δ and $\mathbf{F}$ form the combination $C_B = 2C_\Delta/\mathbf{F}^2$ which does not contain g. The parameter C_B can play an important part in a number of problems in which the weight of water and the parameter g are totally inessential.

The nondimensional quantities, apart from the Froude number, can depend on the loading Δ and on the velocity v only through the combination $C_B = 2\Delta/(\rho B^2 v^2)$. Consequently, the investigation of the influence of the velocity can be replaced by an investigation of the influence of the loading, and conversely. This result is of great value when the velocities attainable are inadequate.

Using this, we can obtain experimentally a number of results which are valid for cases which cannot be covered by direct experiment.

The viscosity property of the water is felt notably only directly at the bottom of the ship (boundary layer) consequently, the Reynolds number does not influence the pressure distribution, the moment of the hydrodynamic forces, the shape of the wetted surface, etc., substantially. The influence of the viscosity of the water on damping disturbances is only felt in practice at very large distances from the ship.

The drag force depends essentially on the frictional force on the bottom surface; consequently, viscosity and the Reynolds number influence the nondimensional drag coefficient.

The hydrodynamic characteristics of steady planing depend on the geometric shape of the ship hull to a strong degree as well as on the mechanical parameters mentioned.

The difference in the nature of the supporting forces causes a distinct difference in the shape of the planing ships from the shape of displacing ships. The outlines of the hull of planing ships are characterized by the flat-bottomed shape of the hull, by the abruptly drawn cheek-bones and by the presence of projections on the bottom of the boat. The flat-bottomed shape is necessary to absorb the large vertical forces on a small wetted surface. During planing the sharp cheek-bones and projections cause the breakaway of jets of water; consequently, the side surface of the boat and a significant part of the lower section of the bottom are not wetted by the water, which decreases the frictional force.

The various peculiarities of the bottom can be characterized by certain nondimensional parameters. The influence of the variation of these parameters can be determined by systematically testing a series of profiles.

In the problem of the planing plate shaped like a plane wedge, we encounter a very interesting property, the substance of which is closely related to dynamical similarity and dimensional analysis. Consider a plane-keeled, prismatic plate planing on the surface of the water. We suppose that the keel of the plate possesses a vertical plane of symmetry and that the motion is parallel to this plane. The rear part of the plate is a plane perpendicular to the plane of symmetry. Let us consider the case when the plate length and the width of the side-piece of the wedge are large enough so that the boundaries of all the wetted surfaces for comparable motions are not related to the structural width and length of the plate. The geometrical width and length of the plate can be assumed to be infinite for all comparable motions. The geometric shape of the plate is determined completely by the angle between the side-pieces $\pi-2\beta$ (β is the careening angle) and by the angle between the keel and the plane surface. These angles can be taken as the geometric shape parameters. For simplicity, we shall analyse a class of motions in which these angles are fixed.

It is not difficult to see that the number of characteristic parameters is reduced in this case since the linear dimension characterizing the plate is absent.

The steady planing of a careening plate with an incompletely wetted width is defined by the parameters

$$\Delta, \theta, v, \rho, g, \mu.$$

The wetted width of the plate along the rear part and the wetted length along the keel are defined completely by the parameters mentioned.

The class of similar motions and the state of motion are characterized by three nondimensional quantities:

$$\theta, \quad \frac{v}{\sqrt{\left[g \sqrt[3]{\left(\frac{\Delta}{\rho g}\right)} \right]}} = \mathbf{F}_1, \quad \frac{\rho v \sqrt[3]{\left(\frac{\Delta}{\rho g}\right)}}{\mu} = \mathbf{R}_1, \tag{10.3}$$

The numbers $\mathbf{F}_1$ and $\mathbf{R}_1$ can be considered as the Froude and Reynolds numbers associated with the loading.

The Reynolds number for the planing regime (larger $\mathbf{F}_1$) only affects quantities which depend on the nature of fluid motion in a boundary layer. In particular, it can be assumed that the wetted length along the keel l is independent of viscosity, i.e.

$$l = f(\theta, \Delta, v, \rho, g). \tag{10.4}$$

This relation can be written in nondimensional form,

$$l = \sqrt{\left(\frac{\Delta}{\rho v^2}\right)} f(\theta, \mathbf{F}_1). \tag{10.5}$$

If it is assumed that the effect of weight does not influence the magnitude of l, i.e. that the acceleration g in (10.5) is not essential, then we obtain

$$l = \sqrt{\left(\frac{\Delta}{\rho v^2}\right)} \cdot f(\theta). \tag{10.6}$$

Under these assumptions, it is evident that all the linear dimensions (for example, the wetted width along the rear end, etc.) are proportional to $\sqrt{(\Delta/\rho v^2)}$. The wetted geometric surfaces will be similar for fixed angle of incidence θ but with different loads on the water Δ and different planing velocities v.

Experiments confirm the validity of these conclusions for large values of the Froude number when $\mathbf{F}_1 > 2$ (Figs. 17, 18, 19) (Sedov and Vladimirov, 1941a).

We took the load on the water Δ and the velocity of the motion v as natural parameters in the system of characteristic parameters since

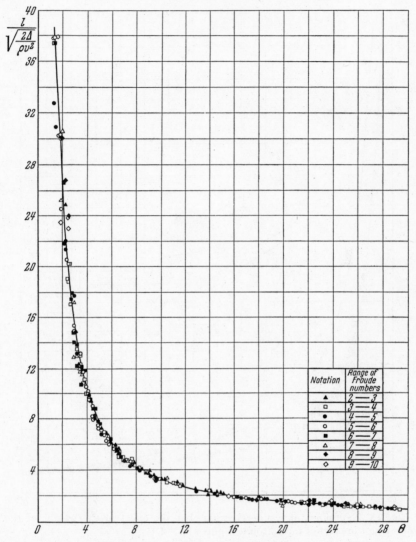

FIG. 17. Gliding of a careening plate.

precisely these quantities are given beforehand in experiments while the rest vary. The wetted length l or the wetted width along the rear end b can be taken as decisive parameters instead of the load Δ. For example, the following system of dimensional characteristic parameters can be taken

$$\theta, \, l, \, v, \, \rho, \, g, \, \mu$$

3*

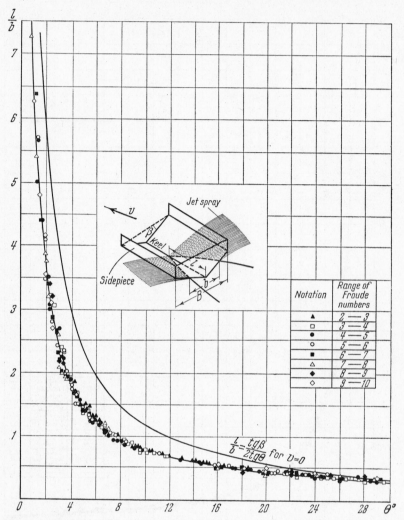

FIG. 18. Experimental results showing that the effect of the weight of the water is not important for a Froude number $F_1 > 2$.

and the corresponding system of nondimensional parameters is

$$\theta, \quad \frac{v}{\sqrt{(gl)}}, \quad \frac{\rho v l}{\mu}.$$

In this case, the relations (10.4) and (10.6) can be considered as relations determining the load on the water Δ.

11. IMPACT ON WATER

The phenomenon of impact on water is encountered in a number of cases; in particular, in the landing of a seaplane on water. Of particular interest in the study of this phenomenon are the explanation of the properties of the reactive force of the water and the investigation of ricocheting on water.

Many readers will have probably observed the skimming of flat stones on water. If the stones are thrown with velocities having a large horizontal component and with rotation which guarantees the conservation of a small angle between the plane of the stone and the horizontal, they

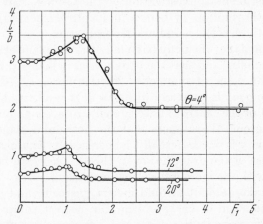

FIG. 19. Gliding of a careening plate. Effect of weight important for Froude numbers $F_1 > 2$.

easily recoil upward upon contact with the water and may bounce several times. Obviously, the horizontal velocity plays a fundamental part in this phenomenon of ricocheting. A flat heavy stone cannot recoil from the water if the horizontal velocity is lacking. Multiple ricochets testify to the small loss in horizontal velocity during contact with the water. The ricocheting of shells is another well-known phenomenon. For example, a circular cannonball of $0 \cdot 16$ m. diameter with an initial velocity of 455 m./sec. can perform more than twenty-two ricochets in water (Jonquières, 1883).

At the present time, ricochet firing is sometimes used deliberately in artillery.

The phenomenon of water ricochet can occur during the landing of a seaplane on water. Water ricochet of an airplane is a source of anxiety in the air transport industry and is considered to be an extremely undesirable phenomenon.

Let us use dimensional analysis and similarity theory to analyse the problem of impact on water as applied to seaplane landings and to experiments with plates—simulated boat models and airplane floats.

We suppose that the geometric shape of the moving body is fixed. The scale, however, which can be chosen as the value of a certain linear dimension, is kept variable. The width of the boat is naturally taken as the characteristic linear dimension. For simplicity, we limit ourselves to the case when the body is absolutely rigid. Differences in aerodynamic behaviour can be reflected strongly in the course of the whole phenomenon. An experimental investigation of the problem can start with an analysis of the motion of bodies of identical geometric shape and, therefore, with identical aerodynamic properties.

Let us assume that the body has a longitudinal plane of geometric and dynamic symmetry and that the body moves in a longitudinal direction parallel to the plane of symmetry. Furthermore, we assume that the motion of the body is unsteady with two degrees of freedom: a vertical displacement of the centre of gravity and a rotation about the centre of gravity can occur. We suppose that the horizontal plane through the centre of gravity is fixed. Actually, a decrease in the horizontal velocity occurs because of the drag; however, this decrease can be neglected in comparison with the more interesting processes.†

We shall take account of fluid inertia, viscosity and gravity in the general formulation of the problem. We neglect compressibility and capillarity. The wave motion of the water can exert substantial influence on the phenomenon being studied but we shall assume that the water adjacent to the body is at rest.

From the assumptions made, it follows that the motion of the body-water system is defined by the following system of parameters which can be given arbitrarily within known limits:

(1) The scale parameter: the boat width B (the plate width in the case of a model).

(2) The kinematic parameters: the instant of time under consideration t (the initial instant $t = 0$ corresponds to the moment the body touches the water surface), the horizontal velocity U and the initial vertical

---

† It is not essential to assume that the horizontal velocity is constant in order to use dimensional analysis in the general formulation of the water impact problem. The necessity to fix the horizontal velocity arises in simulating this phenomenon. Constancy of the horizontal component can be guaranteed under laboratory conditions by the motion of a tow-cart; the model centre of gravity can slide along the vertical guide fastened to the cart which moves with a given constant velocity. The frictional forces arising in the guiding instrumentation can be made negligibly small.

velocity v_0, the initial angle of incidence θ_0, the initial angular velocity Ω_0.†

(3) The dynamic parameters of the body: the coordinates of the centre of gravity ξ, η in a certain coordinate system fixed in the body; the moment of inertia J relative to the transverse axis passing through the centre of gravity; the mass m; the vertical component of the given external forces A ($A = mg$ for a free body). Under laboratory conditions, the quantities A and mg can be made independent by using an artificial counterpoise.‡

(4) Physical constants: acceleration due to gravity g, density ρ and viscosity of the water μ.

All the nondimensional quantities related to our problem are functions of the following system of nondimensional parameters defining the regime and the state of the motion:

$$\tau = \frac{Ut}{B}, \frac{v_0}{U}, \theta_0, \frac{\Omega_0 B}{U}, \frac{\xi}{B}, \frac{\eta}{B}, \frac{J}{\rho B^5}, \frac{m}{\rho B^3}$$

$$C_B = \frac{2A}{\rho B^2 U^2}, \quad \mathbf{F} = \frac{U}{\sqrt{(gB)}}, \quad \mathbf{R} = \frac{\rho U B}{\mu}.$$

The influence of the Reynolds number $\mathbf{R}$ is felt through the frictional forces depending on the viscosity of the water, which are not generally large in comparison with the lift and which are usually directed approximately in a horizontal direction; however, in certain cases the frictional forces can influence the magnitude of the rotational moment significantly. But if the comparatively weak dependence of the friction on the Reynolds number is taken into account, then, apparently, it is completely valid to neglect the influence of the Reynolds number on the characteristics of the vertical and angular motions and, in particular, on the phenomenon of water ricochet.

The influence of the Froude number $\mathbf{F}$ on the hydrodynamic forces, the shape of the wetted surface, etc., is related to the influence of the weight of the water on the perturbed motion of the water near the body. At high horizontal speeds, the phenomenon is of an impact character, consequently, the water reaction forces can be considered to be independent

---

† We assume that the initial state of the motion at the moment of landing is completely defined in practice by the parameters

$$U, v_0, \theta_0, \Omega_0.$$

‡ We assume that the external, additionally assigned, forces are applied at the centre of gravity. The total aerodynamic forces and moments can be considered to be quantities defined by the shape and motion of the body.

of the Froude number. Also it should be remembered that the minimum value of Froude number, at which the influence of this number begins to be insignificant, depends on the character of the mechanical quantities under consideration and is related to the values of the other character-istic parameters.

If $A = mg$, then the parameters $m/(\rho B^2)$, C_B and $\mathbf{F}$ become dependent. In this case, it is sufficient to retain only the two parameters $m/(\rho B^2)$ and C_B since the influence of the weight of the model and the water is taken into account simultaneously by the C_B coefficient.

Let us denote the vertical component of the total hydrodynamic forces by Y. From the above, we deduce the following formula

$$\frac{Y}{A} = f\left(\tau, \frac{v_0}{U}, \theta_0, \frac{\Omega_0 B}{U}, \frac{\xi}{B}, \frac{\eta}{B}, \frac{J}{\rho B^5}, \frac{m}{\rho B^3}, C_B\right). \tag{11.1}$$

The parameters ξ/B, η/B, $J/(\rho B^5)$ and $m/(\rho B^3)$ are constant for motions of a specific model. If the angle of incidence is fixed and only the vertical velocity is variable, then we have motion with only one degree of freedom, progressive vertical motion. In this case $\Omega_0 = 0$ and the parameters ξ, η, J are insignificant; (11.1) becomes

$$\frac{Y}{A} = f\left(\tau, \frac{v_0}{U}, \theta, \frac{m}{\rho B^3}, C_B\right). \tag{11.2}$$

The angle of incidence is retained in this formula as a constant which can vary in different experiments.

The parameter τ is defined if we consider the maximum value of the ratio Y/A in different experiments, or in general, the value of Y/A for certain characteristic times

$$\frac{Y_{\max}}{A} = f^*\left(\frac{v_0}{U}, \theta, \frac{m}{\rho B^3}, C_B\right). \tag{11.3}$$

It is evident that the average values of all the quantities after the period of contact with the water, the nondimensional time $\tau_1 = (Ut_1/B)$, the nondimensional maximum depth of submersion $h_{\max}/B$, etc., are not related to the parameter τ.

The whole set of motions can be subdivided into two parts corre-sponding to the cases when the body does or does not emerge from the water (the presence or absence of ricochets). The boundary between these two regions is defined by the relation

$$\Phi\left(\frac{v_0}{U}, \theta_0, \frac{\Omega_0 B}{U}, \frac{\xi}{B}, \frac{\eta}{B}, \frac{I}{\rho B^5}, \frac{m}{\rho B^3}, C_B\right) = 0, \tag{11.4}$$

and in the case of forward motion with one degree of freedom, by the relation

$$\Phi\left(\frac{v_0}{U}, \theta, \frac{m}{\rho B^3}, C_B\right) = 0. \tag{11.5}$$

At the present time experimental data exist on the form of the function for a plane plate landing on water (Sedov, 1942).

In the problem of a wedge (with a fixed angle) of the kind described in the preceding paragraph making impact with the water, the following formulas hold in addition to (11.3) and (11.5):

$$\frac{Y_{\max}}{A} = f^*\left(\frac{v_0}{U}, \theta, \frac{2A}{\rho^{1/3} m^{2/3} U^2}\right) \tag{11.6}$$

and

$$\Phi\left(\frac{v_0}{U}, \theta, \frac{2A}{\rho^{1/3} m^{2/3} U^2}\right) = 0. \tag{11.7}$$

The equivalent parameters v_0/U and $2A/(\rho^{1/3} m^{2/3} U^2)$ which have symmetry and which represent the influence of the vertical and horizontal velocities separately, for constant A and m, can be used instead of the parameters

$$\frac{v_0}{\sqrt{\left(\frac{A}{m}\sqrt[3]{\frac{m}{\rho}}\right)}} \quad \text{and} \quad \frac{U}{\sqrt{\left(\frac{A}{m}\sqrt[3]{\frac{m}{\rho}}\right)}}.$$

Above, we noted the system of characteristic parameters and the form of certain important relations. Some of these parameters are not essential in a number of cases; this is clarified by means of special investigations outside the scope of dimensional analysis and similarity theory.

The phenomenon of ricocheting on a water surface is closely related to the phenomenon of the longitudinal instability of planing. We encounter the phenomenon of the longitudinal gliding instability in actual motion of seaplanes and hydrofoils and in experiments with models. It is well known at the present time that unstable regions of motion exist for every seaplane and for every model. Strong longitudinal oscillations which are extremely unfavourable and dangerous arise in these regions. Just as in the problem of landing on water, the investigation of the planing instability phenomenon is complicated by the large number of parameters whose influence must be clarified.

It is not difficult to see that a system of nondimensional parameters defining planing instability is obtained from the system of parameters determining the water impact phenomenon if we put $v_0 = \Omega_0 = 0$. A similar concept of a boundary separating the stable and unstable planing regions corresponds to the concept of the ricochet boundary.

If the influence of the weight of water is absent, then the stability boundaries depend on the load and on the planing velocity only through the coefficient $C_B = 2A/(\rho B^2 U^2)$. This is in good agreement with experimental results for a number of regimes important in practice (Sedov, 1940).

As in the water impact problem, the number of parameters defining the planing stability of a careening plate of indefinite width is decreased (Sedov and Vladimirov, 1943).

Let us now concentrate on the problem of vertical entry into water.

The phenomenon of vertical water impact when the body is moving forward† is defined by the parameters

$$t, v_0, m, B, A, g \text{ and } \rho.$$

We take the following four quantities as nondimensional parameters defining the regime and state of the motion:

$$\frac{tv_0}{\sqrt[3]{\dfrac{m}{\rho}}}, \frac{m}{\rho B^3}, \frac{v_0}{\sqrt{\left(g\sqrt[3]{\dfrac{m}{\rho}}\right)}}, \frac{v_0}{\sqrt{\left(\dfrac{A}{m}\sqrt[3]{\dfrac{m}{\rho}}\right)}}. \tag{11.8}$$

If $A = mg$, only three parameters remain.

It is evident that the nondimensional quantities adopted as characteristic times (maximum or average values in time) are determined by only two parameters for $A = mg$:

$$\frac{m}{\rho B^3}, \frac{v_0}{\sqrt{\left(g\sqrt[3]{\dfrac{m}{\rho}}\right)}}.$$

For example, the expressions for the maximum impact and for the momentum of water acting upon a body in a certain characteristic time interval are

$$P_{\max} = f_1\left(\frac{m}{\rho B^3}, \frac{v_0}{\sqrt{\left(g\sqrt[3]{\dfrac{m}{\rho}}\right)}}\right)\rho\left(\frac{m}{\rho}\right)^{2/3}v_0^2, \tag{11.9}$$

$$I = f_2\left(\frac{m}{\rho B^3}, \frac{v_0}{\sqrt{\left(g\sqrt[3]{\dfrac{m}{\rho}}\right)}}\right)mv_0. \tag{11.10}$$

† If the body is not symmetrical, vertical forward motion can be accomplished by using special guides.

If the velocity of the water v_0 on contact is large, and the shape of the wetted surface of the body is almost a horizontal plane, then the phenomenon of submersion in the water has a definite impact character. In this case, the weight of water and the body weight are not essential. Consequently, for entry into water when the quantity

$$\frac{v_0}{\sqrt{\left(g \sqrt[3]{\frac{m}{\rho}}\right)}}$$

is large enough, we must have

$$P_{max} = f_1\left(\frac{m}{\rho B^3}\right) \rho^{1/3} m^{2/3} v_0^2, \qquad (11.11)$$

$$I = f_2\left(\frac{m}{\rho B^3}\right) m v_0. \qquad (11.12)$$

It is evident in the case of constant v_0 and m that the fluid reaction will increase as the body dimensions (the parameter B) increase, consequently, the functions $f_1(m/\rho B^3)$ and $f_2(m/\rho B^3)$ must increase when the parameter $m/\rho B^3$ tends to zero. Formulas (11.11) and (11.12) show that the maximum force is proportional to the square of the velocity but the momentum is proportional to the first power of the incident velocity.†

The parameter $m/(\rho B^3)$ is eliminated in the problem of cone entry into a water surface (the conical section by a horizontal plane can be arbitrary) since there is no pre-assigned characteristic linear dimension in this case.

In this case, we obtain, instead of (11.11) and (11.12), the relations

$$P_{max} = c_1 \rho^{1/3} m^{2/3} v_0^2, \qquad (11.13)$$

$$I = c_2 m v_0. \qquad (11.14)$$

These formulas determine the variation of P_{max} and I with mass. The constants c_1 and c_2 depend on the cone shape. It is interesting to note that the influence of the mass is independent of the cone shape in (11.13) and (11.14). The cone shape only influences the values of the constants c_1 and c_2.

However, the general conclusion that the maximum force is proportional to $m^{2/3}$ for bodies of any shape would be incorrect, in general. In fact, let us consider a long plane wedge with a small careening angle incident on water. Let the plane of symmetry of the wedge be vertical and let the incident velocity be large. Neglecting the weight of the water

† Experimental data bear out this conclusion well; see R. L. Kreps (1939); and R. L. Kreps (1940).

and the wedge, we obtain the following system of characteristic parameters

$$t, v_0, m/L = m_1, L, \rho,$$

where L is the length of the wedge along the keel, m_1 is the wedge mass per unit length.

If L is very large, then we can consider the limiting case $L = \infty$. The linear dimension is eliminated in the limiting case of a plane infinitely long wedge when the wetted surface does not reach the wedge edges; consequently, the phenomenon is defined by only four dimensional quantities

$$t, v_0, m_1, \rho.$$

All the nondimensional characteristics are defined by the one nondimensional quantity

$$\frac{t v_0}{\sqrt{\left(\dfrac{m}{\rho L}\right)}}.$$

The maximum and average values of the nondimensional mechanical characteristics will be nondimensional constants.

Hence, the following formula will be correct for the maximum impact force in this case

$$\frac{P_{\max}}{L} = c_3 \sqrt{(\rho m_1)}\, v_0^2,$$

or

$$P_{\max} = c_3 \sqrt{(\rho m)}\, L v_0^2.$$

Therefore, the maximum force for a very long body is proportional to the square root of the body mass. The constant c_3 depends on the careening and it can be considered as a function of the careening angle β.

It is not difficult to see that c_3 increases as the careening angle decreases. For small careening angles, it is possible to put $c_3 = c_4/\beta$ where c_4 can be considered as independent of the careening angle for small β. In the general case of a finite wedge, the following formula can be given for c_3

$$c_3 = \frac{f\left(\dfrac{m}{\rho L^3}, \beta\right)}{\beta}.$$

It is convenient to express the experimental result in terms of the function $f[(m/\rho L^3), \beta]$ since this function will vary slightly for small β and for very elongated wedges or small m.

12. ENTRY OF A CONE AND WEDGE AT CONSTANT SPEED INTO A FLUID

Let us consider the problem of the unsteady motion of an incompressible fluid caused by the entry of a solid body in the shape of a cone or a wedge into the fluid. The shape of the cone in the three-dimensional case and the shape of the plane wedge of infinite span in the plane case are of interest in that their surfaces are fixed completely by the one requirement of geometrical similarity. The set of geometrically similar cones reduces to a single unique cone. The surface of the cone and the surface of the plane wedge are defined completely by nondimensional geometric quantities.

Let us assume that the fluid occupies the whole lower half-space bounded by the horizontal plane and that the properties of fluid weight and viscosity can be neglected. Therefore, we shall assume that the fluid is incompressible, homogeneous and ideal.

Let $t = 0$ be the initial instant at which the body makes contact with the fluid at rest. The body penetrating the fluid moves forward with a velocity v which is constant in magnitude and direction.

Let us assume that the pressure p on the free surface has the constant value p_0. Since the fluid is incompressible the value p_0 on the free surface cannot influence the perturbed fluid motion. We can consider the difference $p - p_0$ instead of the pressure p; in this case the parameter p_0 is not essential. Consequently, the mechanical properties of the fluid are determined by a single parameter, the density ρ.

It follows from the above, that all the mechanical characteristics of the fluid motion are defined at each point by the quantities

$$\rho, t, v, \alpha, \beta, x, y, z$$

where α and β are angles defining the direction of the velocity v relative to the body and x, y, z are the coordinates of the point under consideration either in a fixed coordinate system with the origin at the point of contact of the cone apex and the fluid level, or in a moving coordinate system which is fixed in the body with its origin at the cone apex.

It is evident that all the relevant nondimensional quantities are defined by the parameters

$$\alpha, \beta, \frac{x}{vt}, \frac{y}{vt}, \frac{z}{vt},$$

in which the parameters $x/(vt)$, $y/(vt)$, $z/(vt)$ can influence only the quantities dependent on position in the fluid. The total nondimensional parameters (for example, the total fluid reaction, etc.) or the parameters which are independent of position depend only on the angles α, β. If the direction of the velocity is fixed (for example the velocity is vertical) then all the nondimensional total characteristics can be considered as absolute constants dependent only on the cone shape.

Let us denote the velocity potential of the disturbed fluid motion by $\phi(x, y, z, t)$ where the cone velocity is given in magnitude and direction. Because the fluid motion is unsteady, the problem of determining the disturbed motion reduces to the determination of the velocity potential as a function of the four independent variables x, y, z, t.

The four independent variables can easily be reduced to three by dimensional analysis.

Actually, the quantity $\phi/(v^2 t)$ is nondimensional, consequently, we have

$$\phi = v^2 tf\left(\frac{x}{vt}, \frac{y}{vt}, \frac{z}{vt}\right). \tag{12.1}$$

The velocity of the fluid particles $\bar{\mathfrak{v}}$ is given by

$$\bar{\mathfrak{v}} = v\bar{f}_1\left(\frac{x}{vt}, \frac{y}{vt}, \frac{z}{vt}\right). \tag{12.2}$$

The magnitudes of the total water reaction P and of the wetted area S are given by the formulas

$$\left.\begin{aligned} P &= c_1\rho v^4 t^2, \\ S &= c_2 v^2 t^2, \end{aligned}\right\} \tag{12.3}$$

in which the coefficients c_1, c_2 and the direction of the force P depend only on the cone shape and on the direction of the velocity of the cone motion. The first formula of (12.3) shows that the reaction of the water is proportional to the fluid density, to the fourth power of the velocity and to the time squared. The wetted area is proportional to the velocity squared and to the time squared. Evidently, the two different states of the motion are dynamically similar.

The following formulas for the velocity potential and for the velocity distribution are correct in the two-dimensional case of entry of a plane wedge (the xy plane is the plane of motion)

$$\phi = v^2 tf\left(\frac{x}{vt}, \frac{y}{vt}\right),$$

$$\bar{\mathfrak{v}} = v\bar{f}_1\left(\frac{x}{vt}, \frac{y}{vt}\right).$$

For the force per unit length of the wedge and for the wetted length along the wedge face, we have

$$\left.\begin{aligned} P_1 &= c_1'\rho v^3 t, \\ l &= c_2' vt. \end{aligned}\right\} \tag{12.4}$$

It is seen from (12.3) and (12.4) that the relation between velocity and the water reaction when a body enters the water at a constant velocity is different for bodies of different shapes.

The constants c_1' and c_2' depend on the careening angle of the wedge, on the angles of inclination of the wedge plane of symmetry and on the wedge velocity relative to the undisturbed level of the free surface.

Approximate theoretical solutions exist for the plane problem of vertical entry and for the penetration of a plate slightly inclined to the fluid level when the horizontal component of the plate velocity is large (Wagner, 1932; Sedov, 1936).

13. SHALLOW WAVES ON THE SURFACE OF AN INCOMPRESSIBLE FLUID

Considering the Cauchy-Poisson problem of waves on the surface of a heavy incompressible fluid, N. E. Kochin (1935) used the reasoning of dimensional analysis and he gave the solution of this classical problem a new, elegant mathematical form.

Amplifying the reasoning of dimensional analysis, an entire class of new solutions of the wave problem can be found in explicit and simple form (Sedov, 1948). The N. E. Kochin solution is a particular case in the class of solutions obtained.

This method and the solutions found can be extended and generalized to the case of the three dimensional problem.

The plane problem of potential waves of infinitely small amplitude on the surface of a heavy incompressible fluid occupying the whole lower half-space can be formulated as follows.

Let us take a Cartesian coordinate system; let the x axis coincide with the unperturbed fluid level; let the y axis be directed vertically upward. The velocity potential $\phi(x, y, t)$ at $y < 0$ is a regular harmonic function, i.e.

$$\frac{\partial^2 \phi}{\partial x^2} + \frac{\partial^2 \phi}{\partial y^2} = 0 \tag{13.1}$$

when $y < 0$. Furthermore, we shall consider cases in which the fluid motion attenuates as the depth of submersion increases, i.e.

$$|\text{grad } \phi| \to 0 \tag{13.2}$$

when $y \to -\infty$.

The condition of constant pressure on the free surface can be represented in the linearized form

$$\frac{\partial^2 \phi}{\partial t^2} + g\frac{\partial \phi}{\partial y} = 0 \tag{13.3}$$

when $y = 0$ and $t \geqslant 0$, where g is the acceleration due to gravity. Certain requirements, in addition to conditions (13.1), (13.2) and (13.3), must be imposed in order to determine the velocity potential $\phi(x, y, t)$.

To provide these additional data we assume that the shape of the free surface and the distribution of the impulsive pressures at $t = 0$ are given.

The initial conditions can be formulated in linearized theory by starting from the relations

$$\zeta = -\frac{1}{g}\frac{\partial\phi}{\partial t}\bigg|_{y=0} \quad \text{and} \quad \frac{p_t}{\rho} = -\phi, \tag{13.4}$$

where $\zeta(x, t)$ is the height of a point of the free surface above the undisturbed level, p_t is the impulsive pressure and ρ is the fluid density.

The initial conditions can be formulated as follows: we have at $t = 0$

$$-\frac{1}{g}\frac{\partial\phi}{\partial t}\bigg|_{y=0} = f(x), \quad \phi\bigg|_{y=0} = F(x). \tag{13.5}$$

If the functions f and F are not simultaneously zero nor infinite and if $f \neq kx$, then obviously, these functions must depend on certain dimensional constants in addition to the variable x. Since the problem is formulated in kinematic terms, not more than two dimensional constants with independent dimensions entering in the functions f and F can exist.

We shall satisfy the Laplace equation if we put

$$\phi(x, y) = \operatorname{Re} w(x + iy) \tag{13.6}$$

where $w(z)$ ($z = x + iy$) is the complex potential. Furthermore, we assume that the function $w(z)$ is single-valued, finite and regular for $t > 0$ and finite x and $y < 0$.

It follows from condition (13.2) that the derivative $\partial w/\partial z$ vanishes as $y \to -\infty$.

Boundary condition (13.3) can be written

$$\operatorname{Re}\left(\frac{\partial^2 w}{\partial t^2} + ig\frac{\partial w}{\partial z}\right) = 0 \tag{13.7}$$

when $y = 0$. This condition permits the combination

$$G(z) = \frac{\partial^2 w}{\partial t^2} + ig\frac{\partial w}{\partial z}$$

to be continued into the upper half-plane; we then find that the function $G(z)$ is single-valued in the whole complex plane. It follows from the assumption made about the general character of the fluid motion that the singular points of $G(z)$ lie on the real axis.

Let us look for solutions in which the characteristic function $w(z)$ depends linearly on the dimensional constants entering in the additional conditions determining the velocity potential; we will not make the form of these conditions concrete.

It follows from the linearity of the problem that it is sufficient to consider the case when we have only one dimensional constant a, on which the characteristic function $w(z)$ depends linearly (the constant a can be complex).

Let the dimensions of the constant a be represented by the formula

$$[a] = L^p T^q.$$

From the assumptions made above, we deduce that the complete system of characteristic parameters is represented by the table

$$z = x + iy, t, g, a.$$

Now, let us put

$$w = a z^\alpha g^\beta \chi(z, t, g);$$

where the exponents α and β are selected so that χ is an abstract quantity. Since $[a] = L^p T^q$, then we must have

$$p + \alpha + \beta = 2, \quad q - 2\beta = -1,$$

from which

$$\beta = \frac{1+q}{2}, \quad \alpha = \frac{3 - 2p - q}{2}. \tag{13.8}$$

It follows from dimensional analysis that the function $\chi(z, t, g)$ depends only on the combination

$$\lambda = \frac{gt^2}{z}$$

that is

$$w = a g^\beta z^\alpha \chi(\lambda). \tag{13.9}$$

From (13.9) we obtain the following equations:

$$\frac{\partial w}{\partial t} = a g^\beta z^\alpha \chi'(\lambda) \frac{\partial \lambda}{\partial t},$$

$$\frac{\partial^2 w}{\partial t^2} = a g^\beta z^\alpha \left[\chi''(\lambda) \left(\frac{\partial \lambda}{\partial t} \right)^2 + \chi'(\lambda) \frac{\partial^2 \lambda}{\partial t^2} \right],$$

$$\frac{\partial w}{\partial z} = a g^\beta z^\alpha \left[\frac{\alpha}{z} \chi(\lambda) + \chi'(\lambda) \frac{\partial \lambda}{\partial z} \right]$$

and

$$G(z) = ag^\beta z^\alpha \left[\chi'' \left(\frac{\partial \lambda}{\partial t} \right)^2 + \chi' \left(\frac{\partial^2 \lambda}{\partial t^2} + ig \frac{\partial \lambda}{\partial z} \right) + \frac{ig\alpha}{z} \chi \right];$$

since

$$\frac{\partial \lambda}{\partial t} = 2 \frac{\lambda}{t}, \quad \frac{\partial^2 \lambda}{\partial t^2} = 2 \frac{\lambda}{t^2}, \quad \frac{\partial \lambda}{\partial z} = -\frac{\lambda}{z},$$

then

$$G = 4ag^{\beta+\alpha} t^{2\alpha-2} \lambda^{2-\alpha} \left[\chi'' + \left(\frac{1}{2\lambda} - \frac{i}{4} \right) \chi' + \frac{i\alpha}{4\lambda} \chi \right]. \qquad (13.10)$$

The function $G(\lambda)$ is a single-valued function of the complex variable λ in the whole plane. Singular points occur only on the real axis.

Since the function $G(\lambda)$ is purely imaginary for real λ, then evidently the coefficients of the Laurent series are purely imaginary at any singular point of this function.

Let us denote: $a = a_0 e^{i\delta}$ and $G = 4a_0 g^{\beta+\alpha} t^{2\alpha-2} G_1(\lambda)$. We obtain from (13.10)

$$e^{-i\delta} C_1(\lambda) = \lambda^{2-\alpha} \left[\chi'' + \left(\frac{1}{2\lambda} - \frac{i}{4} \right) \chi' + \frac{i\alpha}{4\lambda} \chi \right]. \qquad (13.11)$$

We can satisfy the conditions within the fluid and on the free surface if we take any single-valued function which is purely imaginary on the real axis and has singularities only on the real axis as our function $G_1(\lambda)$.

In order to determine the characteristic function of the appropriate wave motion, it is necessary to integrate the differential equation (13.11) for the function $\chi(\lambda)$. In the most general case, wave motions possessing singularities on the free surface of the fluid are obtained.

If we assume that the fluid motion at the free boundary is regular for $t > 0$, then since the combination $(\partial^2 w / \partial t^2) + ig(\partial w / \partial z)$ is finite the function $G_1(\lambda)$ reduces to an imaginary constant which we must put equal to zero by virtue of condition (13.2) and the additional condition on the pressure is independent of time as $y \to -\infty$.

Hence, we arrive at the problem of integrating the ordinary differential equation

$$\chi'' + \left(\frac{1}{2\lambda} - \frac{i}{4} \right) \chi' + \frac{i\alpha}{4\lambda} \chi = 0. \qquad (13.12)$$

In this equation α is an arbitrary constant. It is not difficult to see that the solution of (13.12) also gives a certain wave motion for complex

α. The fundamental solution, considered by N. E. Kochin, corresponds to the particular value $\alpha = -\frac{1}{2}$.

After the change of variable

$$\lambda = \frac{4}{i}\mu$$

Equation (13.12) reduces to

$$\mu\frac{d^2\chi}{d\mu^2} + (\tfrac{1}{2} - \mu)\frac{d\chi}{d\mu} + \alpha\chi = 0. \tag{13.13}$$

The solution of (13.13) is expressed in terms of the confluent hypergeometric functions $y = M(k, \gamma, x)$ which satisfy the differential equation (Jahnke and Emde, 1938):

$$xy'' + (\gamma - x)y' - ky = 0.$$

The general solution of (13.13) is

$$\chi = C_1 M(-\alpha, \tfrac{1}{2}, \mu) + C_2 \mu^{1/2} M(-\alpha + \tfrac{1}{2}, \tfrac{3}{2}, \mu).$$

Using this solution for the characteristic function of the wave motion, we can write

$$w = A_1 w_1 + A_2 w_2, \tag{13.14}$$

where

$$w_1(z, t, \alpha) = z^\alpha M\left(-\alpha, \tfrac{1}{2}, \frac{igt^2}{4z}\right), \tag{13.15}$$

$$w_2(z, t, \alpha) = z^\alpha \sqrt{\left(\frac{igt^2}{4z}\right)} M\left(-\alpha + \tfrac{1}{2}, \tfrac{3}{2}, \frac{igt^2}{4z}\right). \tag{13.16}$$

The arbitrary constants A_1 and A_2 can be complex.

The particular solutions of the wave problem (13.15) and (13.16) which depend on the single arbitrary constant α, can be generalized by replacing t by $t - t_0$ and z by $z - z_0$. The constants t_0 and z_0 (z_0 is real) determine the initial instant and the displacement of the singular point corresponding to the origin in the plane of flow. More general solutions can be constructed by superposition of these solutions: the constants A_1 and A can be considered as functions of the parameters α, t_0 and z_0 while the summation is being carried out.

From formulas (13.15) and (13.16), it is easy to deduce the following properties of the functions w_1 and w_2.

At $t = 0$, we have

$$w_1(z, 0, \alpha) = z^\alpha, \quad \left(\frac{\partial w_1}{\partial t}\right)_{t=0} = 0 \tag{13.17}$$

and

$$w_2(z, 0, \alpha) = 0, \quad \left(\frac{\partial w_2}{\partial t}\right)_{t=0} = \sqrt{\left(\frac{ig}{4}\right)} \cdot z^{\alpha - 1/2}. \tag{13.18}$$

Now, let us consider the wave motions for which the characteristic functions are determined by the formulas

$$\Omega_1(z, t) = -\frac{e^{\pi i \alpha}}{\pi i} \int_{-\infty}^{+\infty} f(x_0)\, w_1(z - x_0, t, \alpha)\, dx_0, \tag{13.19}$$

$$\Omega_2(z, t) = -\frac{2e^{\pi i (\alpha - \frac{1}{2})}}{\sqrt{(ig)} \cdot \pi i} \int_{-\infty}^{+\infty} F(x_0)\, w_2(z - x_0, t, \alpha)\, dx_0, \tag{13.20}$$

where $f(x_0)$ and $F(x_0)$ are certain functions for which the integrals in formulas (13.19) and (13.20) converge.

At $t = 0$, we have

$$\Omega_1 = -\frac{1}{\pi i} \cdot \int_{-\infty}^{+\infty} \frac{f(x_0)\, dx_0}{(x_0 - z)^{-\alpha}}, \quad \frac{\partial \Omega_1}{\partial t} = 0, \tag{13.21}$$

$$\Omega_2 = 0, \quad \frac{\partial \Omega_2}{\partial t} = -\frac{1}{\pi i} \int_{-\infty}^{+\infty} \frac{F(x_0)\, dx_0}{(x_0 - z)^{1/2 - \alpha}}. \tag{13.22}$$

The first of Equations (13.21) yields for $\alpha = -1$ and $z = x$

$$\Omega_1(x) = \Phi_1 + i\Psi_1 = f(x) - \frac{1}{\pi i} \int_{-\infty}^{+\infty} \frac{f(x_0)\, dx_0}{x_0 - x}.$$

where the Cauchy principal value of the integral is taken. Hence, if $f(x)$ is real then (13.19) gives the solution of the problem of the wave motion for the following initial conditions when $\alpha = -1$:

$$\Phi_1(x_0) = f(x), \quad \frac{\partial \Phi_1}{\partial t} = 0 \quad \text{for } t = 0.$$

The following initial conditions are obtained for the function $\Psi(x, y)$

$$\Psi_1(x, 0) = \frac{1}{\pi} \int_{-\infty}^{+\infty} \frac{f(x_0)\, dx_0}{x_0 - x} \quad \text{and} \quad \frac{\partial \Psi}{\partial t} = 0 \quad \text{for } t = 0.$$

If $f(x)$ is a purely imaginary, then the initial conditions have analogous form but the roles of the functions Φ_1 and Ψ_1 are interchanged.

The second of Equations (13.22) yields for $t = 0$ and $z = x$ when $\alpha = -\frac{1}{2}$:

$$\frac{\partial \Omega_2}{\partial t} = \frac{\partial (\Phi_2 + i\Psi_2)}{\partial t} = F(x) - \frac{1}{\pi i} \int\limits_{-\infty}^{+\infty} \frac{F(x_0)\, dx_0}{x_0 - x}.$$

Hence, it follows that if $F(x)$ is real, the following initial conditions are obtained

$$\Phi_2 + i\Psi_2 = 0, \quad \frac{\partial \Phi_2}{\partial t} = F(x) \quad \text{for } t = 0 \text{ and } y = 0.$$

This case was investigated by N. E. Kochin. Now, let us clarify the character of the initial conditions in the general case when $\alpha \neq -1, -\frac{1}{2}$.

Let us consider the initial conditions for the function $\Omega_1(z, t)$. We have

$$\Omega_1(z, 0) = -\frac{1}{\pi i} \int\limits_{-\infty}^{+\infty} (x_0 - z)^\alpha f(x_0)\, dx_0.$$

If $\alpha > 0$ and is an integer, then evidently $\Omega_1(z, 0)$ is a polynomial of degree α. In general, if $\alpha > 0$, then the function $\Omega_1(z, t)$ will become infinite as $z \to \infty$, consequently, the condition that the velocity vanish as $y \to -\infty$ is not satisfied for $\alpha \geqslant 1$.

If $\alpha = -(1 + s)$, where s is a positive integer, then we have

$$\Omega_1(z, 0) = -\frac{1}{\pi i} \int\limits_{-\infty}^{+\infty} \frac{f(x_0)\, dx_0}{(x_0 - z)^{1+s}}. \tag{13.23}$$

Let us assume that the integral $\int\limits_{-\infty}^{\infty} f(x_0)\, dx_0$ is finite,† and introduce the following function

$$\omega(z) = -\frac{1}{\pi} \int\limits_{-\infty}^{+\infty} \frac{f(x_0)}{x_0 - z}\, dx_0. \tag{13.24}$$

Under the assumptions accepted for the function $f(x_0)$, it is easy to see the validity of the relations

$$\frac{d^s \omega}{dz^s} = \omega^s(z) = \Gamma(s+1)\, \Omega_1(z, 0), \tag{13.25}$$

$\omega(z) =$

$$\Gamma(s+1) \int\limits_{-i\infty}^{z} dz_s \int\limits_{-i\infty}^{z_s} dz_{s-1} \cdots \int\limits_{-i\infty}^{z_2} \Omega(z_1)\, dz_1 = \int\limits_{-i\infty}^{z} (z-u)^s\, \Omega_1'(u)\, du. \tag{13.26}$$

† It is evident that the expression for $\Omega_1(z)$ can have meaning in a number of cases when the integral in (13.24) diverges.

Therefore, we have the following relation to determine $f(x)$:

$$\int_{-i\infty}^{x} (x-u)^s \, \Omega_1'(u) \, du = f(x) - \frac{1}{\pi i} \int_{-\infty}^{+\infty} \frac{f(x_0) \, dx_0}{x_0 - x}. \tag{13.27}$$

We have derived formula (13.27) from formulas (13.23) and (13.24) for integral $s > 0$. This relation also remains valid for any real $s > -1$.

Actually, from (13.23) we have

$$\Omega'(u) = -\frac{s+1}{\pi i} \int_{-\infty}^{+\infty} \frac{f(x_0) \, dx_0}{(x_0 - u)^{s+2}}.$$

Multiplying by $(z - u)^s$ and integrating, we obtain

$$\int_{-i\infty}^{z} \Omega'(u) \, (z-u)^s \, du = -\frac{s+1}{\pi i} \int_{-\infty}^{+\infty} f(x_0) \, dx_0 \int_{-i\infty}^{z} \frac{(z-u)^s \, du}{(x_0 - u)^{s+2}}.$$

We make the following change of variable in the inner integral:

$$u = x_0 + (z - x_0)\frac{1}{\lambda};$$

from which

$$x_0 - u = \frac{x_0 - z}{\lambda}, \quad z - u = \frac{x_0 - z}{\lambda}(1 - \lambda), \quad du = (x_0 - z)\frac{d\lambda}{\lambda^2}.$$

Substituting in the inner integral, we obtain

$$\frac{1}{x_0 - z} \int_{0}^{1} (1 - \lambda)^s \, d\lambda = \frac{1}{(s + 1)(x_0 - z)}.$$

Using this, we find

$$\int_{-i\infty}^{z} (z-u)^s \, \Omega'(u) \, du = -\frac{1}{\pi i} \int_{-\infty}^{+\infty} \frac{f(x_0) \, dx_0}{x_0 - z}.$$

Hence, formula (13.27) follows as $z \to x$.

Formula (13.25) can be generalized to fractional s if the fractional derivative $\omega^s(z)$ is defined by the formula

$$\omega^s(z) = \frac{1}{\Gamma(1-s)} \int_{-i\infty}^{z} (z-u)^{-s} \, \omega'(u) \, du.$$

Actually

$$\omega^s(z) = -\frac{1}{\pi i \Gamma(1-s)} \int\limits_{-\infty}^{+\infty} f(x_0)\,dx \int\limits_{-i\infty}^{z} \frac{(z-u)^{-s}\,du}{(x_0-u)^2}.$$

In order to evaluate the inner integral, let us make the change of variable

$$u = x_0 + (z-x_0)\frac{1}{\lambda},$$

after which we obtain

$$\int\limits_{-i\infty}^{z} \frac{(z-u)^{-s}}{(x_0-u)^2}\,du = \frac{1}{(x_0-z)^{s+1}} \int\limits_{0}^{1} \lambda^s(1-\lambda)^{-s}\,d\lambda = \frac{B(s+1,\,1-s)}{(x_0-z)^{s+1}};$$

since

$$B(s+1,\,1-s) = \frac{\Gamma(s+1)\,\Gamma(1-s)}{\Gamma(2)} \quad \text{and} \quad \Gamma(2) = 1,$$

then

$$\omega^s(z) = -\frac{\Gamma(s+1)}{\pi i} \int\limits_{-\infty}^{+\infty} \frac{f(x_0)\,dx_0}{(x_0-z)^{s+1}} = \Gamma(s+1)\,\Omega(z).$$

14. THREE DIMENSIONAL SELF-SIMILAR MOTIONS OF COMPRESSIBLE MEDIA

The formulation of the problem of motion of an incompressible fluid in §§ 12 and 13, which led to a reduction in the number of independent variables, can be extended and generalized.

The motion of a compressible medium, in which the dimensionless parameters depend only on the combination

$$\frac{x}{bt^\delta}, \quad \frac{y}{bt^\delta}, \quad \frac{z}{bt^\delta}$$

where x, y, z denote Cartesian coordinates, t is the time and b is a constant with dimensions $LT^{-\delta}$, will be called self-similar with a centre of similarity at the origin of the coordinate system. It is easy to discover the general character of all problems for which self-similarity exists. It is sufficient for the system of dimensional characteristic parameters, prescribed in part by supplementary conditions and in part by boundary or initial conditions, to contain not more than two constants with independent dimensions other than length or time.

In other words the system of characteristic parameters should consist of the following,

$$a, b, x, y, z, t, \alpha_1, \alpha_2, \dots$$

where $\alpha_1, \alpha_2, \ldots$ are arbitrary combinations of dimensional constants. There can be any number of these and the constants a and b have dimensions of the form

$$[a] = ML^k T^s, \quad [b] = LT^{-\delta}.$$

Here $\delta \neq 0$ and k and s can be arbitrary. Without loss of generality the constant a can always be replaced by $A = ab^{s/\delta}$ with dimensions

$$[A] = ML^{\omega-3}$$

where ω can be arbitrary.

Generally speaking, for self-similarity to exist in the motion of a compressible fluid it is necessary that the formulation of the problem should not contain a characteristic length or time (wedge, cone, etc.).

We shall now give some examples of self-similar motion.

(1) Problem of dispersion of an infinite mass of incompressible fluid, initially at rest, which expands from a point under conditions preserving geometric similarity.

Suppose that the prescribed radial velocities of the interior boundary of the fluid are determined by the equation

$$v(t, \theta, \psi) = bf(\theta, \psi) t^{\delta-1}$$

where θ, ψ are polar coordinates and b is a constant with dimensions: $[b] = LT^{-\delta}$, for $\delta > 0$. The interior surface extends continuously and under conditions of similarity from the origin. The resulting motion of the incompressible fluid is potential and determined by the system of parameters,

$$\rho, b, r, \theta, \psi, t, f(\theta, \psi)$$

where ρ is the density; the initial pressure and the pressure at infinity p_0 are indeterminate so that only pressure differences $p-p_0$ can be considered. It is evident that in this case the velocity potential is of the form

$$\phi = rbt^{\delta-1} \Phi\left(\theta, \psi, \frac{r}{bt^\delta}\right).$$

The problem can be formulated in a different way; instead of prescribing the velocity of the interior volume we can give the pressure acting at the surface of this volume:

$$p - p_0 = \rho b^2 t^{2(\delta-1)} F(\theta, \psi).$$

One such problem has already been solved in § 11, Chapter IV.

(2) The above considerations concerning self-similarity of a fluid carry over to the case when the motion of a fluid with a free surface is considered, on condition that this surface is initially a plane and that the centre of symmetry for motion of the prescribed surface lies in that plane. In this way we can generate problems on penetration by a cone or wedge, in the cases of arbitrary but not hard or rigid surfaces, which change their shape under the conditions of similarity. We can also treat bodies displacing a fluid with a free surface.

In particular it is possible to consider self-similarity conditions when the wedge or cone moves into the fluid with a velocity varying with time like a step function.

If forces due to gravity in the liquid are taken into account it is necessary to include, among the characteristic parameters, the acceleration due to gravity. To preserve self-similarity it is necessary to have $[g] = [b]$, i.e. $\delta = +2$.

Consequently, for a wedge or cone moving with uniform acceleration into an incompressible heavy fluid, the induced motion will be self-similar.

Conditions of self-similarity in a fluid with a free surface are also preserved in cases when the free surface has a conical or wedge-like shape, and when the centre of similarity coincides with the vertex of the cone or wedge.

As an example of a self-similar solution for an elastic medium we should like to point to Boussinesq's problem on the distribution of stress and strain in the elastic semi-space bounded by a plane near the point of application of a prescribed concentrated force P.

In problems of Statics, the properties of the elastic body are completely determined by Young's modulus E and Poisson's ratio σ. The external force is determined by the magnitude of P and by the abstract parameters determining its direction.

We take polar coordinates r, θ based on the point of application of the concentrated load, in a plane which contains the force vector and which is perpendicular to the boundary. The system of characteristic parameters is

$$P, E, r, \theta, \psi, \sigma, \theta_0;$$

θ_0 is the angle defining the direction of the force P. Owing to the linear character of the problem all stresses and strains are linear functions of P and hence the dependence on P is known beforehand; the dependence of all quantities on E and r can be determined immediately from dimensional considerations: this leads to only two independent variables θ and ψ. In the case of axial symmetry (P perpendicular to the bounding plane) the variable ψ will disappear and therefore it is easy to obtain a full solution to the problem, and only the integration of one differential equation is required.

In some cases, which are obvious whenever a particular example is considered, the preceding considerations can be extended to include various problems of unsteady motion of compressible media.

Chapter III

Application to the Theory of Motion of a Viscous Fluid and to the Theory of Turbulence

§ 1. Diffusion of Vorticity in a Viscous Fluid

In this and succeeding paragraphs we shall give examples of the use of dimensional analysis in the mathematical solution of certain physical problems.

We begin with the problem of diffusion of vorticity in the one dimensional unsteady motion of a viscous incompressible fluid of infinite extent (Kochin *et al.*, 1948). We suppose that at the initial instant $t = 0$ potential motion exists everywhere except in the strip O which is the trace on the plane of the motion of an infinite, rectilinear, point vortex with associated circulation Γ.

We assume that the motion has axial symmetry, and denote the angular velocity of the fluid by Ω.

As is known, the circulation around a circle of radius R centre 0 is

$$\Gamma_R = 2 \int\limits_0^R \int\limits_0^{2\pi} \Omega r \, dr \, d\theta = 4\pi \int\limits_0^R r \Omega(r) \, dr. \qquad (1.1)$$

At the initial instant, for any circle, no matter how small, we have

$$\Gamma_R = \Gamma. \qquad (1.2)$$

In the case under consideration, the equation governing the propagation of vorticity will be:

$$\frac{\partial \Omega}{\partial t} = \nu \left(\frac{\partial^2 \Omega}{\partial r^2} + \frac{1}{r} \frac{\partial \Omega}{\partial r} \right), \qquad (1.3)$$

where ν is the coefficient of kinematic viscosity ($\nu = \mu/\rho$). The problem is to determine Ω as a function of the radius r and the time t.

It follows from the formulation of the problem that

$$\Omega = f(\Gamma, \nu, r, t).$$

Since Equation (1.3) is linear the initial condition that Ω is proportional to Γ, gives

$$\Omega = \Gamma f_1(\nu, r, t). \qquad (1.4)$$

The nondimensional combination $\Omega \nu t / \Gamma$ must be expressed as a function of the single independent nondimensional quantity $r^2 / \nu t = \xi$, which must be composed of the parameters ν, r, t. Therefore

$$\Omega = \frac{\Gamma}{\nu t} \psi(\xi). \tag{1.5}$$

It is evident from (1.5) that the partial differential equation (1.3) for Ω, involving two independent variables r and t, reduces to an ordinary differential equation in the one unknown ξ.

Substituting the expression for Ω from (1.5) into (1.3), we have:

$$\psi(\xi) + \xi \psi'(\xi) + 4[\psi'(\xi) + \xi \psi''(\xi)] = 0.$$

Integrating, we obtain

$$\xi \psi + 4 \xi \psi' = C.$$

The constant C is zero for the solution in which $\psi(0)$ and $\psi'(0)$ are finite. Integrating the equation

$$4 \frac{d\psi}{d\xi} + \psi = 0,$$

we find

$$\psi = A \, e^{-\xi/4}.$$

This yields for the magnitude of the vorticity Ω:

$$\Omega = \frac{\Gamma}{\nu t} A \, e^{-r^2/4\nu t}.$$

We determine the constant A from the initial conditions. The circulation around a circle of radius R is

$$\Gamma_R = 4\pi \frac{A\Gamma}{\nu t} \int\limits_0^R r \, e^{-r^2/4\nu t} dr = 8\pi A \Gamma (1 - e^{-R^2/4\nu t}). \tag{1.6}$$

For $t = 0$ and any $R > 0$, we have:

$$\Gamma_R = 8\pi A \Gamma.$$

The initial condition $\Gamma_R = \Gamma$ yields

$$A = \frac{1}{8\pi}.$$

The final solution of the problem is

$$\Omega = \frac{\Gamma}{8\pi \nu t} e^{-r^2/4\nu t}. \tag{1.7}$$

Let us denote the fluid velocity by $\mathfrak{v}(r, t)$. The motion has axial symmetry, hence, the fluid velocity is directed normal to a radius-vector drawn from the pole 0 to the point under consideration.

Taking into account the direction of the velocity vector, we obtain the following relation between Γ_R and $\mathfrak{v}$;

$$\Gamma_R = 2\pi r \mathfrak{v}.$$

Using (1.6), we find that the velocity distribution over the radius r for the time t is

$$\mathfrak{v} = \frac{\Gamma}{2\pi r}(1 - e^{-r^2/4\nu t}).$$

A velocity distribution corresponding to a point vortex in an ideal fluid is obtained for $t = 0$. The fluid motion is potential for $r > 0$ and $t = 0$ and vortices are absent: the fluid motion is rotational at each point of the fluid for $r > 0$ and $t > 0$. Formula (1.7) yields the law of the propagation or diffusion of vorticity: this formula shows that the strength of the vortex at each point increases with time from zero to a maximum equal to $\Gamma/2\pi r^2 e$ and then again approaches zero.

Since (1.3) is linear, then, starting from the solution obtained for the propagation of a point vortex, a solution of the problem of symmetric motion for any initial velocity distribution can be constructed by the principle of superposition.

§ 2. Exact Solutions of the Equations of Motion of a Viscous Incompressible Fluid

We consider the steady motion of a viscous incompressible fluid of infinite extent.

The Navier-Stokes equation and the continuity equation can be written

$$\bar{\mathfrak{v}}.\nabla\bar{\mathfrak{v}} = -\operatorname{grad}\left(\frac{p}{\rho} - U\right) + \nu\Delta\bar{\mathfrak{v}}, \qquad \left.\begin{array}{c} \\ \\ \end{array}\right\} \quad (2.1)$$
$$\operatorname{div}\bar{\mathfrak{v}} = 0.$$

In what follows we use spherical polar coordinates. The independent variables and the characteristic parameters will be

$$r, \theta, \lambda, \nu,$$

where r is the distance of the point under consideration from the pole, θ is the amplitude, λ is the longitude and ν is the coefficient of kinematic viscosity. The quantities to be determined will be the velocity components v_r, v_θ, v_λ and the dynamic pressure referred to the density and equal to $(p/\rho) - U$.

Let us study the solutions of (2.1) which are completely determined by the parameters r, θ, λ, ν and just one constant parameter A. Let the dimensions of A be of the form

$$[A] = L^p T^q$$

where p and q are certain constants.

Under this assumption, it is evident that all the nondimensional combinations of the quantities introduced will be functions of only the three abstract parameters

$$\theta, \lambda, \pi = \frac{r^{p+2q} \nu^{-q}}{A}.$$

Then the desired functions can be represented as

$$v_r = \frac{\nu}{r} f(\pi, \lambda, \theta), \qquad v_\theta = \frac{\nu}{r} \phi(\pi, \lambda, \theta),$$

$$v_\lambda = \frac{\nu}{r} \psi(\pi, \lambda, \theta), \quad U - \frac{p}{\rho} = \frac{\nu^2}{r^2} F(\pi, \lambda, \theta).$$

The most general solution of (2.1) can be represented and considered in such a form.

If we assume that

$$p + 2q = 0$$

the number of independent variables is reduced.

This condition means that the dimensions of the constant A are a certain power of the dimensions of the kinematic viscosity coefficient ν.

In addition to the above assumption, let us assume that the motions to be studied have axial symmetry so that the variable λ is not essential.

The following formulas result from these assumptions:

$$v_r = \frac{\nu}{r} f(\theta), \quad v_\theta = \frac{\nu}{r} \phi(\theta), \quad v_\lambda = \frac{\nu}{r} \phi(\theta), \quad U - \frac{p}{\rho} = \frac{\nu^2}{r} F(\theta). \quad (2.2)$$

These formulas give the velocity and pressure as functions of the variable r.

In this case, a system of ordinary nonlinear equations is obtained for the four functions f, ϕ, ψ, F:

$$\left.\begin{aligned}
f'' + f'(\operatorname{ctg}\theta - \phi) + f^2 + \phi^2 + \psi^2 - 2F &= 0, \\
\phi\phi' - \psi^2 \operatorname{ctg}\theta - f' - F' &= 0, \\
\psi'' - \phi\psi' - \phi\psi \operatorname{ctg}\theta + \psi' \operatorname{ctg}\theta - \frac{\psi}{\sin^2\theta} &= 0, \\
f + \phi' + \phi \operatorname{ctg}\theta &= 0.
\end{aligned}\right\} \quad (2.3)$$

After elimination of F and after certain simple transformations, we obtain:

$$\left.\begin{array}{r} f''' + 2\psi(\psi' + \psi\,\mathrm{ctg}\,\theta) + (f'\,\mathrm{ctg}\,\theta)' - (\phi f')' + 2ff' + 2f' = 0, \\ \phi(\psi' + \psi\,\mathrm{ctg}\,\theta) = (\psi' + \psi\,\mathrm{ctg}\,\theta)', \\ f = -(\phi' + \phi\,\mathrm{ctg}\,\theta). \end{array}\right\} \qquad (2.4)$$

The general solution of this system of equations depends on six arbitrary constants.

Before proceeding to the study of the solution of the system of Equations (2.4) we note certain general properties of the viscous fluid motion under consideration.

The differential equations for the projections of the streamlines on the meridian plane can be written as

$$\frac{dr}{v_r} = \frac{r\,d\theta}{v_\theta} \quad \text{or} \quad \frac{dr}{f(\theta)} = \frac{r\,d\theta}{\phi(\theta)},$$

from which

$$\ln\frac{r}{a} = \int \frac{f(\theta)\,d\theta}{\phi(\theta)}$$

where a is a constant of integration. From the last equation of the system (2.4), we obtain

$$\frac{r}{a} = \frac{1}{\phi\sin\theta}. \qquad (2.5)$$

From general considerations of dimensional analysis and also directly from (2.5), it follows that the streamlines of the flow are similar curves.

Let us denote the mass flow and the momentum flux through the closed surface S by Q and J respectively, i.e.

$$Q = \int_S v_n\,d\sigma, \quad \bar{J} = \int_S \bar{v}v_n\,d\sigma.$$

The dimensions of Q and J are given by the formulas

$$[Q] = L^3\,T^{-1}, \quad [J] = L^4\,T^{-2}.$$

When the surface S shrinks to a point at the pole, we find that Q and J can only depend on the coefficient ν and on the constant A, the dimensions of which can be expressed in terms of the dimensions of ν. Since the dimensions of Q and ν are independent, it is evident that the mass flow Q equals either zero or infinity. The dimensions of J are expressed in terms of the dimension of the coefficient ν; consequently, J can be finite.

Another situation holds in two dimensional motions. If the velocity field over the whole plane depends only on the point coordinates and on a constant with dimensions dependent on the dimensions of the coefficient of kinematic viscosity ν for plane motion, then formulas analogous to (2.2) would be valid in polar coordinates. (In this case, r is a radius-vector in the plane of motion.)

The mass flow and the momentum flux for plane motion can be determined by

$$Q = \int_L v_n \, dS, \quad \bar{J} = \int_L \bar{v} v_n \, dS,$$

where L is a certain closed contour enclosing the origin. In this case, the dimensions of Q and J are represented by

$$[Q] = L^2 T^{-1}, \quad [J] = L^3 T^{-1}.$$

Therefore, plane motions of the kind considered can be characterized by a finite mass flow but the corresponding momentum will be equal to zero or infinity.

This result enabled Hamel and a number of other authors to obtain exact solutions of the Navier-Stokes equations in the problem of the fluid motion in an angle between two planes by reducing them to ordinary differential equations.†

We now consider the solution of the system of Equations (2.4). The first of Equations (2.4) can be written

$$\left[\frac{1}{\sin\theta}\left(\frac{\Phi'}{\sin\theta}\right)\right]' = \frac{(\psi^2\sin^2\theta)'}{\sin^2\theta}, \tag{2.6}$$

where

$$\Phi = (\phi' - \tfrac{1}{2}\phi^2)\sin^2\theta - \phi\sin\theta\cos\theta. \tag{2.7}$$

The following integral can be derived from Equation (2.6):

$$\psi^2\sin^2\theta - \sin\theta\left(\frac{\Phi'}{\sin\theta}\right)' + 2\Phi + 2\operatorname{ctg}\theta \cdot \Phi' = D, \tag{2.8}$$

where D is a constant of integration.

The function $\psi(\theta)$ determines the velocity components perpendicular to the meridian plane.

It is not difficult to see that the system of Equations (2.4) yields the same single relation to determine the functions f and ϕ for $\psi = 0$ or for $\psi\sin\theta = \text{const.}$

The condition $\psi = c/\sin\theta$ yields $v_\lambda = c\nu/r\sin\theta$; the velocity field for v_λ corresponds to a vortex line coincident with the axis of symmetry.

Therefore the equations of motion will be satisfied if we add the velocity field of a vortex line to the velocity field of the kind considered.

† These solutions are explained in detail in Kochin *et al.*, (1948).

If we assume that

$$\psi \sin \theta = \text{const}, \qquad (2.9)$$

then Equation (2.6) can be integrated three times, and the solution of the problem reduces to the integration of the Riccati equation

$$(\phi' - \tfrac{1}{2}\phi^2) \sin^2 \theta - \phi \sin \theta \cos \theta = M \cos 2\theta + N \cos \theta + R, \qquad (2.10)$$

where M, N, R are arbitrary constants of integration.[†]

If we put $M = N = R = 0$, then (2.10) is integrated easily and yields:

$$\phi = \frac{2 \sin \theta}{A + \cos \theta}, \quad \text{whence } f = -2 + \frac{2(A^2 - 1)}{(A + \cos \theta)^2}, \qquad (2.11)$$

where A is an arbitrary constant. The solution (2.11) was obtained by Landau (Landau and Lifshitz, 1959).

It is not difficult to confirm that the mass flow through a surface enclosing the origin equals zero for this solution when $|A| > 1$ and is infinite for $|A| < 1$.

The flow corresponding to the component of momentum along the axis of symmetry through a sphere with centre at the origin is given by

$$J = \rho \nu^2 \left[8(A^2 - 1) \ln \frac{A-1}{A+1} + 8A - \frac{32A}{3(A^2 - 1)} + \frac{8A(3A^2 - 1)}{A^2 - 1} \right]. \qquad (2.12)$$

Therefore, the magnitude of the momentum J is independent of the radius of the sphere and has the physical characteristics of a singular point at the origin.

The equations of the streamlines in this flow have the form

$$\frac{r}{a} = \frac{A + \cos \theta}{2 \sin^2 \theta}.$$

The shape of the streamline for $A > 1$ is shown in Fig. 20.

At infinity, the streamlines approach parabolic form. The radius-vector r, for motion along a streamline, attains a minimum value at a certain value $\theta = \theta^*$ defined by the relation:

$$\cos \theta^* = -A + \sqrt{(A^2 - 1)}.$$

This corresponds to the motion of an infinite viscous fluid issuing from

[†] N. A. Slezkin obtained Equation (2.10) by other means. See Slezkin (1934).

a fine jet at the origin which is oscillating at the end of an infinitely slender pipe with finite momentum in the x direction.

A whole series of solutions of the Riccati equation (2.10) can be obtained corresponding to particular values of the constants M, N and R.

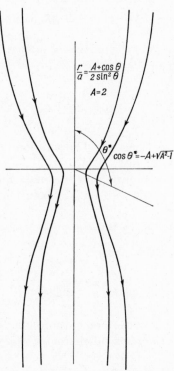

$$\frac{r}{a} = \frac{A + \cos\theta}{2\sin^2\theta}$$

$$A = 2$$

$$\cos\theta^* = -A + \sqrt{A^2 - 1}$$

FIG. 20. Streamline for a source of zero intensity and infinite momentum in a viscous fluid.

For example, for $R = 1$, $N = 0$, $M = \frac{1}{2}$, we have the solution

$$\phi = -\left(\operatorname{ctg}\theta + \coth\frac{\theta + \theta_0}{2}\right)$$

where θ_0 is an arbitrary constant. For $N = 0$, $M = \frac{1}{4} - m^2$ and $R = \frac{9}{4} - (m+1)^2$ we have the solution

$$\phi = (2m - 1)\operatorname{ctg}\theta - \frac{2\sin^{2m}\theta}{\int \sin^{2m}\theta \, d\theta} \quad \text{etc.}$$

The Riccati equation (2.10) can be solved in the general case by using hypergeometric functions (Iatseev, 1950).

Using the substitution

$$\phi = -2\frac{y'(\theta)}{y(\theta)}, \quad \mu = \cos^2\frac{\theta}{2}$$

Equation (2.10) is reduced to

$$\frac{d^2y}{d\mu^2} + \frac{\dfrac{M+R-N}{2} + (N-4M)\,\mu + 4\mu^2\,M}{4\mu^2(\mu-1)^2}\,y = 0. \tag{2.13}$$

This equation is easily integrated in terms of hypergeometric functions; its general integral can be written†:

$$Y(\theta) = \left(\cos\frac{\theta}{2}\right)^{\gamma}\left(\sin\frac{\theta}{2}\right)^{1+\alpha+\beta-\gamma}\left\{PF\left(\alpha, \beta, \gamma, \cos^2\frac{\theta}{2}\right)\right.$$

$$\left. +Q\left(\cos^2\frac{\theta}{2}\right)^{1-\gamma}F\left(\alpha+1-\gamma, \beta+1-\gamma, 2-\gamma, \cos^2\frac{\theta}{2}\right)\right\}, \tag{2.14}$$

where α, β, γ are related to M, N and R by the formulas

$$M = \frac{1-(\alpha-\beta)^2}{4},$$

$$N = 1-(\alpha+\beta)^2 + 2\gamma(\alpha+\beta-1),$$

$$R = \frac{3[1-(\alpha+\beta)^2]}{4} - \alpha\beta - 2\gamma^2 + 2\gamma(\alpha+\beta+1).$$

The parameters α, β, γ can be used as arbitrary constants instead of M, N, R.

The solution obtained for ϕ depends on four arbitrary constants, on three parameters α, β, γ and on the ratio P/Q.

Equation (2.13) degenerates into the equation $y''(\mu) = 0$ for $M = N = R = 0$, which has only one regular singular point at $\mu = \infty$. The corresponding solution was considered above.

If $N = -4M = -\frac{4}{3}R$, then the factor $(\mu-1)^2$ in the denominator of (2.13) cancels and only two singular points $\mu = 0$ and $\mu = \infty$ remain. In

---

† If γ is an integer, then the solution can also be written in another way. To do this, the representation of the solution of the hypergeometric equation in the form cited in Sedov (1950) can be used.

4*

this case, (2.13) transforms into the Euler equation

$$\mu^2 \frac{d^2 y}{d\mu^2} + My = 0,$$

which is integrated easily.

§ 3. Boundary Layer in the Flow of a Viscous Fluid Past a Flat Plate

Consider the flow of an incompressible viscous fluid past an infinitely thin plate. We suppose that the fluid moves forward with the constant velocity U_0 far in front of the plate; the plate is of infinite length and is parallel to the undisturbed velocity direction. The problem is one of plane steady motion and the fluid occupies the whole plane beyond the plate. It is one of the simplest problems concerning the motion of a viscous fluid; nevertheless, it has not been solved as an exact solution of the Navier-Stokes equations because of considerable mathematical difficulties. We analyse this problem by using the Prandtl equations, which are obtained from the general equations of the motion of a viscous fluid by using certain approximations (Prandtl, 1905).

The Prandtl boundary layer equations in the case under consideration are:

$$\left.\begin{aligned} u \frac{\partial u}{\partial x} + v \frac{\partial u}{\partial y} &= \nu \frac{\partial^2 u}{\partial y^2}, \\ \frac{\partial u}{\partial x} + \frac{\partial v}{\partial y} &= 0, \end{aligned}\right\} \quad (3.1)$$

where u and v are the components of the fluid velocity along the co-ordinate axes and ν is the coefficient of kinematic viscosity. The x axis is directed along the plate, the y axis is perpendicular to the plate. Besides (3.1), we have the boundary conditions

$$\left.\begin{aligned} x > 0, y = 0 \quad & u = v = 0; \\ y = \pm \infty \quad & u = U_0 \end{aligned}\right\} \quad (3.2)$$

to determine $u(x, y)$ and $v(x, y)$.

The characteristic parameters will be

$$U_0, \nu, x, y.$$

Since the plate is flat and infinitely long, it is impossible to choose a characteristic linear dimension. It follows from the general considerations of dimensional analysis that all the nondimensional quantities are functions of the two independent combinations:

$$\frac{y}{x}, \ \frac{y}{\sqrt{\left(\dfrac{vx}{U_0}\right)}} \ .$$

Consequently, formulas of the type

$$u = U_0 f\left(\frac{y}{x}, \ \frac{y}{\sqrt{\left(\dfrac{vx}{U_0}\right)}}\right), \tag{3.3}$$

$$v = \sqrt{\left(\frac{vU}{x}\right)} \, \Phi\left(\frac{y}{x}, \ \frac{y}{\sqrt{\left(\dfrac{vx}{U_0}\right)}}\right), \tag{3.4}$$

apply in this problem.

We now show that the first parameter y/x in (3.3) and (3.4) is not essential because of the singularity of (3.1). To do this, we make the transformation of variables:

$$x = l\xi, \quad y = \sqrt{\left(\frac{vl}{U_0}\right)} \cdot \eta, \quad u = U_0 u_1, \quad v = \sqrt{\left(\frac{vU_0}{l}\right)} v_1, \tag{3.5}$$

where l is a certain constant larger than zero. If we attribute the dimensions of a length to l, then the quantities ξ, η, u_1, v_1 can be considered as nondimensional.

When the new variables defined by (3.5) are introduced, Equations (3.1) become

$$\left.\begin{aligned} u_1 \frac{\partial u_1}{\partial \xi} + v_1 \frac{\partial u_1}{\partial \eta} &= \frac{\partial^2 u_1}{\partial \eta^2}, \\[2mm] \frac{\partial u_1}{\partial \xi} + \frac{\partial v_1}{\partial \eta} &= 0. \end{aligned}\right\} \tag{3.6}$$

The boundary conditions (3.2) become, in terms of the new variables:

$$\left.\begin{aligned} \xi > 0, \eta = 0 \quad & u_1 = v_1 = 0; \\[2mm] \eta = \pm\infty \quad & u_1 = 1. \end{aligned}\right\} \tag{3.7}$$

Equations (3.6) and boundary conditions (3.7) give the formulation of the boundary layer problem in nondimensional form. The solution of this problem cannot depend on $U_0 l/v = \mathbf{R}$ which appears nowhere in

Equations (3.6) and the boundary conditions (3.7).† On the other hand, the general formulas (3.3) and (3.4) show that

$$u_1 = \frac{u}{U_0} = f\left(\frac{\eta}{\xi\sqrt{\mathbf{R}}}, \frac{\eta}{\sqrt{\xi}}\right), \tag{3.8}$$

$$v_1 \sqrt{\xi} = \frac{v}{\sqrt{\left(\dfrac{\nu U_0}{x}\right)}} = \Phi\left(\frac{\eta}{\xi\sqrt{\mathbf{R}}}, \frac{\eta}{\sqrt{\xi}}\right). \tag{3.9}$$

Since the solution need not depend on **R**, it follows that the first argument y/x need not enter into the right side of formulas (3.8) and (3.9).

Hence, we have proved that the solution of the problem posed must have the form‡

$$u = U_0 f\left(\frac{y}{\sqrt{\left(\dfrac{\nu x}{U_0}\right)}}\right), \tag{3.10}$$

$$v = \sqrt{\left(\frac{\nu U_0}{x}\right)} \,\Phi\left(\frac{y}{\sqrt{\left(\dfrac{\nu x}{U_0}\right)}}\right). \tag{3.11}$$

Now we introduce the new variable $\lambda = \eta/\sqrt{\xi} = y/\sqrt{(\nu x/U_0)}$ and put $f(\lambda) = \phi'(\lambda)$. Substituting u and v in the continuity equation, we express $\Phi(\lambda)$ in terms of $\phi(\lambda)$ as follows:

$$\Phi'(\lambda) = \tfrac{1}{2}\lambda\phi''(\lambda) = \tfrac{1}{2}(\lambda\phi' - \phi)'.$$

Using this relation, formulas (3.10) and (3.11) can now be written

$$u = U_0 \phi'(\lambda), \tag{3.10'}$$

$$v = \sqrt{\left(\frac{\nu U_0}{x}\right)} \cdot \tfrac{1}{2}[\lambda\phi'(\lambda) - \phi(\lambda)]. \tag{3.11'}$$

Substituting the expressions obtained for u and v in the first of Equations (3.1), we obtain an ordinary third order differential equation for $\phi(\lambda)$:

$$2\phi''' + \phi\phi'' = 0. \tag{3.12}$$

---

† This is a property of the Prandtl equations (3.1). If the transformation (3.5) is applied to the Navier-Stokes equations, then we obtain nondimensional equations containing the parameter **R**, consequently subsequent conclusions lose their validity when applied to the Navier-Stokes equations.

‡ Proof of the validity of formula (3.10) is made by another method in Loitsianskii (1941).

From the boundary conditions (3.2) for the desired function $\phi(\lambda)$ which satisfies (3.12), the following boundary conditions are obtained

$$\phi'(0) = \phi(0) = 0 \quad \text{and} \quad \phi'(\infty) = 1. \tag{3.13}$$

The solution of the nonlinear differential equation (3.12) under the boundary conditions (3.13) can be obtained approximately (Blasius, 1908). A very general property of (3.12) used by Töpfer (1912) in an approximate method of solution is the following.

If $\phi_0(\lambda)$ is a certain solution of (3.12), then the function

$$\phi(\lambda) = a\phi_0(a\lambda),$$

where a is any constant, is also a solution of (3.12). It is easy to see by a direct check that this property is correct.

We take as the initial solution $\phi_0(\lambda)$ that solution of (3.12) which satisfies the boundary conditions

$$\phi_0(0) = \phi_0'(0) = 0 \quad \text{and} \quad \phi_0''(0) = 1.$$

The functions $\phi_0(\lambda)$ can be constructed by the usual approximate methods. The following limit can be calculated from the approximate solution

$$\lim_{\lambda \to +\infty} \phi_0'(\lambda) = k.$$

Numerical calculations yield $k = 2 \cdot 0854$. The solution of (3.12) given by the formula

$$\phi(\lambda) = \alpha^{1/3} \phi_0(\alpha^{1/3} \lambda),$$

satisfies the boundary conditions

$$\phi(0) = \phi'(0) = 0 \quad \text{and} \quad \phi''(0) = \alpha$$

in which

$$\lim_{\lambda \to +\infty} \phi'(\lambda) = k \cdot \alpha^{2/3}.$$

Hence, it is clear that it is sufficient to put

$$\alpha = \frac{1}{k^{3/2}} = 0 \cdot 332,$$

to obtain the desired solution.

Defining $\alpha = \phi''(0)$, we easily find the friction drag acting on the plate by using (3.10').

We have for the friction stress τ at the plate:

$$\tau = \mu\left(\frac{du}{dy}\right)_{y=0} = \mu U_0 \frac{\phi''(0)}{\sqrt{\left(\frac{\nu x}{U_0}\right)}} = 0\cdot332\sqrt{\left(\frac{\rho\mu U_0^3}{x}\right)}. \qquad (3.14)$$

Using this, we calculate the drag W of a section of a plate of width b and of length l:

$$W = b\int_0^l \tau\,dx = 0\cdot664b\cdot\sqrt{(\rho\mu l U_0^3)}. \qquad (3.15)$$

The friction stress τ and the drag W are seen to be proportional to the three-halves power of the stream velocity.

The following formula for the friction coefficient c_f is obtained from relation (3.15):

$$c_f = \frac{2W}{\rho b l U_0^2} = \frac{1\cdot328}{\sqrt{\mathbf{R}}}$$

where

$$\mathbf{R} = \frac{U_0 l}{\nu}.$$

Experimental data (Hansen, 1928, Fage, 1934) on plane smooth flat plates are in good agreement with the velocity distribution and drag laws in the laminar flow region characterized by low values of the Reynolds number

$$\mathbf{R} = \frac{U_0 l}{\nu} < 3\times10^5.$$

The laminar, steady motion considered above is unstable at high Reynolds numbers. Turbulent motion arises which basically alters the drag and velocity distribution near the plate.

§ 4. Isotropic Turbulent Motion of an Incompressible Fluid

1. averaging of turbulent motion

Many fluid motions observed in nature and the majority of motions with which we deal in engineering are characterized by the presence of disorderly, unsteady, fluid motions superposed on a basic fluid motion which can be represented as a certain statistically average motion. Fluid motions of such a kind are called turbulent.

The velocity, pressure and other quantities at each point of the flow in turbulent fluid motion undergo irregular fluctuating variations about certain average values. Consequently, it is feasible to use probability theory concepts to investigate turbulent flows; in this case, the instantaneous values of the mechanical characteristics are considered as random variables and the average values are defined as the mathematical means (Millionshchikov, 1939). Often, however, the average values are determined as the usual time averages. The time interval within which the averaging is carried out must be sufficiently large in comparison with the time of the separate fluctuations and must be small in comparison with the time for a noticeable change to take place in the average quantities, if the average motion is not stationary (Kochin et al., 1948).

The average values of the pressure, the velocity components, the products of the fluctuating velocity components, evaluated at the same point or at neighbouring points (the correlation coefficient of the velocity) etc., depend to a great degree on the presence of turbulent mixing. This causes the variations in the average quantities to be equalized and smoothed out, to an extent which varies with position in space.

Experiment shows that steady, laminar motions of a fluid at large values of Reynolds number, i.e. at high speeds and large scales, become unstable and change to unsteady turbulent motions. In a number of cases the mean of such motion is steady.

The investigation of turbulent flow is very important in modern hydro and aerodynamics and affects many basic problems including the calculation of profile drag.

In all theoretical investigations of viscous fluid motion, it is assumed that the Navier-Stokes equations apply to the actual unsteady, pulsating motion. However, when turbulence is present individual fluid particles execute very tortuous, complex motions and to obtain the solution of the Navier-Stokes equations would be an extremely awkward and complicated problem, comparable to that of describing the motion of the individual molecules of a large volume of gas. Consequently, the fundamental problems of turbulent fluid motion in hydromechanics are posed, exactly as in kinetic gas theory, as problems of finding functional relations between average quantities.

The equations of motion for the average quantities can be obtained by averaging the equations for the quantities describing the instantaneous state of the motion. Because these equations are non-linear, after the averaging we are left with more unknowns than equations. The average values of non-linear terms, for example, the product of two or of several quantities, are new unknowns (Keller, 1925). Hence, when averaging the Navier-Stokes equations for an incompressible fluid,[†] average values of the products $\overline{u_i u_k}$ (i, $k = 1$, 2, 3) must be taken into account in addition to the average values[‡] of the velocity components $\overline{u}_1$, $\overline{u}_2$, $\overline{u}_3$.

Therefore, the one vector momentum equation which is sufficient for studying real motions is inadequate for a mathematical study of average turbulent motion. Consequently, a complete, theoretical investigation of average turbulent motions is possible only if we make certain additional hypotheses the validity of which must be established by experiment.[§]

The essence of a number of papers on turbulence reduces to the study of the validity of various, plausible hypotheses. These are simple and natural in concept, can be verified experimentally, and permit fundamental problems on turbulent fluid motion to be formulated and solved theoretically.

No general mathematical formulation of the problem of arbitrary average turbulent motions yet exists and, in general, the possibility of giving a mathematical formulation of the problem similar to that of the problem of the real motion of a viscous fluid has still to be realized.

Dimensional analysis and similarity concepts are frequently used as the fundamental means of investigating turbulent fluid motions.

2. PROPERTIES OF HOMOGENEITY AND ISOTROPY

Let us consider the turbulent motion of a viscous fluid of infinite extent.[‖]

The state of the motion at each instant t is determined by the initial disturbances (the disturbances in the fluid at $t = 0$) and by the inertial and viscous properties of the fluid, i.e. by the quantities ρ and μ.

Consider a system of initial disturbances which are kinematically similar. Then each individual disturbed state can be singled out by assigning a length and time scale and by selecting a certain characteristic velocity u_0 and a certain characteristic quantity l_0 with the dimensions of length.

Therefore, the turbulent state of the fluid motion for a system of kinematically similar initial disturbances is defined by the parameters

$$\rho, \ \mu, \ l_0, \ u_0, \ t, \ x_1, \ x_2, \ x_3,$$

where x_1, x_2, x_3 are the coordinates of a point in space.[¶]

[†] In what follows we shall assume that the fluid is always incompressible.

[‡] As is customary, we shall denote the average values of quantities by means of a bar.

[§] We are discussing the formulation of mathematical problems with a finite number of unknowns.

[‖] The following analysis also refers to fluid motion in a finite region if the influence of boundaries can be neglected.

[¶] The coordinate systems are located similarly with respect to the distribution of the initial disturbances.

The set of turbulent motions so obtained contains motions which are not dynamically similar. For two turbulent motions to be similar, it is necessary and sufficient that the Reynolds number should have the same value in both motions:

$$\frac{u_{01} l_{01} \rho_1}{\mu_1} = \frac{u_{02} l_{02} \rho_2}{\mu_2}.$$

The instants of time and the coordinates of the points corresponding to similar states are determined from the relations

$$\frac{u_{01} t_1}{l_{01}} = \frac{u_{02} t_2}{l_{02}}; \quad \frac{x_{i1}}{l_{01}} = \frac{x_{i2}}{l_{02}} \quad (i = 1, 2, 3).$$

The values of u and l are determined by functions such as

$$\left. \begin{aligned} \frac{u}{u_0} &= f\left(\frac{u_0 t}{l_0}, \frac{u_0 l_0 \rho}{\mu}, \frac{x_1}{l_0}, \frac{x_2}{l_0}, \frac{x_3}{l_0}\right), \\[2mm] \frac{l}{l_0} &= \phi\left(\frac{u_0 t}{l_0}, \frac{u_0 l_0 \rho}{\mu}, \frac{x_1}{l_0}, \frac{x_2}{l_0}, \frac{x_3}{l_0}\right). \end{aligned} \right\} \quad (4.1)$$

In general, formulas of a similar kind will hold for any nondimensional mechanical variables evaluated at one point of a fluid. These functions depend on nondimensional parameters defining the distributions of the initial perturbations as well as on the parameters mentioned above.

Variables depending on the state of the motion at two or more points must be considered when studying turbulent motion; these variables can depend on the coordinates of several points. For example, the average values of the product of the velocity components at m points $M_1(x_1', x_2', x_3')$, $M_2(x'', x_2'', x_3'')$, ..., $M_m(x_1^m, x_2^m, x_3^m)$

$$\tau_{k_1 k_2 \dots k_n}(M_r, M_s, \dots, M_d) = \overline{u_{k_1}(M_r) \, u_{k_2}(M_s) \dots u_{k_n}(M_d)} \quad (4.2)$$
$$(k_1, k_2, \dots, k_n = 1, 2, 3)$$

generate a tensor whose components depend on $3m$ coordinates $x_i^k (i = 1, 2, 3; \quad k = 1, 2, \dots m)$ for a system of similar initial disturbances. The subscripts $r, s, \dots, d$ are functions of all the coordinates in the general case.

A turbulent flow is called homogeneous if all the average quantities at each point are independent of the position of the point and if the average values for quantities depending on the positions of several points only depend on their relative location, i.e. only on the differences of the coordinates $x_i^p - x_i^q$. The functions (4.1) in a homogeneous turbulent velocity field are independent of x_1/l_0, x_2/l_0, x_3/l_0. The averaged characteristics of the fluid motion near any two points are identical.

Evidently, the initial perturbations in the case of a homogeneous turbulent flow must be uniformly distributed over the region occupied by the fluid.

A homogeneous turbulent flow is called isotropic if the tensor relations formed from the velocity components, defined by (4.2), for arbitrary n and m, are independent of the orientation of the polygon $M_1 M_2 \ldots M_m$ in space and of reflections of this polygon in the coordinate planes.[†]

The components of the tensor relation depend on the positions of the coordinate axes relative to the polygon $M_1 M_2 \ldots M_m$ and on the choice of the positive direction along the axes. The quantities $\tau_{k_1 k_2 \ldots k_n}$, for an isotropic turbulent flow, have the same value in all coordinate systems oriented identically relative to various positions of the same polygon.

By definition, isotropic turbulent motion has the property of symmetry in the mean. It is assumed that all velocity directions are equally probable at each point of space during the sufficiently long time interval over which the average is taken; this is combined with the assumption that the fluid motion at each instant is continuous and has approximately the same velocity at neighbouring points.

Homogeneous, isotropic turbulence can be considered as the simplest kind of turbulent motion. The perturbed fluid is moved by inertia; and dissipation of kinetic energy occurs under the effect of internal viscous forces. The motion is characterized by damping, and decay of the turbulent disturbances. The fundamental problem in the study of isotropic turbulence is the determination of the damping law.[‡]

Turbulent motions are characterized in the general case by the equalization and diffusion of disturbances. Isotropic turbulent motion can be considered as homogeneous isotropic turbulent motion. Turbulent motion of air blown through a fixed grating also reduces to this case if a constant velocity translation is added to the whole system. We encounter such a problem in investigating turbulence in wind tunnels.

3. SYMMETRY PROPERTY OF THE VELOCITY CORRELATION TENSORS

The isotropy assumption leads to a number of relations between the components of the tensor $\tau_{k_1 k_2 \ldots k_n}$. For example, if the components of the tensor $\tau_{k_1 k_2 \ldots k_n}$ are formed from the average values of the velocity components of the common point $M_1 = M_2 = \ldots = M_m$, then, obviously, we shall have, for odd values of n,

$$\tau_{k_1 k_2 \ldots k_n} = 0.$$

[†] If this condition is valid only for $2 \leqslant n < N$, $2 \leqslant m < M$, where N and M are certain integers, then such a flow can be considered to be approximately isotropic.

[‡] At the present time, the question of the existence and of the possible different kinds of isotropic turbulent motion of a viscous fluid has still not been examined.

The only non-vanishing components are those in which each velocity component occurs to an even power, for even values of n. In particular, we have for $n = 1$:

$$\bar{u}_1 = \bar{u}_2 = \bar{u}_3 = 0$$

for $n = 2$, we shall have

$$\overline{u_1 u_2} = \overline{u_1 u_3} = \overline{u_2 u_3} = 0, \quad \overline{u_1^2} = \overline{u_2^2} = \overline{u_3^2} = \tfrac{1}{3}\overline{\mathfrak{v}^2}, \tag{4.3}$$

and for $n = 3$, all the components are zero, etc.

Certain symmetry conditions hold if the points M_1, M_2, ..., M_m are different, in particular, if we have just two points. In the latter case, the correlation tensor depends only on the distance r between the points in question. The components of $\tau_{k_1 k_2 \ldots k_n}$ depend on r and on the orientation of the coordinate axes relative to the segment $M_1 M_2$.

The relations

$$\overline{u_i(M_1) u_j(M_2)} = \overline{u_i(M_2) u_j(M_1)} \quad (i = 1, 2, 3; \quad j = 1, 2, 3) \tag{4.4}$$

can be deduced from the property of isotropy of the motion.

Let us take the point M_1 at the origin and the point M_2 on the x_1 axis; then we have

$$\tau_{11} = b_d^d \neq 0, \quad \tau_{22} = \tau_{33} = b_n^n \neq 0, \quad \tau_{12} = \tau_{13} = \tau_{23} = 0. \tag{4.5}$$

The superscripts refer to the M_2 point and the subscripts to the M_1 point. It is evident that the quantities b_d^d and b_n^n depend on x_1 and t and are even functions of x_1.

If the position of the point M_2 is referred to the coordinate axes, then all the τ_{ik} components can easily be expressed in terms of b_d^d and b_n^n. The formulas expressing τ_{ik} in terms of b_d^d and b_n^n are found from the formulas for the tensor components when the special coordinate system in which the x_1 axis passes through the point M_2 is transformed to an arbitrarily assigned coordinate system. These formulas are

$$\tau_{ik} = (b_d^d - b_n^n) l_{i1} l_{k1} + b_n^n \delta_{ik} \left(\delta_{ik} \begin{array}{l} = 1; \quad i = k \\ = 0; \quad i \neq k \end{array} \right) \tag{4.6}$$

where l_{ik} are the direction cosines of the axes of the assigned coordinate system with respect to the special system.

We have for $r = 0$:

$$b_d^d = b_n^n = \tfrac{1}{3}\overline{\mathfrak{v}^2} = b. \tag{4.7}$$

Now, let us consider the symmetry conditions satisfied by a third order correlation tensor. The transformation formulas of third order tensor components for a transformation from one coordinate system to another are

$$\Pi'_{ijk} = \sum_{\alpha, \beta, \gamma} \Pi_{\alpha\beta\gamma} l_{\alpha i} l_{\beta j} l_{\gamma k} \qquad (4.8)$$

where l_{sm} are the direction cosines of the new coordinate system. Let us take the transformation

$$x'_1 = -x_1, \quad x'_2 = x_2, \quad x'_3 = x_3$$

which reduces simply to a reversal in the direction of the x_1 axis. We have in this case

$$l_{11} = -1, \quad l_{22} = l_{33} = 1, \quad l_{sm} = 0 \quad (s \neq m)$$

consequently, if the subscript 1 occurs among the subscripts i, j, k an odd number of times, it follows from (4.8) that

$$\Pi'_{ijk} = -\Pi_{ijk}.$$

Let the point M_1 coincide with the origin and let the point M_2 lie on the x_1 axis. It then follows from the isotropy properties that the components of the third order correlation tensor formed for the velocity component of the points M_1 and M_2 are independent of the directions of the x_2 and x_3 axes.

Hence it follows that the components containing the subscripts 2 or 3 an odd number of times are zero since they cannot change sign with a change in the direction of the x_2 or x_3 axes.

Hence, if the point M_2 lies on the x_1 axis, then the correlation tensor (formed from two velocity components at the point M_1 and one velocity component at the point M_2) has the following five non-vanishing components

$$\tau_{111} = b_{dd}^d,$$

$$\tau_{122} = \tau_{133} = b_{dn}^n,$$

$$\tau_{221} = \tau_{331} = b_{nn}^d.$$

The third order correlation tensor components[†] in any coordinate system can be expressed in terms of b_{dd}^d, b_{dn}^n and b_{nn}^d by using (4.8). In the general case, the distance between the M_1 and M_2 points, which we shall subsequently denote by the letter r, must be taken instead of the variable x_1. As the distance between the points increases, their velocities become

---

[†] Analogous symmetry properties of the velocity projection coupling tensors can be explained for higher order tensors. The symmetry conditions for fourth order correlation tensors are given in Millionshchikov (1941).

more and more independent statistically, consequently, the components of the velocity correlation tensors must approach zero as r increases without limit.

The points M_1 and M_2 coincide at $r = 0$; in this case, we have,

$$b_{dd}^d = b_{dn}^n = b_{nn}^d = 0.$$

Interchanging the roles of the points M_1 and M_2 is equivalent to reversing the direction of the coordinate axes. Hence, it follows that

$$\tau_{ijk}(M_1, M_1, M_2) = -\tau_{ijk}(M_2, M_2, M_1). \tag{4.9}$$

These relations can be written in the alternate form,

$$\left.\begin{aligned}
b_{dd}^d &= -b_d^{dd}, \\
b_{dn}^n &= -b_n^{dn}, \\
b_{nn}^d &= -b_d^{nn}.
\end{aligned}\right\} \tag{4.10}$$

Therefore, the correlation coefficients b_{dd}^d, b_{dn}^n and b_{nn}^d are odd functions of x_1 or r.

If the fluctuating velocity components are regular functions of the coordinates expanded in a Taylor series, then it is evident that the correlation coefficients can also be expanded in a Taylor series in r. The Taylor series for the second order moments b_d^d and b_n^n will contain only even powers of r and the series for the third order moments b_{dd}^d, b_{dn}^n and b_{nn}^d will only contain odd powers of r.

We shall show that the series for b_{dd} does not contain a linear term in r. We have

$$b_{dd}^d = \overline{u_1^2(0, 0, 0)\, u_1(r, 0, 0)} = \overline{u_1^2\left(\frac{\partial u_1}{\partial r}\right)}_{r=0} r + \frac{1}{6}\overline{u_1^2\left(\frac{\partial^2 u_1}{\partial r^2}\right)}_{r=0} r^3 + \cdots$$

The first term becomes zero since

$$\overline{u_1^2\left(\frac{\partial u_1}{\partial r}\right)}_{r=0} = \frac{1}{3}\left[\frac{\partial \overline{u_1^3}}{\partial r}\right]_{r=0},$$

and $\overline{u_1^3}(r) = 0$ because of the isotropy of the turbulent motion. Thus, the power series expansion of the moment b_{dd}^d must start with a term at least of order r^3.

4. INCOMPRESSIBILITY CONDITION AND DYNAMIC RELATIONS

Using the averaging operation, relations between the independent components of the velocity correlation tensors can be established from the incompressibility and the Navier-Stokes equations (Friedman and

Keller, 1924; Kármán and Howarth, 1938). The following relations are obtained from the incompressibility equations:

$$b_n^n = b_d^d + \frac{r}{2} \frac{\partial b_d^d}{\partial r},$$ (4.11)

$$b_n^{nd} = -b_d^{nn} - \frac{r}{2} \frac{\partial b_d^{nn}}{\partial r},$$ (4.12)

$$b_d^{dd} = -2b_d^{nn}.$$ (4.13)

The correlation coefficients b_d^d and b_n^n can be measured directly and independently in experiments. Using Simmons' experimental results,

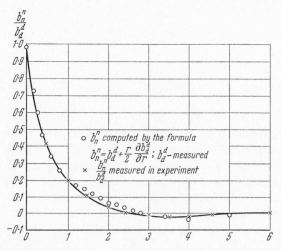

FIG. 21. Experimental results in good agreement with the theoretical formula relating b_d^d and b_n^n.

obtained in a wind tunnel, Taylor (1937) showed that the experiments are in very good agreement with (4.11) (Fig. 21).

Relations (4.12) and (4.13) show that the power series in r for b_d^{nn} and b_n^{nd} must start with a term at least of order r^3 for small r since the series for b_d^{dd} has this property. Equations (4.11), (4.12) and (4.13) show that the determination of the velocity component correlation tensors of second and third order reduces to the determination of two functions $b_d^d(r, t)$ and $b_d^{nn}(r, t)$. Only one equation is obtained for these quantities from the Navier-Stokes equations: namely,

$$\nu\left(\frac{\partial^2 b_d^d}{\partial r^2} + \frac{4}{r} \frac{\partial b_d^d}{\partial r}\right) - \frac{1}{2} \frac{\partial b_d^d}{\partial t} = \frac{\partial b_d^{nn}}{\partial r} + \frac{4}{r} b_d^{nn},$$ (4.14)

where $\nu = \mu/\rho$. This equation was obtained by Kármán and Howarth in another form for the functions $f = b_d^d/b$ and $h = b_d^{nn}/b^{3/2}$. Loitsianskii (1939) and Millionshchikov (1941) analysed (4.14) in the form described.

Multiplying (4.14) by r^4, we obtain:

$$\frac{1}{2}\frac{\partial b_d^d r^4}{\partial t} = \frac{\partial}{\partial r}\left[\nu r^4 \frac{\partial b_d^d}{\partial r} - r^4 b_d^{nn}\right]. \tag{4.14'}$$

Assuming that the order to which the functions $\partial b_d^d/\partial r$ and b_d^{nn} vanish as $r \to \infty$ is higher than $1/r^4$ and that the expression in the square brackets becomes zero at $r = 0$, we find

$$\frac{d}{dt}\int\limits_0^\infty b_d^d r^4 \, dr = 0$$

or

$$\Lambda = \int\limits_0^\infty b_d^d r^4 \, dr = \text{const.} \tag{4.15}$$

The existence of the invariant quantity Λ was discovered by Loitsianskii (1939). However, we may note that the properties of boundedness and invariance of Λ are established on the assumption about the order to which b_d^{nn} and the derivative $\partial b_d^d/\partial r$ vanish.

The single equation (4.14) is inadequate to determine b_d^d and b_d^{nn}. Transformation to fourth (Millionshchikov, 1941) and higher order correlation coefficients also does not yield a closed system of equations. The equation for the third order moments contains a fourth order moment; the next equations contain fifth order moments, etc. Besides having to contend with the inconsistency of these equations, we have also to assign initial conditions; consequently, additional hypotheses of a mechanical nature are required for a theoretical study of isotropic turbulence.[†]

Similar conclusions were reached by Kármán and Howarth (1938) in analysing the cases of small and large Reynolds numbers.

5. CONCLUDING STAGES IN THE DECAY OF ISOTROPIC TURBULENCE

When the turbulent motion is damped, the quantity $b = \overline{u_1^2}$ approaches zero as t approaches infinity.

The components of the third order correlation tensor are small in comparison with the components of the second order correlation tensor

[†] An explanation of these questions as well as the results contained in paragraphs 5, 7, 8 were published in the first edition in 1944. A brief explanation is also given in Sedov (1944).

when the fluctuating velocities are very small. This affords a basis for neglecting third order moments in (4.14). Replacing the right side of (4.14) by zero, we obtain

$$\nu\left(\frac{\partial^2 b_d^d}{\partial r^2}+\frac{4}{r}\frac{\partial b_d^d}{\partial r}\right)-\frac{1}{2}\frac{\partial b_d^d}{\partial t} = 0. \tag{4.16}$$

To find a particular solution of (4.16), the function $b_d^d(r, 0)$ which gives the initial form of the correlation function of the longitudinal velocities of two points, must be known. We propose to obtain the solution of (4.16) for large values of the time t. The singularities of the function $b_d^d(r, 0)$ play a secondary role for large t, consequently, the initial conditions can be put into a simplified form when investigating the asymptotic behaviour of the function $b_d^d(r, t)$ as $t \to +\infty$.

Since the dimensions of b_d^d and r are different the function $b_d^d(r, 0)$ involves dimensional constants, as well as the distance r.

Now let us assume that the influence of the initial distribution of the correlation coefficients is represented in the asymptotic behaviour of b_d^d as $t \to +\infty$ by means of the single constant factor A with the dimensions $L^p T^q$. This constant represents the total properties of the initial distribution or the coupling moment b_d^d. In particular, the value of the constant A can be determined from the formula

$$A = \left[\int_0^\infty b_d^d(r, t)\frac{r^{p+2q+1}}{\nu^{2+q}}\,\Phi\left(\frac{r^2}{\nu t}\right)dr\right]_{t=0},$$

where $\Phi(r^2/\nu t)$ is a certain function.

Let us put

$$b_d^d = A\nu^{1-p/2}\frac{f(r, t, \nu)}{t^{1+q+p/2}}.$$

By hypothesis, the nondimensional quantity $f(r, t, \nu)$ depends only on three dimensional parameters r, t, ν, consequently, the function $f(r, t, \nu)$ depends only on the combination $\xi = r^2/\nu t$. We may therefore write

$$b_d^d = \frac{A\nu^{1-p/2}}{t^{1+q+p/2}} f\left(\frac{r^2}{\nu t}\right). \tag{4.17}$$

Later, we shall consider such motions for which $f(\infty) = 0$ and $f(0) = 1$; the first condition corresponds to the statistical independence of the velocities of two points as $r \to +\infty$ and the second condition is easily satisfied by a choice of the numerical value of the constant A.

We have from (4.17) for $r = 0$:

$$b = \tfrac{1}{3}\overline{\mathfrak{v}}^2 = \frac{A\nu^{1-p/2}}{t^{1+q+p/2}}. \tag{4.18}$$

Because of the damping of the turbulent motion, we must have $1+q+p/2 > 0$, consequently, since $f(0) = 1$, the distribution of the disturbances defined by (4.17) is characterized by irregularity at $t = 0$.

It is evident that the function

$$f\left(\frac{r^2}{\nu t}\right) = \frac{\overline{u_1(0, 0, 0, t)\, u_1(r, 0, 0, t)}}{\overline{u}_1^2}$$

is the correlation coefficient between the velocity components of two points on the segment connecting these points. The correlation coefficient equals zero in the case under consideration at $t = 0$ if $r \neq 0$ and equals unity if $r = 0$.

Using (4.17), we obtain from (4.16)

$$f'' + \left(\frac{1}{8} + \frac{5}{2\xi}\right)f' + \frac{5\alpha}{4\xi}f = 0 \tag{4.19}$$

where

$$\alpha = \frac{1+q+p/2}{10}.$$

This equation has been obtained by Kármán and Howarth (1938) directly, from the assumption that the correlation coefficient depends only on the combination $r^2/\nu t$. The number α is an arbitrary constant introduced by Kármán.

The solution of (4.19) depends only on the constant α. For $\alpha = \mathrm{const}$ the dimensions of the constant A are fixed solely by the difference between the values of p and q. Formula (4.17) shows that the constant $A\nu^{1-p/2}$ must have the dimensions

$$L^2\, T^{q+p/2-1} = L^2\, T^{10\alpha-2}.$$

The general solution of (4.19) is regular for all $\xi \neq 0$, ∞. The regular solution at $\xi = 0$ which satisfies the condition $f(0) = 1$ is

$$f(\xi) = M\left(10\alpha, \frac{5}{2}, -\frac{\xi}{8}\right)$$

$$= 1 - \alpha\xi + \frac{\alpha(10\alpha+1)}{4.7.2!}\xi^2 - \frac{\alpha(10\alpha+1)(10\alpha+2)}{4^2.7.9.3!}\xi^3 + \dots \tag{4.20}$$

where $M(\alpha, \gamma, x)$ is a confluent, hypergeometric function (Jahnke and Emde, 1938). The asymptotic expansion

$$f(\xi) = \frac{\Gamma\left(\frac{5}{2}\right) 8^{10\alpha}}{\Gamma\left(\frac{5}{2} - 10\alpha\right)} \frac{1}{\xi^{10\alpha}} \left[1 + 8 \frac{10\alpha\left(10\alpha - \frac{3}{2}\right)}{\xi} \right.$$

$$\left. + 8^2 \frac{10\alpha(10\alpha + 1)\left(10\alpha - \frac{3}{2}\right)\left(10\alpha - \frac{1}{2}\right)}{2! \, \xi^2} + \ldots \right], \qquad (4.21)$$

where Γ is the Euler Gamma function, gives the solution for very large values of ξ.

The behaviour of the moment b_d^d as $r \to +\infty$ or as $t \to 0$ is easily explained by use of (4.17) and (4.21). For all $\alpha > 0$, we have

$$\lim_{r^2/\nu t \to \infty} b_d^d r^{20\alpha} = A\nu^{2+q} \frac{\Gamma\left(\frac{5}{2}\right) 8^{10\alpha}}{\Gamma\left(\frac{5}{2} - 10\alpha\right)}. \qquad (4.22)$$

If $10\alpha - 5/2 = k$, i.e., if $\alpha = 1/4 + 0 \cdot 1k$ where k is a positive integer, then the required solution of (4.19) finally takes the simple form

$$f = M\left(k + \frac{5}{2}, \frac{5}{2}, -\frac{\xi}{8}\right) = \frac{\Gamma\left(\frac{5}{2}\right)}{\Gamma\left(k + \frac{5}{2}\right)} \left[\frac{d^k}{d\lambda^k} \left(\lambda^{k+3/2} e^{-\xi\lambda/8} \right) \right]_{\lambda = 1}$$

$$= \Gamma\left(\frac{5}{2}\right) \left[\frac{1}{\Gamma\left(\frac{5}{2}\right)} - \frac{k}{\Gamma\left(\frac{7}{2}\right)} \frac{\xi}{8} + \frac{k(k-1)}{\Gamma\left(\frac{9}{2}\right)} \frac{\xi^2}{2!} \frac{1}{8^2} - \ldots + (-1)^k \frac{\xi^k}{\Gamma\left(k + \frac{5}{2}\right) 8^k} \right] e^{-\xi/8}.$$

In this case, the moment b_d^d for $t = 0$ is a source like function. We have $b_d^d = 0$ for $r \neq 0$ and $t = 0$, and $b_d^d \to \infty$ for $r = 0$ as $t \to 0$.

Equation (4.16) can be interpreted as the heat conduction equation in a five-dimensional space with symmetry relative to the origin. The solution corresponding to $\alpha = 1/4$ (when $k = 0$) can be considered as the analogue of a heat source in a five-dimensional space (Loitsianskii, 1939; Millionshchikov, 1939). In this case, the solution is

$$b_d^d = A\nu^{1-p/2} \frac{e^{-r^2/8\nu t}}{\sqrt{t^5}}. \qquad (4.23)$$

The constant $A\nu^{1-p/2}$ for $\alpha = 1/4$, has the dimensions $L^2 T^{0\cdot5}$.

It is easy to verify that the above solutions for $\alpha = 1/4 + 0 \cdot 1k$, where k is a positive integer, can be obtained from the solution (4.23) for a simple source, by differentiation with respect to time, viz.

$$b_d^d = \frac{A\nu^{1-p/2}}{(-1)^k \cdot \frac{5}{2} \cdot \frac{7}{2} \cdot \frac{9}{2} \cdots \left[\frac{5}{2} + (k-1)\right]} \frac{\partial^k}{\partial t^k} \left(\frac{e^{-r^2/8\nu t}}{t^{5/2}}\right). \qquad (4.24)$$

This is the solution for an unsteady dipole of order k. The behaviour of the correlation coefficient $f(r/\sqrt{(\nu t)})$ in this case is shown in Fig. 22.

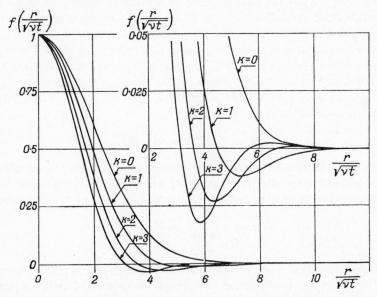

FIG. 22. Correlation coefficient for source type motions with different values of k.

It can be shown that the parameter $\varDelta$ is finite and non-zero at $\alpha = 1/4$. From (4.17), we obtain

$$\varLambda = \int_0^\infty b_d^d r^4 \, dr = \frac{A\nu^{1-p/2}}{2} \frac{(\nu t)^{5/2}}{t^{10\alpha}} \int_0^\infty f(\xi)\xi^{3/2} \, d\xi. \qquad (4.25)$$

This equation shows that constant values of $\varLambda$ other than zero or ∞ are incompatible with the inequality $\alpha \neq 1/4$. The expansion (4.21) shows that $\varLambda = \infty$ for $0 < \alpha < 1/4$. We have $\varLambda = 0$ for $\alpha > 1/4$; in this case $f(\xi)$ changes sign as ξ varies between zero and infinity.

The solutions for $b_d^d(r, t_0)$ determined by (4.17) and (4.20), are continuous functions for all α when $t_0 > 0$. Clearly in these particular cases and, therefore, in the general case, the damping law is strongly influenced by the properties of the initial disturbances. *Consequently, in order to obtain asymptotic damping laws from the solutions considered, we require either additional hypotheses of a mechanical character or experimental data.*

6. PROBLEM OF TURBULENT MOTION IN A WIND-TUNNEL

We have already pointed out that the investigation of isotropic turbulence is related to the study of turbulence produced by the guide vanes of wind-tunnels.

We now discuss the development of turbulent motion of an incompressible fluid behind a grating which moves forward with constant velocity u along the x axis. For simplicity, we assume that the fluid is infinite in extent and that the grating is formed by a doubly-periodic system of congruent cells lying in a plane perpendicular to the x axis. We consider the family of motions behind gratings of fixed shape.

The fluid motion in a plane perpendicular to the x axis is determined by the system of parameters

$$\rho, \mu, u, M, x = u(t - t_0)$$

where M is the characteristic grating dimension, and x is the coordinate of the plane in question; the constant t_0 is determined by the origin for x.

The nondimensional variables of the motion depend on two parameters

$$\frac{x}{M} \quad \text{and} \quad \frac{\rho u M}{\mu}.$$

We assume that the turbulent motion is isotropic for large enough x/M and that the development of isotropic turbulence in different planes perpendicular to the x axis only differs in phase. In this case, the characteristics of the turbulent fluid motion are determined by the parameters ρ, μ, u, M, t.

The correlation coefficients $f = b_d^d/b$ and $h = b_d^{nn}/b^{3/2}$ depend on the nondimensional parameters

$$\frac{\rho u M}{\mu}, \frac{u t}{M}, \frac{r^2}{\nu t}.$$

Formula (4.17) is obtained on the assumption that the parameter ut/M becomes negligible for sufficiently large values.

It follows from (4.18) that

$$\frac{1}{\sqrt{b}} = \frac{1}{\sqrt{b_0}} \left(\frac{t}{t_0} \right)^{5\alpha}.$$

Putting $t = t_0 + x/u$, we obtain

$$\frac{u}{\sqrt{b}} = \frac{u}{\sqrt{b_0}}\left(1 + \frac{x}{ut_0}\right)^{5\alpha}. \qquad (4.26)$$

Formula (4.26) gives the law of decay of turbulent eddies along the axis of a tube.

Taylor (1935, 1936) proposed the empirical formula

$$\frac{u}{\sqrt{b}} = A + B\frac{x}{M}, \qquad (4.27)$$

where A and B are constants, based on a number of experimental results obtained in wind-tunnels.

To ensure agreement between (4.26) and (4.27) we must put $\alpha = 1/5$. We have $[A\nu^{1-p/2}] = L^2$ for $\alpha = 1/5$.

7. TURBULENT MOTIONS WITH LARGE EDDIES

If the parameter ut/M is negligible, then the correlation coefficients remain constant for $\chi = r/\sqrt{(\nu t)} = r/l = $ const. The influence of the time reduces to a variation in the scale of r.

The scale variation in the solution considered is determined by the relation

$$l = \sqrt{(\nu t)} = \sqrt{\left(\frac{\nu(x+x_0)}{u}\right)}. \qquad (4.28)$$

It follows from (4.28) that the scale l is constant in time for $x = $ const, i.e. at a point fixed with respect to the grating. In addition, (4.28) shows that this scale depends on the velocity u.

Taylor (1935, 1936) gives certain experimental results which do not confirm the last conclusion; because of this it was necessary to modify and improve the theory dealing with the large eddy case. The exchange of momentum between masses of displaced fluid is very important for large eddy motion. The fluid inertia property plays a leading role in these processes.

Viscosity plays a large part in the development of motions with very small fluctuations when the basic process is one of kinetic energy dissipation. To study large eddies, we assume, following Kármán and Howarth (1938), that for sufficiently large values of r, the relations

$$\frac{b_d^d}{b} = f\left(\frac{r}{l}\right) \qquad (4.29)$$

$$\frac{b_d^{nn}}{b^{3/2}} = f\left(\frac{r}{l}\right) \qquad (4.30)$$

are valid, where l is a certain linear quantity depending on the time and on constant parameters defining isotropic turbulent motion.† Formulas (4.29) and (4.30) show that the influence of the time on the correlation coefficient reduces to a change in scale for the distance l. Viscosity influences the energy dissipation phenomenon implicitly by means of the quantities b and l.

We shall not attribute any specific geometric or physical meaning to l. The definition of l can be considered as an additional hypothesis. In particular, if it is assumed that the invariant $\Lambda \neq 0, \infty$, i.e. that an appropriate integral exists and conditions formulated on p. 119 are satisfied, then we obtain a relation between l and b. In fact

$$\Lambda = b \int_0^\infty f\left(\frac{r}{l}\right) r^4 \, dr = bl^5 \int_0^\infty f(\chi)\,\chi^5 \, d\chi.$$

Hence, we find

$$l = a\left(\frac{\Lambda}{b}\right)^{1/5} \tag{4.31}$$

where a is a constant. If we still assume that $l = \sqrt{(\nu t)}$, as in the solution for small fluctuations analysed above, then we again obtain the damping law corresponding to $\alpha = 1/4$,

$$b = \frac{a^5 \Lambda}{(\nu t)^{5/2}}.$$

Using (4.29) and (4.30), (4.14) yields:

$$h'(\chi) + \frac{4h(\chi)}{\chi} - \tfrac{1}{2}\chi f'(\chi)\frac{1}{b^{1/2}}\frac{dl}{dt} + \tfrac{1}{2}f(\chi)\frac{l}{b^{3/2}}\frac{db}{dt} = \frac{\nu}{b^{1/2}l}\left(f''(\chi) + \frac{4f'(\chi)}{\chi}\right). \tag{4.32}$$

This relation can be satisfied by using various assumptions. Following Kármán and Howarth (1938) we consider the motion corresponding to high values of Reynolds number $\sqrt{(bl/\nu)}$ and, consequently, we neglect the right side of (4.32). Then we obtain

$$h' + \frac{4h'}{\chi} = \tfrac{1}{2}\chi f' kc + \tfrac{1}{2}fc, \tag{4.33}$$

where

$$\frac{1}{\sqrt{b}}\frac{dl}{dt} = kc, \tag{4.34}$$

$$\frac{1}{\sqrt{b^3}}\frac{dl}{dt} = -c. \tag{4.35}$$

† Further arguments will be based on the assumptions included in (4.29) and (4.30). Whether or not these assumptions are realistic is a point to be clarified.

The nondimensional quantities kc and c do not depend on χ. It follows from (4.33) and from the general properties of $f(\chi)$ that the quantities kc and c are constants.

Integrating (4.34) and (4.35), we obtain

$$\frac{l}{l_0} = \left(\frac{b_0}{b}\right)^k. \tag{4.36}$$

We have for† $k \geqslant -\frac{1}{2}$

$$\frac{b}{b_0} = \left(\frac{t_0+t}{2t_0}\right)^{-2/(2k+1)} \quad \text{and} \quad \frac{l}{l_0} = \left(\frac{t_0+t}{2t_0}\right)^{2k/(2k+1)} \tag{4.37}$$

where the time corresponding to the values of l_0 and b_0 is denoted by t_0.

If we combine the assumption that an invariant $\varLambda \neq 0$ exists with the formula (4.37), then it follows from (4.31) that $k = 1/5$. Substituting $k = 1/5$ into (4.37) we obtain Kolmogorov's (1941) result

$$\frac{b}{b_0} = \left(\frac{t_0+t}{2t_0}\right)^{-10/7}, \quad \frac{l}{l_0} = \left(\frac{t_0+t}{2t_0}\right)^{2/7} \tag{4.37'}$$

which he found by using a number of assumptions including (4.29) and (4.31).‡

The exponent k above is arbitrary. If we assume that the scale l is constant for a wind-tunnel and proportional to the dimensions of the grating cell M, i.e. $l = \text{const } M$, then we find $k = 0$ and

$$\frac{u}{\sqrt{b}} = \frac{u}{2\sqrt{b}}\left(1+\frac{t}{t_0}\right) = A + B\frac{x}{M}.$$

This formula agrees with the Taylor formula (4.27).

8. LAWS OF DECAY OF TURBULENCE TAKING THIRD ORDER MOMENTS INTO ACCOUNT

The laws of turbulent growth, represented by (4.37), are based on the Kármán and Howarth (1938) assumption that the right side of (4.32) is zero. If the right side of (4.32) is put equal to zero then we obtain for $f(\chi)$,

$$f'' + \frac{4f}{\chi} = 0$$

which has no solution satisfying the condition $f(0) = 1$ except the solution $f(\chi) = 1$ and does not satisfy the physical condition $f(\infty) = 0$.

---

† The condition $k \geqslant -1/2$ must be satisfied to ensure that the mean velocity of the turbulent motion does not vanish at any time.

‡ It will be shown under sub-heading 8 that assumptions (4.29), (4.30) and (4.31) are contradicted if $h \neq 0$ and if $h = 0$, laws are obtained which differ from (4.37).

Therefore, the Kármán-Howarth solution is approximate; moreover, it does not enable us to determine the functions $f(\chi)$ and $h(\chi)$. Consequently, this solution is unsatisfactory.

Furthermore, we can find all the physically admissible exact solutions of (4.14) *simply by using the two basic Kármán-Howarth assumptions included in the formulas*

$$\frac{b_d^d}{b} = f\left(\frac{r}{l}\right) \tag{4.29}$$

and

$$\frac{b_d^{nn}}{b^{3/2}} = h\left(\frac{r}{l}\right) \tag{4.30}$$

where l and b are certain functions of time.

Because of these assumptions (4.14) reduces to (4.32). Now let us consider the exact solutions of this equation in more detail. Differentiating (4.32) with respect to time for constant χ, we obtain:

$$\tfrac{1}{2}\chi f'(\chi)\frac{d}{dt}\left(\frac{1}{b^{1/2}}\frac{dl}{dt}\right) - \tfrac{1}{2}f(\chi)\frac{d}{dt}\left(\frac{l}{b^{3/2}}\frac{db}{dt}\right) + \left(f''(\chi) + \frac{4f'(\chi)}{\chi}\right)\frac{d}{dt}\left(\frac{\nu}{\sqrt{(b)}\,l}\right) = 0. \tag{4.38}$$

We shall analyse the following possible cases associated with (4.38):

(1) The functions $\chi f'(\chi)$, $f(\chi)$ and $f''(\chi) + 4f(\chi)/\chi$ are linearly independent.

(2) There exists only one independent linear relation with constant coefficients of the type

$$c_1\chi f' + c_2 f + c_3\left(f'' + \frac{4f'}{\chi}\right) = 0 \tag{4.39}$$

in which not all the c_1, c_2, c_3 equal zero.

(3) Two linearly independent relations with constant coefficients exist between the functions $\chi f', f, f'' + 4f'/\chi$. Since $f \neq 0$, these relations can always be written in the form

$$f'\chi = c_1'f \quad \text{and} \quad f'' + \frac{4f'}{\chi} = c_2'f.$$

It is not difficult to see that in the last case

$$c_1' = c_2' = 0 \quad \text{and} \quad f = \text{const.}$$

This solution is of no physical interest, and the third case can be ignored.

In the first case, all the coefficients in (4.38) equal zero, consequently:

$$\frac{1}{\sqrt{b}}\frac{dl}{dt} = kc, \quad \frac{l}{\sqrt{b^3}}\frac{db}{dt} = -c, \quad \frac{\nu}{\sqrt{b}\,.\,l} = \frac{1}{mc} \qquad (4.40)$$

where k, c and m are certain constants. The system (4.40) gives:

$$l = c\sqrt{[m\nu(t+t_0)]}, \quad b = \frac{m\nu}{t+t_0}, \quad k = \frac{1}{2}. \qquad (4.41)$$

Thus, we have obtained l and b as specific functions of time. These[†] agree with (4.37) for $k = 1/2$. In this case, only one equation is obtained for the two functions $f(\chi)$ and $h(\chi)$.

Now, let us investigate the second case. The coefficient c_3 in (4.39) is clearly not zero, since otherwise, the function $f(\chi)$ would have to satisfy the equation

$$c_1 f + c_2 \chi f' = 0.$$

This has no solution satisfying the condition $f(0) = 1$ for $c_1 \neq 0$ and for $c_1 = 0$, we obtain $f' = 0$ or $f = \text{const} \neq 0$, in conflict with the condition $f(\infty) = 0$.

Since $c_3 \neq 0$, (4.39) can be written

$$f'' + \frac{4}{\chi}f' + \frac{a_1}{2}\chi f' + \frac{a_2}{2}f = 0 \qquad (4.42)$$

where a_1 and a_2 are constant coefficients.

We find from (4.38) and (4.42)

$$\tfrac{1}{2}\chi f'\left[\frac{d}{dt}\frac{1}{\sqrt{b}}\frac{dl}{dt} - a_1\frac{d}{dt}\frac{\nu}{\sqrt{(b)}\,l}\right] - \tfrac{1}{2}f\left[\frac{d}{dt}\frac{l}{\sqrt{b^3}}\frac{db}{dt} + a_2\frac{d}{dt}\frac{\nu}{\sqrt{(b)}\,l}\right] = 0.$$

Since the functions $\chi f'$ and f are linearly independent, the expressions in the square brackets must vanish, so that

$$\frac{1}{\sqrt{b}}\frac{dl}{dt} = a_1\frac{\nu}{\sqrt{(b)}\,l} + p, \quad \frac{l}{\sqrt{b^3}}\frac{db}{dt} = -a_2\frac{\nu}{\sqrt{(b)}\,l} + q, \qquad (4.43)$$

where p and q are constants of integration.

Substituting (4.43) into (4.32) and using (4.12), we find, as the equation determining $h(\chi)$,

$$h' + \frac{4h}{\chi} - \tfrac{1}{2}\chi f'p + \tfrac{1}{2}fq = 0. \qquad (4.44)$$

---

† It is evident that in this case the quantity Λ cannot be finite and cannot be identically zero, otherwise (4.31) would contradict (4.41).

5

We have already shown that the expansion of the function $b_d^{nn}/b^{3/2} = h$ in powers of $r/l = \chi$ must start with terms of order χ^3, consequently, it follows from (4.44) that $q = 0$.

Hence, in case 2 we have a complete system of equations (4.42), (4.43) and (4.44) to determine h and f as functions of the variable χ and the quantities l and b as functions of the time t.

These equations contain three arbitrary constants a_1, a_2 and p. One of these constants is clearly negligible. In fact, a transformation $\chi = \lambda\chi'$ where λ is a certain constant reduces to multiplying the still undefined scale l by a constant $1/\lambda$ ($l = l'/\lambda$). Carrying out this transformation in (4.42), (4.43) and (4.44), we obtain the same equations as before with transformed values a_1', a_2' and p' which are defined by

$$a_1' = a_1\lambda^2, \quad a_2' = a_2\lambda^2, \quad p' = p\lambda.$$

Fixing a particular non-zero value for a_1 or for a_2 is equivalent to fixing the scale for l.

From the second equation in (4.43), it follows from the condition of damping of turbulent fluctuations ($b \to 0$ as $t \to \infty$) that $a_2 > 0$.

Furthermore, we also assume as a physical condition that $l \to +\infty$ as $t \to \infty$ or as $b \to 0$.

Equations (4.43) can be integrated, to give the relation

$$\frac{1}{l} = \frac{2pa_2}{\nu(a_2 - 2a_1)}\sqrt{(b)} + cb^{a_1/a_2},$$

where c is a constant of integration and $2a_1 \neq a_2$. It follows from this relation, from the inequality $a_2 > 0$ and from the condition $l \to \infty$ as $b \to 0$, that $a_1 \geqslant 0$.

Let us consider the case $a_1 = 0$. Equation (4.42) has a unique solution satisfying the condition $f(0) = 1$ for $a_1 = 0$.

This solution is easily determined and is:

$$f(\chi) = 3\left[\frac{\sin\sqrt{\left(\frac{a_2}{2}\right)}\chi}{\left[\sqrt{\left(\frac{a_2}{2}\right)}\chi\right]^3} - \frac{\cos\sqrt{\left(\frac{a_2}{2}\right)}\chi}{\frac{a_2}{2}\chi^2}\right].$$

This function oscillates as $\chi = r/l$ increases, has a finite number of zeros and decreases slowly, like $1/\chi^2$. On this basis, we shall not consider the case $a_1 = 0$ any further.

We are free to choose the parameter λ, so we put $a_1 = 1/2$; then (4.42) takes the form

$$f'' + \left(\frac{4}{\chi} + \frac{\chi}{4}\right)f' + \frac{a_2}{2}f = 0. \tag{4.45}$$

Putting $a_2 = 10\alpha$ and

$$\frac{r^2}{l^2} = \chi^2 = \xi$$

(4.45) transforms into (4.19) which we investigated earlier in studying the damping of very small fluctuations. Therefore, we obtain the same solution for $f(\xi)$ as in the small fluctuation case, namely,

$$f(\xi) = M\left(10\alpha, \frac{5}{2}, -\frac{\xi}{8}\right)$$

but now $l \neq \sqrt{(\nu t)}$ and $h \neq 0$. *Therefore, taking third order moments into account influences the correlation coefficient $f = b_d^d/b$ only by changing the way in which the linear scale l varies with time.*

Using (4.44) ($q = 0$), the function $h(\chi)$ is easily expressed in terms of the function $f(\chi)$. Integrating (4.44) and using the condition $h = 0$ when $\chi = 0$, we obtain

$$h = \frac{p}{2}\frac{1}{\chi^4}\int\limits_0^\chi \chi^5 f'(\chi)\, d\chi. \qquad (4.46)$$

If $\alpha = 1/4$ and $h \neq 0$, then it is evident that the function $h(\chi)$ vanishes like $1/\chi^4$ as $\chi \to \infty$; consequently, the third order moment b_d^{nn} is of order $1/r^4$ as $r \to \infty$ and, therefore, the conditions necessary for the invariance of Λ (see p. 119) are not satisfied.

Hence, we have $\Lambda = \infty$ for $\alpha < 1/4$, $\Lambda = 0$ for $\alpha = 1/4$ and Λ is finite and non-zero but not constant for $\alpha = 1/4$ and $p \neq 0$ (third order moments different from zero). In fact, we find from (4.14′) and (4.46), for $\alpha = 1/4$,

$$\frac{d\Lambda}{dt} = -p\int\limits_0^\infty \chi^5 f'(\chi)\, d\chi\, l^4 b^{3/2} \neq 0.$$

If $p = 0$, then the integral of Λ is finite and invariant but the third order moments are zero in this case.

Hence, if the third order moments are not zero and the hypotheses expressed by (4.29) and (4.30) are fulfilled, then the integral of Λ is either zero or infinity and is therefore unsuitable to determine scales according to (4.31). Alternatively the quantity Λ varies with time for $\alpha = 1/4$ and, consequently, cannot be considered as a characteristic constant.

Multiplying (4.45) by $\chi^4\, d\chi$ and integrating, we find

$$\int\limits_0^\chi \chi^5 f'\, d\chi = -4\chi^4 f' - 20\alpha\int\limits_0^\chi \chi^4 f\, d\chi = -4\chi^4 f' - 4\alpha\chi^5 f + 4\alpha\int\limits_0^\chi \chi^5 f'\, d\chi.$$

Using this relation for $\alpha \neq 1/4$, we obtain

$$h = \frac{2p}{4\alpha - 1}[f'(\chi) + \alpha \chi f(\chi)]. \tag{4.47}$$

Equations (4.43) determine l and b as functions of time. If we put $\alpha_1 = 1/2$, $\alpha_2 = 10\alpha$, $q = 0$, they become

$$\left. \begin{aligned} \frac{1}{\sqrt{b}}\frac{dl}{dt} &= \frac{1}{2}\frac{\nu}{\sqrt{(b)}\,l} + p, \\[2mm] \frac{l}{\sqrt{b^3}}\frac{db}{dt} &= -10\alpha\frac{\nu}{\sqrt{(b)}\,l}. \end{aligned} \right\} \tag{4.48}$$

In the particular case when $p = 0$ the system of Equations (4.48) and (4.45) is seen to have the solution

$$l = \sqrt{(\nu t)}, \quad b = \frac{A'}{(\nu t)^{10\alpha}}, \quad f = M\left(10\alpha, \frac{5}{2}, -\frac{\xi}{8}\right) \quad \text{and} \quad h \equiv 0$$

which corresponds to the Kármán-Howarth small fluctuation solution.

From (4.48), for any p and $\alpha \neq 0 \cdot 1$, we obtain

$$\frac{1}{l} = Cb^{1/20\alpha} + \frac{2p\sqrt{b}}{(10\alpha - 1)\nu}, \tag{4.49}$$

where C is a constant of integration.

For $p \neq 0$ and $\alpha \neq 0 \cdot 1$ the general solution of Equations (4.48) can be written

$$\frac{l}{l^*} = \frac{1}{w^{1/20\alpha} + \sqrt{w}} \tag{4.50}$$

and

$$\frac{\nu(t + t^*)}{l^{*2}} = \frac{1}{10\alpha}\int_w^\infty \frac{dw}{w^{1+1/10\alpha}\,(1 + w^{(10\alpha - 1)/20\alpha})^2} \tag{4.51}$$

where

$$w = \frac{b}{b^*}, \quad l^* = \frac{(10\alpha - 1)\nu}{2p\sqrt{b^*}} \tag{4.52}$$

and b^* and t^* denote constants of integration which have the dimensions of velocity and time squared (l^* has the dimensions of a length).

It is not difficult to see that the law of turbulent decay is determined essentially by the constant α since b^* and l^* play the part of scale constants and the constant t^* depends only on the origin of the time readings.

If we define the Reynolds number by the formula

$$\mathbf{R} = \frac{\sqrt{b}.l}{\nu}$$

it follows from (4.50) that

$$\mathbf{R} = \frac{\sqrt{b^*}.l^*}{\nu} \frac{\sqrt{w}}{\sqrt{w+w^{1/20\alpha}}} = \frac{\sqrt{b^*}.l^*}{\nu} \frac{1}{1+w^{1/2(1/10\alpha-1)}}.$$

Hence, if $\alpha > 0\cdot 1$, then $\mathbf{R} \to 0$ as $w \to 0$ or $t \to +\infty$; if $\alpha < 0\cdot 1$, then we have $\mathbf{R} = \sqrt{b^*}.l^*/\nu$ as $w \to 0$; i.e., the number $\mathbf{R}$ has a finite limit. Clearly the limiting behaviour of the number $\mathbf{R} = \sqrt{(b)} \sqrt{(\nu t)}/\nu$, has the same character (for $\alpha > 0\cdot 1$) if the third order moments are neglected.

The laws of decay of isotropic turbulence which we obtained, (4.50) and (4.51) are consequences of assumptions (4.29) and (4.30). To determine the constant α, additional hypotheses or experimental data are necessary.

From (4.50), (4.51) and (4.52) when $\alpha > 0\cdot 1$, we find that as $t \to +\infty$

$$\frac{b}{b^*} \doteq \left(\frac{l^{*2}}{\nu t}\right)^{10\alpha}, \quad l \doteq \sqrt{(\nu t)} \quad \text{and} \quad h \to 0$$

which show that the laws we found for the variation of b and l as functions of time when the third order moments are taken into account are asymptotic to the laws obtained by Kármán and Howarth by neglecting the third order moments. However, the limiting motion will be attained slowly for small α.

§ 5. STEADY TURBULENT MOTION

In a number of cases, in particular, in fluid motion through pipes and channels, we encounter turbulent motions for which the average motion is steady; these are called steady, turbulent flows.

Consider the problem of the steady turbulent motion of an incompressible fluid through a fixed, smooth, circular pipe of infinite length.

We assume that the average fluid motion is axisymmetric and the average velocities are directed parallel to the pipe axis. It follows from the equations of motion that the magnitudes of the average velocities are independent of the x coordinate along the pipe axis; the average velocity is a variable across the pipe section and depends on the distance r from the pipe centre.

The average motion is evidently defined by the following system of parameters:

$$\rho, \mu, a, \tau_0 = -\frac{1}{2}\frac{\partial \overline{p}}{\partial x}a, \quad r = a-y, \tag{5.1}$$

where a is the pipe radius; $\partial \overline{p}/\partial x$ is the average value of the pressure gradient along the pipe; and τ_0 is the friction stress on the pipe walls. All the nondimensional quantities are functions of the two parameters:

$$\mathbf{R} = \frac{v_* a \rho}{\mu}, \quad \frac{r}{a} = 1 - \frac{y}{a}, \tag{5.2}$$

where $v_* = \sqrt{(\tau_0/\rho)}$ is called the friction velocity. The nondimensional quantities characterizing the properties of the motion as a whole do not depend on the variable r and, therefore, are defined solely by the Reynolds number.

Let us denote the magnitude of the average velocity of the motion by u and the velocity at the centre of the pipe by u_{max}. It follows from dimensional analysis that

$$\frac{u_{max}}{v_*} = f(\mathbf{R}) \tag{5.3}$$

and

$$\frac{u_{max} - u}{v_*} = F\left(\mathbf{R}, \frac{a - y}{a}\right). \tag{5.4}$$

The quantity $u_{max} - u$ is called the velocity defect, and defines the velocity distribution over the pipe cross-section relative to the motion at the centre.

Later, we shall consider fully developed turbulent motion which corresponds to high values of Reynolds number.

The velocity distribution in the pipe is closely connected with the turbulent mixing phenomenon; consequently, an exchange of momentum occurs between adjacent fluid layers. The equalization of velocities as a result of momentum transport is determined by the fluid inertia property.

From the viewpoint of the kinetic theory of matter, the viscosity property is explained by the presence of a random molecular motion which contributes to the equalization of the observed velocities and which leads to transformation of the kinetic energy of the observed motion into thermal energy.

The law of conservation of energy states that the sum of the mechanical energy of the observed motion and of the energy of molecular motion is constant. Both kinds of energy can be considered as components of different kinds of mechanical energy. If the intermolecular forces are neglected, then the viscosity property is determined by the average kinematic characteristics of the state of the molecular motion and by the inertia property of the fluid molecules.

The relation between disordered turbulent mixing and the averaged motion is analogous to that between molecular motion and real turbulent motion. Turbulent fluctuations are analogous to fluctuations of random molecular motion. The difference lies in the different orders of the average quantities which characterize the fluctuating motion. Instead of the motion of individual molecules in the thermal process during turbulent mixing, we have the fluctuating motion of "moles", volumes of fluid which are very large in comparison to the size and mass of molecules. Moreover, the magnitudes of the average fluctuating velocities in turbulent motion are very small in comparison with the magnitude of the average velocity of thermal motion.

The conversion of the energy of the averaged motion into the energy of turbulent molar motion can be analysed in turbulent flow. We can introduce the concept of dissipation of energy of the average motion, where the dissipation is not directly related to the transformation of mechanical energy into thermal and, therefore, is independent of fluid viscosity. The redistribution of the kinetic energy between the average observed motion and the fluctuating motion can also be considered in an ideal fluid. As is known, the transformation of mechanical energy into thermal is impossible in an ideal incompressible fluid. Therefore, the transformation of the energy from the average motion into molar turbulent motion of fluctuations can be determined, basically, only by the inertia property.

A loss of mechanical energy occurs during the motion of a fluid in a pipe, therefore, regions must exist in which the influence of viscosity is essential. The instantaneous and average velocities of the fluid at the walls equal zero because the fluid adheres to the pipe walls. Consequently, intensive fluid mixing cannot exist directly at the pipe wall. This is the basis for the conclusion that the sharp variation in the velocity directly near the wall must be determined by the fluid viscosity and that a laminar layer must exist near the wall. Experimental results are in good agreement with this conclusion.

Let us assume that equalization of the velocities in the main core of turbulent flow near the pipe axis is determined by the molar mixing of the fluid, in which viscosity has a secondary insignificant value. Let us denote the thickness of a fluid layer near the wall in which viscosity cannot be neglected, by δ. Approximately, the quantity δ can be compared to the thickness of the laminar layer at the pipe wall. By our assumption, the viscosity is insignificant for $y > \delta$ and, therefore, the Reynolds number is insignificant in (5.4) for $y > \delta$, i.e.

$$\frac{u_{\max} - u}{v_*} = F\left(\frac{a-y}{a}\right).$$ (5.5)

The equation (5.5) indicates the existence of a universal law of velocity distribution in pipes.

Experiments measuring the velocity distribution in pipes during turbulent motion, carried out for all possible values of the Reynolds number, confirm the validity of this universal law, independent of Reynolds number (Fig. 23). Darcy (1858) proposed the empirical formula

$$\frac{u_{max} - u}{v_*} = 5 \cdot 08 \left(1 - \frac{y}{a} \right)^{3/2}. \tag{5.6}$$

Graphs taken from experimental results showing $(u_{max} - u)/v_*$ as a function of y/a are also given in Stanton (1911), Fritsch (1928a, b) and Nikuradse (1930). These authors noted the existence of a universal law which appeared to be valid in the central part of smooth and rough pipes independently of the roughness, even though the resistance, as well as the ratio u_{max}/v_*, etc., depend substantially on the Reynolds number and on the roughness.

This interpretation of the validity of the law (5.5) was obtained from dimensional analysis as a consequence of neglecting viscosity in the motion near the pipe centre. This explanation is given in the works of Prandtl and Kármán.

Now, let us consider the problem of determining the form of the function defining the velocity distribution in the pipe cross-section for turbulent motion. Keeping in mind the large value of the Reynolds number which is equivalent to a large value of the radius a for given v_* and μ/ρ, let us consider the limiting case $a \to +\infty$. This corresponds to the problem of turbulent motion in the upper semi-infinite region bounded by the plane $y = 0$.

The system of characteristic parameters will be

$$\rho, \mu, \tau_0, y.$$

We obtain the following formula for the averaged velocity distribution

$$\frac{u}{v_*} = \phi(\eta), \tag{5.7}$$

where

$$\eta = \frac{v_* \rho y}{\mu}.$$

The form of the function $\phi(\eta)$ for laminar motion is easily determined theoretically. In fact, all the fluid particles move uniformly and in a straight line in laminar motion in a cylindrical pipe, consequently, the inertia property is negligible and the velocity distribution does not depend on the quantity ρ. Since the characteristic parameters are τ_0,

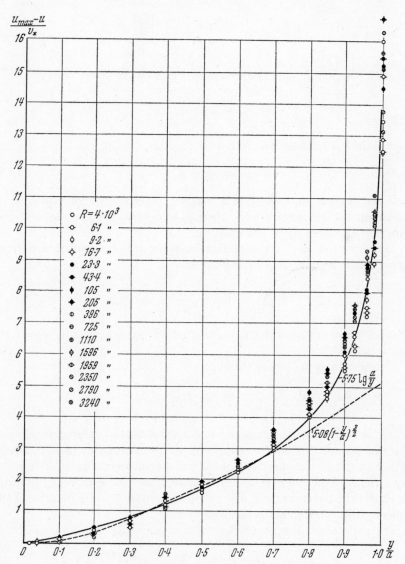

FIG. 23. Experimental confirmation of the universal velocity distribution law along a pipe axis.

μ and y, we obtain from dimensional analysis

$$u = k\tau_0 \frac{y}{\mu} = k \cdot \sqrt{\left(\frac{\tau_0}{\rho}\right)} \frac{\sqrt{\left(\frac{\tau_0}{\rho}\right)} \cdot \rho y}{\mu},$$

where k is a nondimensional constant. From the relation

$$\tau_0 = \mu \left(\frac{\partial u}{\partial y}\right)_{y=0},$$

we obtain $k = 1$. Therefore, in laminar motion, we have

$$\phi(\eta) = \eta. \tag{5.8}$$

If we assume that there is a laminar layer in the general case of turbulent motion near a wall, then (5.8) determines the shape of the function $\phi(\eta)$ directly at the wall.

The function $\phi(\eta)$ for turbulent motion can be determined from experiment. But the Reynolds number

$$\mathbf{R} = \frac{v_* \rho a}{\mu}$$

is involved in addition to the parameter η in experiments in pipes with finite radius.

Empirical power laws of the type

$$\frac{u}{v_*} = A\eta^n, \tag{5.9}$$

where A and n are constant, are used a great deal to determine velocity distributions in turbulent motion. These constants can be determined either by direct measurement of the velocity distribution or, indirectly, by using an experimental determination of the pipe resistance. In order to clarify the latter method, we express the pipe resistance in terms of the velocity distribution along the pipe radius.

The resistance (drag) coefficient of a circular pipe is determined by

$$\psi = \frac{(p_1 - p_2) a}{l \cdot \rho \frac{u^2}{2}} = \frac{4\tau_0}{\rho \bar{u}^2} = 4\left(\frac{v_*}{\bar{u}}\right)^2, \tag{5.10}$$

where

$$\bar{u} = \frac{Q}{\pi a^2}$$

$\overline{u}$ is the mean velocity along the pipe cross-section, and Q is the volume discharge of fluid. The formula

$$\frac{u}{v_*} = \phi\left(\mathbf{R}_1 = \frac{\rho a \overline{u}}{\mu}, \frac{\rho v_* y}{\mu}\right)$$

gives the velocity distribution along the pipe radius in the general case. Averaging over the pipe section, we obtain

$$\frac{u}{v_*} = \frac{1}{\pi a^2}\int\limits_0^a \phi \cdot 2\pi(a-y)\,dy = 2\int\limits_0^1 \phi\left(\mathbf{R}_1, \frac{v_*}{\overline{u}}\mathbf{R}_1\lambda\right)(1-\lambda)\,d\lambda. \quad (5.11)$$

Solving this equation for v_*/u we find ψ as a function of $\mathbf{R}_1$.

It is not difficult to see that the formula for ψ is

$$\psi = \frac{a}{\mathbf{R}_1^m}, \quad (5.12)$$

where a and m are constants, corresponding to a power law for the velocity distribution defined by (5.9) in which A and n are independent of the Reynolds number $\mathbf{R}_1$. In fact, substituting

$$\phi = A\left(\frac{\rho v_* y}{\mu}\right)^n = A \cdot \mathbf{R}_1^n \left(\frac{v_*}{\overline{u}}\right)^n \lambda^n$$

in (5.11), we obtain

$$\frac{\overline{u}}{v_*} = \frac{2A}{(n+1)(n+2)}\mathbf{R}_1^n\left(\frac{v_*}{\overline{u}}\right)^n,$$

hence, we determine $v_*/\overline{u}$ and, subsequently, we find using (5.10)

$$\psi = 4\left[\frac{(n+1)(n+2)}{2A}\right]^{2/(n+1)}\frac{1}{\mathbf{R}_1^{2n/(n+1)}}. \quad (5.13)$$

Comparing (5.12) and (5.13), we obtain simple relations between the constants m and n and the constants a, A and n.

The Blasius empirical formula for the resistance of smooth cylindrical pipes is

$$\psi = \frac{0 \cdot 132}{\mathbf{R}_1^{1/4}}. \quad (5.14)$$

If a power law is used for the velocity distribution, the Blasius formula (5.14) reduces to a "one-seventh" law:

$$\frac{u}{v_*} = 8 \cdot 6 \left(\frac{v_* \rho y}{\mu}\right)^{1/7}. \quad (5.15)$$

Formulas (5.14) and (5.15) are in good agreement with experiment for the Reynolds number $2\mathbf{R}_1$ in the 10^4–10^5 range; a formula of the type (5.9) with exponent $1/6$ is in better agreement with experiment for lower values. The exponent must decrease for $2\mathbf{R}_1 > 10^5$.

Results of experiments (Fig. 24) show that the best agreement with experiment is obtained with an empirical formula of the form

$$\frac{u}{v_*} = \phi(\eta) = 5 \cdot 75 \lg \eta + 5 \cdot 5. \tag{5.16}$$

Formulas (5.9) and (5.16) lose their validity very near the wall where $\eta = 0$. There is a laminar layer near the wall for which $\phi(\eta) = \eta$. If we

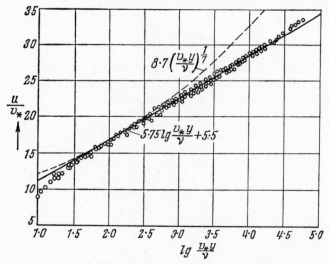

FIG. 24. Velocity distribution in the turbulent boundary layer.

assume that the laminar layer is adjacent to the turbulent flow and if we require that the velocity is continuous on transition from the laminar layer boundary to the turbulent velocity distribution defined by (5.15) and (5.16) then we have a condition to determine the laminar layer thickness either from the equation

$$\eta = 5 \cdot 75 \lg \eta + 5 \cdot 5,$$

or from

$$\eta = 8 \cdot 7 \, \eta^{1/7}.$$

The solution of these equations yields a value close to $\eta = 12$ in both cases. Putting $\eta = 12$, we find the formula for the laminar layer thickness:

$$\frac{\delta}{a} = 12\frac{\mu}{\rho v_* a} = \frac{24}{\mathbf{R}_1 \sqrt{\psi}}. \tag{5.17}$$

Putting $\mathbf{R}_1 = 40{,}000$ and using the Blasius formula, we obtain:

$$\frac{\delta}{a} = \frac{68}{\mathbf{R}_1^{7/8}} = 0\cdot0065.$$

Hence, it can be concluded that the laminar layer thickness is small compared with the pipe radius a.

Now, let us consider the theoretical reasoning of Prandtl and Kármán in determining the form of the function $\phi(\eta)$. We denote the x and y components of the turbulent fluctuation velocities by u' and v'. The

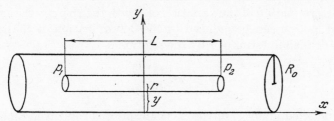

FIG. 25. Diagram for the computation of turbulent motion in a circular pipe.

mean value of the fluid momentum transport along the y axis, referred to unit time and area, is represented as

$$\tau = \rho\overline{u'v'}.$$

Using the theorem on the variation of momentum in a fluid volume enclosed within a circular cylinder of radius r coaxial with the pipe (Fig. 25), we obtain after averaging

$$(p_1 - p_2)\pi r^2 = \tau.2\pi rL + \mu\frac{d\overline{u}}{dy}2\pi rL.$$

Hence, using the relation

$$(p_1 - p_2)a = 2\tau_0 L$$

we find

$$\tau + \mu\frac{d\overline{u}}{dy} = \tau_0\left(1 - \frac{y}{a}\right). \tag{5.18}$$

For laminar motion $\tau = 0$ and we have Poiseuille flow. In this case, a parabolic velocity distribution is obtained from (5.18). There is a laminar layer in which $\tau = 0$ and $y/a \doteq 0$, and at the wall $u = 0$. Consequently, for turbulent motion, (5.18) reduces to

$$\mu \frac{d\bar{u}}{dy} = \tau_0,$$

from which $\bar{u} = (\tau_0 y)/\mu$ or $u/v_* = (\rho v_* y)/\mu$, which is in agreement with (5.8).

In the kinetic theory of gases, the tangential viscous stress $\mu\,(d\bar{u}/dy)$ can be considered as the average value of the momentum transport per unit time and area which is being specified by the random thermal motion of the individual molecules. In this sense, both terms on the left side of (5.18) have the same character.

In the region of a fully developed turbulent flow, $\tau \neq 0$ and is large in comparison with $\mu\,(d\bar{u}/dy)$, consequently, it is permissible to neglect the term $\mu\,(d\bar{u}/dy)$ in comparison with τ.

The determination of the variation of τ with the characteristics of the averaged motion can be reduced to the determination of the length l which is defined in terms of τ by the relation

$$\tau = \rho l^2 \left|\frac{d\bar{u}}{dy}\right|\frac{d\bar{u}}{dy}. \tag{5.19}$$

More detailed analysis of the mechanism of turbulent mixing, leads to the interpretation of the length l as a quantity analogous to the molecular mean free path in thermal gas motion (Prandtl, 1935). Consequently, l is called the mixing length.

The reason for the change from τ to l is related to the physical character of the quantity l. The quantity τ depends on the velocity squared in the absence of viscosity, consequently, l is independent of the velocity, which allows us to use intuitive reasoning to establish the relation with the characteristic dimensions. In a more detailed analysis, it can be shown that l decreases as the wall is approached and l can be related to the roughness characteristics near the wall.

The mixing path for $a = \infty$ is determined by the parameters ρ, μ, v_* and y; consequently, we have in this case

$$l = y \cdot \Phi\left(\frac{\rho v_* y}{\mu}\right).$$

Let us assume that viscosity is insignificant when the origin of the y coordinate is suitably chosen. It then follows that

$$l = ky$$

where k is a certain nondimensional constant. Neglecting the terms $\mu(du/dy)$ and $y/a \doteq 0$ in (5.18) we obtain

$$\tau = \tau_0 = \rho k^2 y^2 \left(\frac{du}{dy}\right)^2.$$

Integrating this equation, we find

$$u = \frac{v_*}{k}[\ln y - \ln y_0]. \tag{5.20}$$

We have $u = -\infty$ for $y = 0$ on the axis of similarity. The integration constant y_0 gives the distance between the point at which $u = 0$ and the axis of similarity. A laminar layer, which is adjacent to the turbulent stream will exist near the wall; if we were to continue the turbulent stream up to the wall on which the condition $u = 0$ is satisfied, we would find that y_0 is equal to the distance of the axis of similarity from the wall. Since y_0 must be determined by the quantities ρ, μ and v_*, we then have

$$y_0 = \beta \frac{\mu}{\rho v_*},$$

where β is a nondimensional constant.

The quantity y_0 is small in fully developed turbulent motion. Substituting the value found for y_0 into (5.20), we find

$$u = \frac{v_*}{k}(\ln \eta - \ln \beta). \tag{5.21}$$

Formula (5.21) can be considered as the theoretical basis of the empirical formula (5.16) in the region of turbulent motion. The constants k and β must be obtained experimentally. As the Reynolds number increases, the assumptions made in deriving (5.21) become more accurate: this leads to the conclusion that (5.16) is in good agreement with reality when the Reynolds number is large enough.

If it is assumed that the logarithmic velocity distribution is correct for turbulent motion in a circular pipe right up to the axis of the pipe, then we obtain

$$\frac{u_{\max} - u}{v_*} = F\left(\frac{y}{a}\right) = 5 \cdot 75 \lg \frac{a}{y}$$

which is in good agreement with experimental results for both smooth and rough pipes; the latter is explained by the fact that the influence of the roughness amounts to a change in y_0, which is excluded in the derivation of this formula (Prandtl, 1935).

We now discuss further applications of dimensional analysis and similarity theory to the problem of turbulent motion.

Let us denote the velocity components of the instantaneous motion by v_x, v_y and the fluctuation velocity components by u' and v'. In the problem of rectilinear averaged motion we have

$$v_x = u + u',$$

$$v_y = v'.$$

Consider the velocity field of the relative motion which is defined by the components $u + u' - u_M$, v' where u_M is the average velocity at a certain point M.

The basic Kármán hypothesis is that the turbulent velocity fields of the relative motion are dynamically similar at different points in the stream. Averaging, we find that the field of mean relative velocities $[u(y) - u_M, 0]$ is also dynamically similar at different points in the stream.

The transformation of values of all the kinematic quantities when we move from one point to another can be expressed in terms of the scales of transformation of two independent kinematic quantities. The magnitudes of these scales can be obtained from an analysis of the average velocity distribution.

The average relative velocity distribution near a given point is defined by the successive derivatives

$$u', u'', u''', u^{IV}, \ldots \tag{5.22}$$

Any two derivatives have independent dimensions and a nondimensional combination can be formed from any three derivatives. The following nondimensional combination can be formed from the first three derivatives

$$\frac{u''^2}{u' u'''} = k.$$

Since all the derivatives are functions of the same variable, it is evident that a functional relation which does not contain dimensional constants, holds between any two nondimensional combinations formed from the sequence (5.22). For example

$$\frac{u'''^2}{u'' u^{IV}} = \Phi\left(\frac{u''^2}{u' u'''}\right).$$

Therefore, the necessary and sufficient condition for similarity of relative motions for all values of y is:

$$\frac{u''^2}{u' u'''} = k = \text{const.} \tag{5.23}$$

Condition (5.23) is a differential equation in $u(y)$. Integrating this, we find for $k \neq 1$ and $k \neq 2$:

$$u = A(y+B)^{(2-k)/(1-k)} + C, \qquad (5.24)$$

where A, B, C are certain dimensional constants of integration. The solution for $k = 1$ is:

$$u = Me^{y/d} + N, \qquad (5.25)$$

and, finally, we have for $k = 2$:

$$u = P \ln (y+Q) + S, \qquad (5.26)$$

where M, d, N, P, Q, S are integration constants.

Thus the assumption of exact similarity leads to completely defined special forms of the average velocity distributions.

Formula (5.26) corresponds to the logarithmic velocity distribution.

The following formulas for the mixing length l and the turbulent momentum transport τ follow from the similarity condition:

$$l = k_1 \frac{u'}{u''} \qquad (5.27)$$

and

$$\tau = \rho l^2 u'^2 = \rho k_1^2 \frac{u'^4}{u''^2}, \qquad (5.28)$$

where k_1 is a nondimensional constant.

Formula (5.27) for the velocity distributions (5.24) and (5.26) leads to

$$l = k_1 y,$$

for a suitable choice of the origin for y, which we obtained earlier as a consequence of the hypothesis that viscosity can be neglected.

The result $\tau = \tau_0 = \text{const}$ corresponds to the velocity distribution (5.26). For the velocity distributions (5.25) and (5.24) τ is obtained as a variable.

Condition (5.23) strongly restricts the average velocity distributions possible. The original exact condition on similarity can be weakened and similarity of relative velocity fields in the vicinity of a point can be assumed with accuracy up to only second order quantities. Only u' and u'' are retained as characteristic quantities in the sequence (5.22) under such an assumption and, consequently, (5.23) drops out while formulas (5.27) and (5.28) are retained.

Using (5.28) and (5.18), theoretical formulas can be obtained for the average velocity distributions (Kármán, 1931).

Chapter IV

One-Dimensional Unsteady Motion of a Gas

§1. Self-similar Motion of Spherical, Cylindrical and Plane Waves in a Gas

1. THE CONCEPT OF SELF-SIMILARITY

The motion of a gas or liquid is said to be one dimensional when all its properties depend on only one geometric coordinate and on the time. It can be shown that the only possible one-dimensional motions are produced by spherical, cylindrical and plane waves (Lipschitz, 1887, Liubimov, 1956). The methods of dimensional analysis can be used to find exact solutions of certain problems of one-dimensional unsteady motion of a compressible fluid (Sedov, 1945a, 1945b).† Many of these problems are of considerable theoretical and practical interest. But even when the formulation of the problem is not itself of interest, the exact solutions obtained can be used as examples to check the validity of various approximate methods of solving problems of gas dynamics.

To distinguish the problems which can be solved by the methods of dimensional analysis, we analyse the dependent variables and the fundamental parameters of one-dimensional motion. The basic physical variables in the Eulerian approach are the velocity v, the density ρ and the pressure p and the characteristic parameters are the linear coordinate r, the time t and the constants which enter into the equations, the boundary and the initial conditions of the problem.

Since the dimensions of the quantities ρ and p contain the mass, at least one constant a, the dimensions of which also contain the mass must be a characteristic parameter. Without loss of generality, it can be assumed that its dimensions are

$$[a] = ML^k T^s.$$

† Similar solutions without using dimensional analysis or group theory considerations and without relation to the formulation of the problems selected below are considered in Bechert (1941), (Bechert considers only polytropic motion); and in Staniukovich (1945),

Use of the methods we developed in the filtration of a fluid in a porous medium and their generalization to develop solutions of motions limiting to self-similar (see §4) are given in Barenblatt (1952) and Barenblatt (1954).

We can then write for the velocity, density and pressure

$$v = \frac{r}{t}V, \quad \rho = \frac{a}{r^{k+3}\,t^s}R, \quad p = \frac{a}{r^{k+1}\,t^{s+2}}P, \tag{1.1}$$

where V, R, and P are arbitrary quantities and, therefore, can depend only on nondimensional combinations of r, t and other parameters of the problem.

In the general case they are functions of two nondimensional variables. But if an additional characteristic parameter b can be introduced with dimensions independent of those of a, the number of independent variables which can be formed by combining a and b is reduced to one.†

Since the dimensions of the constant a contain the mass, we can choose the constant b, without loss of generality, so that its dimensions do not contain the mass, i.e.

$$[b] = L^m\,T^n.$$

The single nondimensional independent variable in this case will be

$$\frac{r^m\,t^n}{b},$$

which can be replaced for $m \neq 0$ by the variable

$$\lambda = \frac{r}{b^{1/m}\,t^\delta} \quad \text{where } \delta = -\frac{n}{m}. \tag{1.2}$$

If $m = 0$, V, R and P depend only on the time t, and the velocity v is proportional to r; the motion corresponding to this special case is studied in §15.

The solution depending on the independent variable may contain a number of arbitrary constants.

This argument shows that, when the characteristic parameters include two constants with independent dimensions in addition to r and t, the partial differential equations satisfied by the velocity, density and pressure in the one-dimensional unsteady motion of a compressible fluid can be replaced by ordinary differential equations for V, R and P. The solution of these ordinary differential equations can sometimes be obtained exactly in closed form and, in other cases, approximately by using numerical integration.

Such motions are called self-similar. We now formulate problems which can easily be solved by the method just described.

To be definite, we assume that the gas is perfect, inviscid and non-heat-conducting, so that the motion does not involve any kind of physical

---

† In general there may be several fundamental constants, but their dimensions must depend on a and b.

or chemical change (we shall consider later the extent to which these assumptions are valid in any given problem). The equations of motion, continuity and energy take the form

$$
\left.
\begin{aligned}
\frac{\partial v}{\partial t} + v\frac{\partial v}{\partial r} + \frac{1}{\rho}\frac{\partial p}{\partial r} &= 0, \\[2mm]
\frac{\partial \rho}{\partial t} + \frac{\partial \rho v}{\partial r} + (\nu - 1)\frac{\rho v}{r} &= 0, \\[2mm]
\frac{\partial}{\partial t}\left(\frac{p}{\rho^{\gamma}}\right) + v\frac{\partial}{\partial r}\left(\frac{p}{\rho^{\gamma}}\right) &= 0,
\end{aligned}
\right\}
\tag{1.3}
$$

where γ is the adiabatic index; $\nu = 1$ for plane flow, $\nu = 2$ for flow with cylindrical symmetry and $\nu = 3$ for flow with spherical symmetry.

These equations do not contain any dimensional constants; consequently, the question of the self-similarity of the motion is determined by the number of parameters with independent dimensions introduced by the remaining conditions in the problem. If there are only two of these the motion is self-similar. We shall now consider examples of self-similar problems.

2. MOTION OF A GAS WITH GIVEN INITIAL DISTRIBUTIONS OF VELOCITY $v_0(r)$, DENSITY $\rho_0(r)$ AND PRESSURE $p_0(r)$ (CAUCHY PROBLEM)

The general forms of the functions $v_0(r)$, $\rho_0(r)$ and $p_0(r)$ at the initial instant $(t = 0)$ are easily determined, when the motion which follows for $t > 0$ is self-similar.

In fact, since the whole motion involves only two constants a and b which have independent dimensions, the initial state must be defined by the three quantities a, b, and r.

Later, we shall assume that the dimensions of b and r are independent† and, therefore, that $n \neq 0$. Without loss of generality, let us put $[b] = LT^{-\delta}$ with $m \neq 0$.

From dimensional considerations, it follows that the initial distributions must be of the form

$$
v_0 = \alpha_1 b^{1/\delta} r^{1-1/\delta}, \quad \rho_0 = \alpha_2 ab^{s/\delta} r^{-(k+3+s/\delta)},
$$
$$
p_0 = \alpha_3 ab^{(s+2)/\delta} r^{-[k+1+(s+2)/\delta]},
\tag{1.4}
$$

† If the dimensions of b and r are connected then $n = \delta = 0$ and, consequently, a nondimensional combination br^{-m} exists.

It then follows from dimensional considerations that the initial velocity can only equal zero or infinity. The same applies to the initial density and pressure of $s \neq 0$ or -2.

If $n = 0$ and $s = 0$, then the initial density is an arbitrary function of r but the pressure and velocity are either zero or infinite. If $n = 0$ and $s = -2$, then the initial pressure is arbitrary but the velocity and density are either zero or infinite.

where α_1, α_2, α_3 are arbitrary constants; in plane wave motion the values of these constants when $r > 0$ are different from those when $r < 0$.

The uniqueness and existence of the solution of the initial value problem must be considered when the r interval is infinite. The physical variables of the motion may be infinite when $r = 0$ or $r = \infty$.

To solve the initial value problems for arbitrary values of k, s and δ, it is necessary to analyse self-similar motions of the most general type.

3. THE PISTON PROBLEM

A gas occupies a long cylindrical tube closed at one end by a piston. The gas is at rest ($v_1 = 0$) at the initial instant and the piston starts to move with constant speed U (Fig. 26).

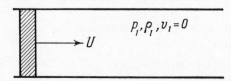

FIG. 26. The piston starts to move with the constant velocity U: the gas ahead of the piston is at rest initially with uniform density ρ_1 and pressure p_1.

The fundamental parameters in this problem, in addition to r and t, are the initial density ρ_1, the initial pressure p_1 and the piston velocity U. Since the dimensions of ρ_1, p_1 and U are connected by the relation

$$[U^2] = \frac{[p_1]}{[\rho_1]},$$

just two constants with independent dimensions enter into the problem.

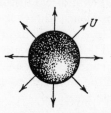

FIG. 27. Expansion of a sphere or a circular cylinder with constant velocity U. Initially the gas is at rest and the radius of the sphere or cylinder is zero. The initial gas density ρ_1 and the initial pressure p_1 are constant.

A similar problem can be formulated for motions with cylindrical and spherical symmetry: an infinite region of gas is set in motion at the initial instant by a cylinder or sphere, respectively, the radius of which increases from zero in proportion to the time (Fig. 27).

If the piston velocity is not constant but, for example, is proportional to some power of the time, so that

$$U = ct^n,$$

then a third fundamental constant c appears, the dimensions of which (for $n \neq 0$) are independent of the dimensions of ρ_1 and p_1

$$[c] = LT^{-n-1},$$

consequently, the resulting disturbance of the gas will not be self-similar.

The motion with this variable piston velocity $(n \neq 0)$ will be self-similar in the limiting case when $p_1 = 0$. In this case, only two independent dimensional constants ρ_1 and c enter into the problem.

4. PROBLEM OF GAS MOTION CONVERGING ON A POINT AND OF DISPERSION FROM A POINT

Consider an infinite homogeneous region of gas in which the pressure and density have constant initial values p_1 and ρ_1 and all particles have the same initial velocity, either directed toward the centre (focusing) or away from the centre (dispersion).

Correspondingly in the case of cylindrical symmetry we consider a motion in which all particles have the same initial velocity directed either toward the axis of symmetry or away from it.

Clearly, the analogous problem in plane wave motion reduces, after the addition of a forward velocity equal and opposite to the initial velocity of the gas to the problem of uniform motion of a piston.

In contrast to the plane flow case the present problems differ from the problems of a spherical and cylindrical piston.

As in the preceding problems, there are just two constants with independent dimensions (p_1, ρ_1) among the dimensional constants appearing in the initial and boundary conditions.

The problems of focusing and of dispersion of a gas in a uniform initial state are particular cases of the more general first initial Cauchy problem. In other cases, when the fundamental constants have dimensions which are different and independent of the dimensions of velocity, pressure and density, the initial radial distributions must be variable if they are not zero.

5. PROPAGATION OF A FLAME FRONT OR DETONATION WAVE

An infinite region of homogeneous combustible mixture with constant density ρ_1 and pressure p_1, is initiated at time $t = 0$ along a plane (plane case), a line (cylindrical symmetry) or at a point (spherical symmetry).

A plane, cylindrical or spherical flame front or detonation wave will then be propagated through the mixture.

As is known, the thickness of the combustion zone is insignificant (of the order of parts of a millimetre).

If the processes taking place in the combustion zone itself are not of interest, then we can take its thickness as zero, so that the gas burns instantaneously at a certain geometric surface. In this case the characteristic parameters will be: the initial density of the mixture ρ_1, the initial pressure p_1, the quantity of heat Q liberated during the combustion of a unit mass of gas, and in the case of flame front propagation, its velocity u, which is a constant determined by the physical chemistry of the given mixture.

The dimensions of Q in mechanical units can be expressed in terms of the dimensions of ρ_1 and p_1:

$$[Q] = \frac{[p_1]}{[\rho_1]},$$

Therefore, again only two of the four characteristic parameters have independent dimensions.

If the initial density ρ_1 is variable, obeying the relation

$$\rho_1 = \frac{A}{r^\omega},$$

then we have three fundamental constants A, p_1, Q and the disturbed gas motion will not be self-similar.

Analysis of the equations of motion and the boundary conditions shows that the initial pressure p_1 only enters into the boundary condition at the shock wave; if the initial pressure is neglected in this condition in comparison with the large pressure at the detonation wave-front, then p_1 is eliminated from the set of fundamental parameters. The problem becomes self-similar in this approximate formulation, with two independent dimensional constants: $[A] = ML^{\omega-3}$ and $[Q] = L^2 T^{-2}$

6. PROBLEM OF THE DECAY OF AN ARBITRARY DISCONTINUITY IN A COMBUSTIBLE MIXTURE

At time $t = 0$ a gas in a uniform state with velocity v_1, density ρ_1 and pressure p_1 occupies the region to the left of the plane $r = 0$ ($v = 1$ case). The region to the right of this plane is occupied by a combustible mixture in which the velocity, density and pressure have constant values and since, in general, the conditions of conservation of mass, momentum and energy will not be satisfied across such a discontinuity, it cannot exist in isolation after the initial instant, and a gas motion with one or more surfaces of discontinuity must develop on each of which the conservation conditions will be satisfied (a flame or detonation front may be propagated through the combustible mixture). The parameters of the problem, namely, Q, the quantity of heat liberated during the combustion of a unit mass of gas, u, the flame front velocity, and v_1, ρ_1, p_1, v_2, ρ_2, p_2, always include two with independent dimensions. Therefore, the motion which occurs will be self-similar.

To solve the Cauchy problem in the plane case for arbitrary values of the constants k, s, δ, and α_1^+, α_2^+, α_3^+ for $r > 0$, α_1^-, α_2^-, α_3^- for $r < 0$ in (1.4), we must solve the more general self-similar problem of the decay of the corresponding singularities of the discontinuity at $r = 0$.

When the solution exists and is unique, the decay of the corresponding discontinuity is a local phenomenon determined solely by the character of the singularity.

7. PROBLEM OF A STRONG EXPLOSION

An explosion occurs at time $t = 0$ in a gas at rest at the centre of symmetry (a point), so that a finite amount of energy E_0 is liberated instantaneously. In this formulation, we shall neglect the mass and dimensions of the substance liberating the energy. The problem of a strong explosion reflects the essential characteristics of the phenomena in an atomic bomb explosion; we give experimental results about this in §11.

Three constants with independent dimensions enter into the problem: the initial gas density ρ_1, the initial pressure p_1 and the explosion energy E_0.

The system of fundamental parameters influencing the motion of the disturbed gas after the explosion, under adiabatic conditions, is represented by the quantities

$$\rho_1, p_1, E_0, r, t, \gamma.$$

General considerations of dimensional analysis then show that all the independent nondimensional quantities can only depend on the three nondimensional parameters:

$$\gamma; \quad \frac{\rho_1^{1/5} r}{E_0^{1/5} t^{2/5}} = \lambda; \quad \frac{p_1^{5/6} t}{E_0^{1/3} \rho_1^{1/2}} = \tau, \quad (1.5)$$

of which λ and τ are variables. Experiment and theory show that an abrupt jump in the characteristics of the motion takes place on the boundaries of the region of disturbed gas motion during an explosion, and a shock wave is formed. This will be a sphere with radius which increases with time. The influence of the initial pressure p_1 and; therefore, of the parameter τ enters only because of the dynamic conditions on the shock wave.

However, if the explosion is strong (E_0 is large), then the pressure behind the shock wave produced by the explosion will be many times larger than the initial pressure in the gas and the gas motion behind the shock wave when its radius is small will be practically independent of the initial pressure p_1; hence, only two dimensional constants are essential: ρ_1 and E_0.

Mathematically, neglecting the initial pressure is equivalent to putting the unperturbed pressure $p_1 = 0$ in the shock wave equations; consequently, the parameter p_1 drops out and, therefore, so does the independent variable τ, with the obvious result that the disturbed gas motion can be considered self-similar.

To neglect the initial pressure, the "counter-pressure" p_1, as the shock wave attenuates further is invalid and, therefore, the problem of the disturbed gas motion due to a point explosion at large distances from the centre of the explosion ceases to be self-similar.

We note that it is sufficient to carry out the computation of just one special case in the numerical solution of the problem of a point explosion when the counter pressure p_1 is not neglected; the relation between all the quantities required and the nondimensional quantities λ, τ can then be obtained and hence, the characteristics of the disturbed explosion field can easily be obtained for a given value of γ, for any values of E_0, initial density ρ_1 and initial pressure p_1.

The explosion occurs along a line† in the cylindrical symmetry case and along a plane‡ in the plane wave case. Here, E_0 denotes the energy liberated per unit length or area, respectively.

Evidently, the explosion problem can be generalized to the case of variable initial density obeying the relation $\rho_1 = A/r^\omega$.

To do this, the constant A must be taken in the dimensions formula $[A] = ML^{\omega-3}$ as the fundamental parameter, instead of ρ_1.

In this case, the following variables can be used instead of the variables of (1.5):

$$\gamma, \quad \left(\frac{A}{E_0}\right)^{1/(5-\omega)} \frac{r}{t^{2/(5-\omega)}} = \lambda, \quad \frac{p_1^{(5-\omega)/6} t}{A^{1/2} E_0^{(2-\omega)/6}} = \tau.$$

The influence of the second parameter can be neglected for small values of the parameter τ because of the high energy E_0 liberated, the small initial pressure p_1 and of the small time interval t, and hence, we

---

† In certain cases, strong explosions along a line can be considered as a high intensity electric discharge in a gas.

‡ The dimensions of E_0 vary in the cylindrical and plane cases and the parameters λ and τ vary correspondingly, see §§ 11, 12, 13 and 14.

154 SIMILARITY AND DIMENSIONAL METHODS IN MECHANICS

obtain a self-similar problem of propagation of a strong explosion in a medium with variable density.

The solution of the problems of strong explosions is explained in §§11, 12 and 14.

8. PROPERTIES OF IDEAL MEDIA AND SELF-SIMILARITY

It has been shown above that if the boundary and initial conditions contain only two independent dimensional constants in the determination of one-dimensional, unsteady, adiabatic† motion of an ideal (inviscid) perfect gas, then self-similarity holds.

It is not difficult to see that when the characteristic constants have dimensions dependent on the density ρ_1 and the pressure p_1, the results about self-similarity of the motion apply to other ideal media (those in which tangential stresses are absent, or in which the thermodynamic state is determined by two parameters, p and ρ, say).

Actually, the usual reasoning of dimensional analysis shows that the internal energy, per unit mass, which enters into the shock conditions, has the form

$$\epsilon = \frac{p}{\rho} F\left(\frac{p}{p^*}, \frac{\rho}{\rho^*}, \alpha_1, \alpha_2, \ldots\right),$$

while the entropy is of the form

$$S = AG\left(\frac{p}{p^*}, \frac{\rho}{\rho^*}, \beta_1, \beta_2, \ldots\right),$$

where p^*, ρ^* are constants with the dimensions of pressure and density. The dimensional constant A is insignificant since it can be taken out of the adiabatic condition: $\alpha_1, \alpha_2, \ldots, \beta_1, \beta_2, \ldots$ are abstract constants. Evidently, the general form of the equation of state is

$$c_v T = \frac{p}{\rho} H\left(\frac{p}{p^*}, \frac{\rho}{\rho^*}, \alpha_1, \beta_1, \ldots\right).$$

Hence, it is clear that the addition of the constants p^* and ρ^* to the table of characteristic parameters does not violate self-similarity if the two characteristic dimensional constants have dimensions dependent on p^* and ρ^*.

The internal properties of the medium may be self-similar when certain of the essential functions F, G, H, are independent of the two parameters p/p^* and ρ/ρ^* and depend only on the single nondimensional parameter $p^{k_1}\rho^{k_2}/C$. In this case, the properties of the medium can be expressed

---

† The adiabatic condition can be replaced by other conditions: for example, the absence of a temperature gradient: $\partial T/\partial r = 0$ (infinite heat conduction).

solely in terms of the dimensional constant C in place of the constants p^* and ρ^*. Generally speaking, the dimensional constant C must be introduced as one of the characteristic constants. Self-similar problems are possible for such media and also arise when the dimensions of one of the characteristic constants are fixed and equal to the dimensions of C.

The functions F, H, reduce to arbitrary constants in the case of an ideal, perfect gas and the additive constant in the function G is insignificant in many formulations of the problem. Consequently, self-similar motions with two arbitrary independent dimensional constants can be constructed for an ideal, perfect gas.

§2. ORDINARY DIFFERENTIAL EQUATIONS AND THE SHOCK CONDITIONS FOR SELF-SIMILAR MOTIONS

1. ORDINARY DIFFERENTIAL EQUATIONS

In order to solve the problems mentioned, let us derive the equations which V, R, P must satisfy.

Substituting the expressions for v, ρ and p in terms of V, R, P in (1.3) from (1.1) and taking (1.2) into account, we obtain

$$\left. \begin{aligned} \lambda\left[(\delta-V)\,V'-\frac{P'}{R}\right] &= V^2-V-(k+1)\frac{P}{R}, \\[2mm] \lambda\left[-V'+(\delta-V)\frac{R'}{R}\right] &= -s-(k-\nu+3)\,V, \\[2mm] \lambda(\delta-V)\left[\frac{P'}{P}-\gamma\frac{R'}{R}\right] &= -s(1-\gamma)-2-[k(1-\gamma)+1-3\gamma]\,V. \end{aligned} \right\}$$

Introducing the new variable† $z = \gamma P/R$, we transform these equations to the following:

$$\frac{dz}{dV} = \frac{z\{[2(V-1)+\nu(\gamma-1)\,V](V-\delta)^2-(\gamma-1) \times V(V-1)(V-\delta)-[2(V-1)+\kappa(\gamma-1)]z\}}{(V-\delta)[V(V-1)(V-\delta)+(\kappa-\nu V)z]}, \quad (2.1)$$

$$\frac{d\ln\lambda}{dV} = \frac{z-(V-\delta)^2}{V(V-1)(V-\delta)+(\kappa-\nu V)z}, \quad (2.2)$$

$$(V-\delta)\frac{d\ln R}{d\ln\lambda} = [s+(k-\nu+3)\,V]-\frac{V(V-1)(V-\delta)+(\kappa-\nu V)z}{z-(V-\delta)^2}, \quad (2.3)$$

† The temperature T and the variable z are related by

$$\mathbf{R}T = \frac{r^2}{\gamma t^2}z,$$

where $\mathbf{R}$ is the gas constant.

where

$$\kappa = \frac{s + 2 + \delta(k+1)}{\gamma}.$$

The dimensions of the constant a can be altered for a given type of self-similar motion by introducing a new constant a_1 where $a_1 = ab^\chi$, and the exponent χ is arbitrary. The modified values of k_1 and s_1 are determined by the formulas

$$k_1 = k + \chi, \quad s_1 = s - \delta\chi.$$

The parameter λ can also be changed by introducing the new constant $b_1 = b^m$. The functions $P(\lambda)$ and $R(\lambda)$ and the parameter λ depend on the choice of the dimensions of the constants a and b; it is evident that the variables z, V as well as the function $z(V)$, are independent of the choice of the exponents k, s, m, but are determined completely by the kind of self-similar motion, which depends essentially on just two parameters κ and δ.

When the quantities k and s are replaced by k_1 and s_1 in the expression for κ, we obtain $\kappa = \kappa_1$.

The marked peculiarity of the function $z(V)$ explains the relation of the field of integral curves of (2.1) in the z, V plane to the type of self-similar motion, independently of the method of introducing the characteristic constants a and b.

It is easy to see that the fundamental problem reduces to the integration of (2.1). If (2.1) is integrated, then the relation of V and R to λ is determined from (2.2) and (2.3) by using quadratures.

The plane of the nondimensional variables z, V can be considered for arbitrary non-self-similar motions. A certain curve in the z, V plane corresponds to the field of one-dimensional unsteady gas motion at each instant. Points of discontinuity will be shocks (jumps) on this curve in the presence of strong explosions. Different curves in the z, V plane will correspond to the gas motion at different instants for non-self-similar motions.

The points corresponding to strong discontinuities in the z, V plane will move with the passage of time. Different curves correspond to different fixed points in space or to different fixed particles in the z, V plane. If the motion is self-similar, then the same integral curve of (2.1) will correspond to motion in the z, V plane of different points or particles at different instants. It follows from the formulation of self-similar problems that the shock coordinate r and the variable $\lambda = r/bt^\delta$ at the shock are functions of the time t and of the characteristic dimensional constants a and b.†

It is impossible to form a nondimensional combination of the three quantities a, b and t, consequently, the surface of discontinuity is given by

$$\lambda = \lambda_0 = \text{const}, \quad r = \lambda_0 b t^\delta.$$

Therefore, fixed values of the variables λ, R, z, P, V correspond to a shock in self-similar motions. Fixed points correspond to jumps in the z, V plane.

The magnitude of the shock velocity c is given by

$$c = \frac{dr}{dt} = \delta \frac{r}{t}. \tag{2.4}$$

Evidently, δ is a constant for self-similar motion. The shock velocity for $r > 0$, $t > 0$ is directed away from the centre when $\delta > 0$ and toward the centre when $\delta < 0$. Therefore, the shock wave diverges when $\delta > 0$ and converges when $\delta < 0$, where the velocity of the phase motion is slowed down for $\delta < 0$. If $r > 0$, the time t increases, but $t < 0$, so the motion at the shocks is reversed in character. The particle velocity on the parabola $z = (\delta - V)^2$ equals the sound speed; above this parabola this velocity is subsonic, below supersonic. The abstract quantity δ is a certain function of time in the general case of non-self-similar motions.

2. CONDITIONS AT COMPRESSION SHOCKS

In the majority of the problems mentioned above, discontinuities occur in the flow (shock waves, detonation fronts, flame fronts); consequently, we consider the general relations between V, z and R on both sides of a surface of strong discontinuity.

The conditions for conservation of mass, momentum and energy must be satisfied on crossing a surface of discontinuity. Denoting the quantities on one side of the discontinuity by the subscript 1 and on the other side by the subscript 2, we can write:

$$\left.\begin{aligned} \rho_1(v_1-c) &= \rho_2(v_2-c), \\ \rho_1(v_1-c)^2+p_1 &= \rho_2(v_2-c)^2+p_2; \end{aligned}\right\} \tag{2.5}$$

$$\tfrac{1}{2}(v_1-c)^2+\frac{\gamma}{\gamma-1}\frac{p_1}{\rho_1} = \tfrac{1}{2}(v_2-c)^2+\frac{\gamma}{\gamma-1}\frac{p_2}{\rho_2}. \tag{2.6}$$

---

† Sometimes, the gas motion is self-similar but the motion of the boundaries, the shock wave say, is determined by additional constants and, consequently, the shock coordinate r depends not only on a, b and t but also on other dimensional constants; the formula $\lambda = \text{const}$ at the shock is not true in these cases. In conformity with the definitions used, such motions, considered as a whole, will be called non-self-similar although the self-similarity is violated only on the boundary.

These equations are written for a perfect gas, in which it is assumed that the enthalpy i per unit mass of gas is defined as

$$i = c_p T + \text{const} = \frac{\gamma}{\gamma - 1} \frac{p}{\rho} + \text{const.}$$

We replace the quantities v, ρ, p in (2.5) and (2.6) by their expressions in terms of V, R, P from (1.1), the velocity c by $\delta(r/t)$ according to (2.4), and we introduce the variable $z = \gamma P/R$.

The relations at the shock become

$$R_1(V_1 - \delta) = R_2(V_2 - \delta),$$

$$V_1 - \delta + \frac{z_1}{\gamma(V_1 - \delta)} = V_2 - \delta + \frac{z_2}{\gamma(V_2 - \delta)},$$

$$(V_1 - \delta)^2 + \frac{2z_1}{\gamma - 1} = (V_2 - \delta)^2 + \frac{2z_2}{\gamma - 1}.$$

Solving these equations for V_2 and z_2 in terms of V_1 and z_1 we find that

$$\left.\begin{aligned}
V_2 - \delta &= (V_1 - \delta)\left[1 + \frac{2}{\gamma + 1}\frac{z_1 - (V_1 - \delta)^2}{(V_1 - \delta)^2}\right], \\
z_2 &= \left(\frac{\gamma - 1}{\gamma + 1}\right)^2 \frac{1}{(V_1 - \delta)^2}\left[(V_1 - \delta)^2 + \frac{2z_1}{\gamma - 1}\right]\left[\frac{2\gamma}{\gamma - 1}(V_1 - \delta)^2 - z_1\right].
\end{aligned}\right\} \quad (2.7)$$

Knowing the point (V_1, z_1) in the (V, z) plane, we find from (2.7) the point (V_2, z_2) where the (V, z) curve passes through the shock. We assume that the gas particles pass through the shock from state 1 into state 2. It is clear from the symmetry of (2.5), (2.6) that the subscripts 1 and 2 in (2.7) can be interchanged.

The points of the parabola

$$z = (V - \delta)^2 \tag{2.8}$$

transform into themselves. On this parabola, weak discontinuities, i.e. surfaces of discontinuity of the derivatives, can arise. Actually, Equation (2.8), written in terms of dimensional quantities, yields:

$$\frac{\gamma p}{\rho} = (v - c)^2,$$

i.e. the square of the particle velocity on the shock equals the square of the speed of sound. Points lying above the parabola (2.8) transform into points lying beneath it and conversely.

Since z is always positive in the physical sense, then only those cases when the points of the upper half-plane transform into points of the upper half-plane, have physical meaning.

Points on the V axis, corresponding to the limiting case when $z = 0$, transform into points of the parabola

$$z = \frac{2\gamma}{\gamma - 1}(V - \delta)^2.$$

Therefore, the transformation (2.7) maps the region between the V axis and the parabola $z = (V - \delta)^2$ into the region between the parabola $z = \{2\gamma/(\gamma - 1)\}(V - \delta)^2$ and the parabola $z = (V - \delta)^2$ and conversely.

Furthermore, since

$$\frac{\rho_2}{\rho_1} = \frac{R_2}{R_1} = \frac{V_1 - \delta}{V_2 - \delta} > 0, \tag{2.9}$$

it is evident that points on different sides of the shock in the z, V plane lie on the same side of the line $V = \delta$. Relation (2.9) also follows from (2.7) for any $z_1 \geqslant 0$.

For points above the parabola (2.8)

$$z > (V - \delta)^2 \quad \text{or} \quad a^2 = \frac{\gamma p}{\rho} > (v - c)^2,$$

and for points below the parabola (2.8)

$$z < (V - \delta)^2 \quad \text{or} \quad a^2 = \frac{\gamma p}{\rho} < (v - c)^2.$$

In other words, we have: the gas particle velocity relative to the shock is subsonic for $z > (V - \delta)^2$ and supersonic for $z < (V - \delta)^2$. Consequently,[†] the region between the parabola $z = (V - \delta)^2$ and the line $z = 0$ corresponds to states ahead of compression shocks and the region between the parabolas $z = \{2\gamma/(\gamma - 1)\}(V - \delta)^2$ and $z = (V - \delta)^2$ corresponds to states behind the compression shocks.

Points lying above the parabola $z = \{2\gamma/(\gamma - 1)\}(V - \delta)^2$ transform into points of the lower half-plane according to (2.7) and, therefore, cannot correspond to a state of the gas either in front of or behind the shock.

The regions in Fig. 28 corresponding to the state of a gas in front of the shock are shaded vertically and the regions corresponding to the state behind the shock are shaded horizontally.

The direction of the possible transformations from the (V_1, z_1) point to the (V_2, z_2) point are shown by arrows.

† An analysis of the conditions on strong discontinuities is explained in the general case, for example, in Sedov (1950).

We note that if the shock wave is propagated into gas at rest, i.e. if the point (V_1, z_1) is on the z axis ($V = 0$), then the point (V_2, z_2) must lie on the parabola

$$z_2 = -\delta(V_2 - \delta)\left(1 + \frac{\gamma - 1}{2\delta} V_2\right). \qquad (2.10)$$

This parabola is shown on the left of Fig. 28.

The phase velocities $\lambda = \text{const}$ (in particular, the shock velocity) $c = \delta(r/t)$ to the left of the line $V = \delta$ are larger and to the right of this line are less than the gas particle velocity $v = V(r/t)$ at the same point in space. The directions of both velocities in space are identical in the first and second cases. The relative particle phase velocity in the first case

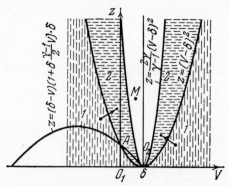

Fig. 28. The transformation of points from the vertically shaded region into the horizontally shaded region corresponds to shocks in the $z = (\gamma p t^2)/(\rho r^2)$, $V = vt/r$ plane. Possible transformations are shown by arrows.

agrees with the direction of the particle velocity and is opposite to the particle velocity in the second.

It is evident that some of the conclusions of this section are not related to self-similarity of the motion.

3. CONDITIONS ON A DETONATION FRONT AND A FLAME FRONT

Let us first consider the general properties of discontinuities with energy addition. Let Q be the energy addition per unit mass when a gas particle passes through a shock front from the state 1 to the state 2. This flow can be realized by chemical reactions (flame front, detonation front), by heat conduction, heat emission or any other process.

The mechanical conditions (2.5) still apply when there is energy addition at the shock front but the energy equation (2.6) is changed because of the addition of energy Q to the state 2 (increase of a constant in the heat content formula). The modified energy equation will be

$$\frac{\gamma_1}{\gamma_1-1}\frac{p_1}{\rho_1}+\tfrac{1}{2}(v_1-c)^2+Q = \frac{\gamma_2}{\gamma_2-1}\frac{p_2}{\rho_2}+\tfrac{1}{2}(v_2-c)^2. \tag{2.11}$$

In this equation we take account of the fact that the values of the Poisson coefficient $\gamma = c_p/c_v$ in front of the wave, γ_1, and behind the wave, γ_2, can be different.

Equations (2.5) and (2.11) can be transformed into

$$v_2-c = \frac{\rho_1}{\rho_2}(v_1-c), \tag{2.12}$$

$$p_2 = p_1+\left(1-\frac{\rho_1}{\rho_2}\right)\rho_1(v_1-c)^2, \tag{2.13}$$

$$\left(\frac{\rho_1}{\rho_2}\right)^2-\frac{2\gamma_2}{\gamma_2+1}\left(1+\frac{p_1}{\rho_1(v_1-c)^2}\right)\frac{\rho_1}{\rho_2}$$

$$+\frac{\gamma_2-1}{\gamma_2+1}\left[\frac{2\gamma_1}{\gamma_1-1}\frac{p_1}{\rho_1(v_1-c)^2}+1+\frac{2Q}{(v_1-c)^2}\right] = 0. \tag{2.14}$$

The absolute value of the difference $|c-v_1| = u$ is the particle velocity relative to the shock ahead of the wave front. The velocity u is a given chemical constant in the case of a flame front.

The following relations are self-evident:

$$\frac{a_1^2}{(v_1-c)^2} = \frac{\gamma_1 p_1}{\rho_1(v_1-c)^2} = \frac{z_1}{(V_1-\delta)^2}$$

and

$$\frac{a_2^2}{(v_2-c)^2} = \frac{\gamma_2 p_2}{\rho_2(v_2-c)^2} = \frac{z_2}{(V_2-\delta)^2}.$$

In these formulas, z_1 and z_2 are defined by means of γ_1 and γ_2, respectively; a_1 and a_2 are the corresponding speeds of sound.

The equation $z/(V-\delta)^2 = $ const defines a parabola in the z, V plane. The region between the two parabolas in the z, V plane

$$z = \alpha(V-\delta)^2 \quad \text{and} \quad z = \beta(V-\delta)^2$$

corresponds to the interval

$$\alpha = \frac{\gamma p}{\rho(v-c)^2}, \quad \beta = \frac{\gamma p}{\rho(v-c)^2}.$$

Equation (2.14) is a quadratic equation defining the ratio ρ_1/ρ_2 as a function of the following four parameters: γ_1, γ_2, $p_1/\rho_1 u^2$ and Q/u^2.

6

A flame front is a rarefaction shock on which the following inequalities hold:

$$\frac{\rho_1}{\rho_2} > 1, \quad \frac{\gamma_2 p_2}{\rho_2(v_2-c)^2} = \frac{z_2}{(V_2-\delta)^2} > 1, \quad \frac{\gamma_1 p_1}{\rho_1(v_1-c)^2} = \frac{z_1}{(V_1-\delta)^2} > 1, \quad (2.15)$$

i.e. the density of the combustion products is less than the density of the combustible mixture and the particle shock speeds (v_1-c) and (v_2-c) in front of and behind the flame front are subsonic.

A detonation front is a compression shock across which the following inequalities hold

$$\frac{\rho_1}{\rho_2} < 1, \quad \frac{\gamma_2 p_2}{\rho_2(v_2-c)^2} = \frac{z_2}{(V_2-\delta)^2} \geqslant 1, \quad \frac{\gamma_1 p_1}{\rho_1(v_1-c)^2} = \frac{z_1}{(V_1-\delta)^2} < 1. \quad (2.16)$$

The fluid particle speed at the shock is supersonic ahead of the front and subsonic or exactly sonic behind the front.

A detonation $\rho_1/\rho_2 < 1$ corresponding to supersonic fluid particle speed behind the shock when

$$\frac{a_2^2}{(v_2-c)^2} = \frac{z_2}{(V_2-\delta)^2} < 1$$

can also be considered; however, such regimes are not usually realized in experiments (Zel'dovich and Kompaneets, 1955).

The roots of (2.14) coincide when the relation

$$\gamma_2^2 \left(1+\frac{p_1}{\rho_1 u^2}\right)^2 = (\gamma_2^2-1)\left[\frac{2\gamma_1}{\gamma_1-1}\frac{p_1}{\rho_1 u^2}+1+\frac{2Q}{u^2}\right], \qquad (2.17)$$

is satisfied. Since, from (2.12) and (2.13),

$$\frac{\gamma_2 p_2}{\rho_2(v_2-c)^2} = \gamma_2 \left(1+\frac{p_1}{\rho_1 u^2}\right)\frac{\rho_2}{\rho_1}-\gamma_2, \qquad (2.18)$$

then it is evident from (2.14) that (2.17) is equivalent to the relation

$$\frac{\gamma_2 p_2}{\rho_2(v_2-c)^2} = \frac{z_2}{(V_2-\delta)^2} = 1. \qquad (2.19)$$

Therefore, coincidence of the roots of (2.14) is attained when the particle speed just behind the shock front is exactly sonic.

Condition (2.17), equivalent to (2.19) for compression shocks, is called the Chapman-Jouguet condition. This condition is realized in actual gas motions involving detonation waves. The quantities ρ_1, p_1 as well as the given quantities γ_1, γ_2 and Q, can be calculated independently of the character of the particular value of the front speed $u = v_1 - c$

by using the additional condition (2.17), for a given state ahead of the wave front, and the quantities ρ_2, p_2, $v_2 - c$ can be calculated behind the detonation wave front from (2.12), (2.13) and (2.14).

It should again† be pointed out that in a number of problems the detonation front velocity $u = |v_1 - c|$ is determined by the formulation of the problem and is larger than the velocity u found from (2.17) according to the Chapman-Jouguet rule, where $a_2^2 > (v_2 - c)^2$ in accordance with the sign of the inequality in (2.16).

4. ON THE RELATION BETWEEN TEMPERATURE AND VELOCITY BEHIND A SHOCK

Without loss of generality, let us assume that $v_1 = 0$, i.e. that the motion relative to the gas ahead of the shock is known.

It then follows from the mechanical relations (2.5) that

$$\chi = \frac{v_2^2}{RT_2} = \frac{v_2^2 \rho_2}{p_2} = \frac{\left(1 - \dfrac{\rho_1}{\rho_2}\right)^2}{\dfrac{\rho_1}{\rho_2}\left[\left(1 + \dfrac{p_1}{\rho_1 c^2}\right) - \dfrac{\rho_1}{\rho_2}\right]}. \tag{2.20}$$

The ratio ρ_1/ρ_2 is less than unity for compression shocks. The non-dimensional function $\chi[\rho_1/\rho_2,\, p_1/(\rho_1 c^2)]$ is positive and monotonic in both the arguments for $p_1/(\rho_1 c^2) > 0$ and $\rho_1/\rho_2 < 1$, hence, it follows that the quantity χ has a maximum value, corresponding to $p_1/(\rho_1 c^2) = 0$ and $(\rho_1/\rho_2)_{\min}$

$$\chi_{\max} = \left(\frac{\rho_2}{\rho_1}\right)_{\max} - 1. \tag{2.21}$$

This value is independent of the energy equation.

Evidently, the greatest compression ρ_2/ρ_1 is attained for simple shock waves or for waves with additional energy absorption. The magnitudes of $(\rho_2/\rho_1)_{\max}$ and $\chi_{\max}$ are less for waves with energy addition and, in particular, for flame or detonation fronts. It follows from the shock conditions that the greatest value of the ratio ρ_2/ρ_1 is $(\gamma + 1)/(\gamma - 1)$ for simple shock waves and, therefore,

$$\chi_{\max} = \frac{2}{\gamma - 1}. \tag{22.2}$$

We find from (2.14), that for strong detonation waves $[p_1/(\rho_1 c^2) = 0]$, satisfying the Chapman-Jouguet condition:

$$\left(\frac{\rho_2}{\rho_1}\right)_{\max} = \frac{\gamma_2 + 1}{\gamma_2},$$

† Suitable examples are given in §8.

and correspondingly
$$\chi_{\max} = \frac{1}{\gamma_2}.$$

The upper limits found for χ permit the temperature behind the front to be found in terms of the fluid particle velocity behind the front.

In many cases, the particle velocity is known; the relative gas velocity for a body moving in a fixed gas equals the body velocity; the particle velocity in nebulae or in stellar photo spheres can be determined in astrophysics by using emission of spectra.

For example, if the velocity behind the front in hydrogen is 1,000 km./sec., then the temperature behind the front in this case must be larger than 50 million Celsius degrees.

5. CONDITIONS AT A SHOCK FRONT WITH HEAT ADDITION FOR SELF-SIMILAR MOTIONS

The constant Q in this case is the key parameter; since $[Q] = L^2 T^{-2}$, then we can assume $m = 2$, $n = -2$, i.e. $\delta = 1$ and, therefore,

$$\lambda = \frac{\beta r}{t \sqrt{Q}},$$

where β is a certain constant whose value can be controlled.

We denote the second dimensional constant by A; evidently we can always take its dimensions formula to be: $[A] = M L^{\omega - 3}$.

It follows from self-similarity that the density and the pressure at $t = 0$ will be given by

$$\rho_1 = k_1 \frac{A}{r^\omega}, \quad p_1 = k_2 \frac{AQ}{r^\omega}, \tag{2.23}$$

where k_1 and k_2 are constants. If we assume that the gas is initially in equilibrium with no body forces acting, then it follows that $k_2 = 0$ for $\omega \neq 0$ and, therefore, in an undisturbed medium in equilibrium

$$p_1 = 0. \tag{2.24}$$

If $\omega = 0$, then $p_1 = \text{const}$ and p_1 can be different from zero in equilibrium.

The conditions on the shock become, for self-similar motions with heat addition:

$$\left. \begin{array}{c} R_1(V_1 - 1) = R_2(V_2 - 1), \\[2mm] V_1 - 1 + \dfrac{z_1}{\gamma_1(V_1 - 1)} = V_2 - 1 + \dfrac{z_2}{\gamma_2(V_2 - 1)}, \\[2mm] \tfrac{1}{2}(V_1 - 1)^2 + \dfrac{z_1}{\gamma_1 - 1} + \dfrac{Q}{c^2} = \tfrac{1}{2}(V_2 - 1)^2 + \dfrac{z_2}{\gamma_2 - 1}, \end{array} \right\} \tag{2.25}$$

where $\lambda = \lambda^* = \text{const}$ on the shock.

For the motion of the shock we have:

$$r_2 = \frac{\lambda^*}{\beta}\sqrt{(Q)}t, \quad c = \frac{dr_2}{dt} = \frac{\lambda^*}{\beta}\sqrt{Q} = \frac{r_2}{t}, \quad \frac{\beta^2}{\lambda^{*2}} = \frac{Q}{c^2}.$$

If there are several shocks, then it is always possible to take $\lambda^* = 1$ on one of them. Thus, the constant β is defined.

The Chapman-Jouguet condition (2.19) yields:

$$z_2 = (V_2 - \delta)^2. \tag{2.26}$$

Hence, the detonation front in the z, V plane must correspond to a certain point of the parabola (2.26) if the Chapman-Jouguet condition is satisfied.

If the shock is propagated into a gas at rest, then $V_1 = 0$; we obtain from (2.25)

$$\left.\begin{aligned}
R_2 &= R_1\left[\frac{\gamma_2}{\gamma_2+1}\left(1+\frac{z_1}{\gamma_1}\right)(1-\Lambda)\right]^{-1}, \\[2mm]
V_2 &= 1 - \left[\frac{\gamma_2}{\gamma_2+1}\left(1+\frac{z_1}{\gamma_1}\right)(1-\Lambda)\right], \\[2mm]
z_2 &= \frac{\gamma_2^2}{(\gamma_2+1)^2}\left(1+\frac{z_1}{\gamma_1}\right)^2(1-\Lambda)(1+\gamma_2\Lambda),
\end{aligned}\right\} \tag{2.27}$$

in which

$$\Lambda^2 = 1 - \frac{\gamma_2^2-1}{\gamma_2^2}\frac{\left[\dfrac{2}{\gamma_1-1}z_1 + 1 + \dfrac{2Q}{c^2}\right]}{\left(1+\dfrac{z_1}{\gamma_1}\right)^2}$$

The Chapman-Jouguet condition is equivalent to the condition $\Lambda = 0$; but if the Chapman-Jouguet condition is not satisfied, and $V_1 = z_1 = 0$, then

$$\left.\begin{aligned}
R_2 &= R_1\frac{\gamma_2+1}{\gamma_2(1-\Lambda)}, \quad V_2 = \frac{1+\gamma_2\Lambda}{\gamma_2+1}, \\[2mm]
z_2 &= \frac{\gamma_2^2}{(\gamma_2+1)^2}(1-\Lambda)(1+\gamma_2\Lambda).
\end{aligned}\right\} \tag{2.28}$$

The values V_2, z_2 behind the shock are located on the parabola

$$z_2 = \gamma_2 V_2(1-V_2). \tag{2.29}$$

The Chapman-Jouguet point corresponds to the point of intersection of the parabolas (2.26) and (2.29) at which $\Lambda = 0$.

The parabola (2.29) passes through the origin where $\varLambda = -1/\gamma_2$. The quantity $\varLambda$ increases as we move upward along the parabola (2.29) from the origin and becomes zero on the parabola (2.26) and then it increases as we move along further and approaches $\varLambda = 1/\gamma_2$ corresponding to a strong, simple shock wave ($Q = 0$).

§3. Algebraic Integrals for Self-Similar Motions

Algebraic integrals of the system of ordinary differential equations can be established independently of particular boundary or initial conditions by using dimensional analysis for self-similar motions. In other words, the order of the system of ordinary equations can always be lowered in the general case.

We show below that the number of such integrals can be increased in certain examples when the characteristic constants a and b have particular values.

The conclusions which follow remain valid for cases more general than the gas motion described by the system (1.3).

To be concrete, we consider the unsteady adiabatic motion of a perfect gas with spherical symmetry taking Newtonian gravitation into account. We then have the system of equations:

$$\frac{\partial \rho}{\partial t} + \frac{\partial \rho v}{\partial r} + \frac{(\nu - 1)\rho v}{r} = 0, \tag{3.1}$$

$$\frac{\partial \mathscr{M}}{\partial r} = \sigma_\nu \rho r^{\nu - 1}, \tag{3.2}$$

$$\frac{\partial v}{\partial t} + v\frac{\partial v}{\partial r} + \frac{1}{\rho}\frac{\partial p}{\partial r} + \frac{f\mathscr{M}}{r^2} = 0, \tag{3.3}$$

$$\frac{\partial S}{\partial t} + v\frac{\partial S}{\partial r} = 0, \tag{3.4}$$

where $\sigma_\nu = 2(\nu - 1)\pi + \delta_{\nu 1}$, in which

$$\delta_{\nu 1} = \begin{cases} 1, \nu = 1, \\ 0, \nu \neq 1, \end{cases}$$

$\nu = 3$ in the spherical case; f is the gravitational constant, $[f] = M^{-1}L^3T^{-2}$. We also consider simultaneously the cylindrical wave case† $\nu = 2$ or the plane wave case $\nu = 1$ when $f = 0$; $\mathscr{M}$ is the mass contained between a fixed surface and the surface under consideration, $[\mathscr{M}] = ML^{\nu-3}$; S is the entropy or a certain function of the entropy.

† Gravitational forces can be taken into account in the following analysis for $\nu = 2$.

Let us consider the self-similar motions defined by the two dimensional constants† a, b:

$$[a] = M L^k T^s \quad \text{and} \quad [b] = L^m T^n.$$

For $m \neq 0$, we can assume without loss of generality, that $m = 1$, $n = -\delta$, $k = -3$. It is sufficient to put $a_1 = a b^{(-k-3)/m}$ and $b_1 = b^{1/m}$. It is necessary to assume $[a] = [1/f]$ for $k = -3$ when Newtonian gravitation is taken into account; consequently, $s = 2$ and, therefore, δ will be the single characteristic parameter. The exponent s can be arbitrary for $f = 0$.

In the general case of self-similar motions when $m \neq 0$, it is possible to write

$$\left. \begin{aligned} \lambda &= \frac{r}{b_1 t^\delta}; \quad v = \frac{r}{t} V(\lambda); \quad \rho = \frac{a}{r^{k+3} t^s} R(\lambda); \\[2mm] p &= \frac{a r^2}{r^{k+3} t^{s+2}} P(\lambda); \quad \mathcal{M} = \frac{a r^\nu}{r^{k+3} t^s} M(\lambda). \end{aligned} \right\} \tag{3.5}$$

Substitution of (3.5) into (3.1)–(3.4) leads to a system of four ordinary differential equations for

$$V(\lambda), \ R(\lambda), \ P(\lambda) \text{ and } M(\lambda).$$

We now find integrals in analytical form defining

$$V, \ R, \ P, \ M \text{ in terms of } \lambda$$

for this system of equations. There is no need to write these equations down.

1. MASS INTEGRAL

It follows from (3.2) that

$$\mathcal{M}'' - \mathcal{M}' = \int_{\mathcal{M}'}^{\mathcal{M}''} d\mathcal{M} = \sigma_\nu \int_{r'}^{r''} \rho r^{\nu-1} dr.$$

Let us consider the moving surfaces $r'(t)$ and $r''(t)$ on which the parameter λ takes the constant values λ' and λ''. It is not difficult to verify the following identity, which is true for any function $F(r, t)$:

$$\frac{d}{dt} \int_{r'}^{r''} F \sigma_\nu \rho r^{\nu-1} dr = \frac{\tilde{d}}{dt} \int_{r'}^{r''} F \sigma_\nu \rho r^{\nu-1} dr + \left[F \sigma_\nu \rho r^{\nu-1} \left(\frac{dr}{dt} - v \right) \right]_{r'}^{r''}, \tag{3.6}$$

† The expression of the entropy S in terms of p and ρ must not contain essential dimensional constants independent of a and b. It is evident that multiplicative and additive constants are insignificant.

where the symbol d/dt denotes differentiation with respect to time evaluated for a moving volume of integration composed of identical gas particles (differentiation following the fluid. Ed.).

Since

$$\mathcal{M}'' - \mathcal{M}' = \frac{ab_1^{\nu-k-3}}{t^{s+\delta(k+3-\nu)}}[\lambda''^{\nu-k-3} M(\lambda'') - \lambda'^{\nu-k-3} M(\lambda')],$$

then

$$\frac{d(\mathcal{M}'' - \mathcal{M}')}{dt}\bigg|_{\substack{\lambda'' = \text{const} \\ \lambda' = \text{const}}} = -\frac{s+\delta(k+3-\nu)}{t}(\mathcal{M}'' - \mathcal{M}')$$

From the law of conservation of mass, we have

$$\frac{\tilde{d}(\mathcal{M}'' - \mathcal{M}')}{dt} = 0.$$

Consequently, (3.6) yields for $F = 1$:

$$-\frac{s+\delta(k+3-\nu)}{t}(\mathcal{M}'' - \mathcal{M}') = \sigma_\nu\left[\rho r^{\nu-1}\left(\frac{dr}{dt}-v\right)\right]_{r'}^{r''}.$$

Hence, using (3.5) and the relations $dr'/dt = \delta(r'/t)$ and $dr''/dt = \delta(r''/t)$ we obtain the integral

$$\lambda^{\nu-k-3}\{[s+\delta(k+3-\nu)]M - \sigma_\nu R(V - \delta)\} = \text{const}, \tag{3.7}$$

which is a consequence of the law of conservation of mass and, therefore, always true. The integral (3.7) can be considered independently of the equations of motion in the absence of Newtonian gravitation, as a formula expressing $M(\lambda)$ in terms of λ, V and R in final form. The function $M(\lambda)$ enters in the ordinary differential equation obtained from the momentum equation when gravitation is taken into account. The function $M(\lambda)$ can be eliminated from this equation by using the integral (3.7). If $\mathcal{M} = 0$ for the solution being studied at $\lambda = 0$, i.e. no mass is concentrated at the origin and there is no mass source with a finite discharge, then the constant on the right side of (3.7) is zero.

2. ADIABATIC INTEGRAL

In reversible adiabatic motion of a gas we obtain one integral (Lidov, 1955b), a consequence of the law of conservation of entropy.

Let $\Phi(p, \rho) = f(S)$ be a certain function of the entropy. The relation between the entropy S and p and ρ is arbitrary. The condition (3.4) that entropy is conserved on particle paths is equivalent to a relation of the following type,

$$\Phi(p, \rho) = F(\mathcal{M}, a, b, \alpha_1, \alpha_2, \ldots),$$

where α_1, α_2 are abstract constants.

Consider the dimensions formula for Φ. Let

$$[\Phi] = M^\omega L^\mu T^\kappa.$$

If there is no value of κ for which $[ab_1^\kappa] = [\mathcal{M}]$†), then it is impossible to form a nondimensional combination of the three dimensional parameters $a, b_1, \mathcal{M}$. Consequently, the following equation holds:

$$\Phi(p, \rho) = \mathcal{M}^\omega \left(\frac{ab_1^{s/\delta}}{\mathcal{M}}\right)^{[|\mu - \omega(\nu-3)\delta|]/[s+\delta(k+3-\nu)]} \left(\frac{ab_1^{\nu-k-3}}{\mathcal{M}}\right)^{\chi/[s+\delta(k+3-\nu)]}$$

$$\times f(\alpha_1, \alpha_2, \ldots). \quad (3.8)$$

After $p, \rho, \mathcal{M}$ have been replaced in (3.8) by (3.5) and M has been replaced by (3.7), we obtain the final relation between V, R, P and λ which is an integral of the continuity equation and the equation of conservation of entropy.

If an arbitrary combination $ab^\kappa/\mathcal{M}$ exists for a certain κ, then the quantity f can depend on $ab^\kappa/\mathcal{M}$, according to a relation not known in advance; consequently, the entropy equation, in general, can not be reduced to an integral.

If the gas is perfect, then we can put $\Phi = p/\rho^\gamma$; we have in this case:

$$\omega = 1-\gamma; \quad \mu = 3\gamma-1; \quad \chi = -2.$$

The conservation of entropy integral takes the form

$$\frac{P}{R^\gamma} = M^{\{2-(\gamma-1)\,s+\delta[k+1-\gamma(k+3)]\}/\{s+\delta(k+3-\nu)\}} \lambda^{-\{[2+\nu(\gamma-1)]\,s+2(k+3-\nu)\}/\{s+\delta(k+3-\nu)\}}$$

$$\times f(\alpha_1, \alpha_2, \ldots). \quad (3.9)$$

By means of the integrals derived from the conditions of conservation of mass and entropy, the order of the system of ordinary equations is diminished from four to two.

3. THE ENERGY INTEGRAL

We shall show that this integral occurs (Sedov, 1946a) if a constant with the dimensions $ML^{\nu-1}T^{-2}$, equal to the dimensions of the energy

---

† Since $\mathcal{M} = \dfrac{ab_1^{\nu-k-3}}{t^{s+\delta(k+3-\nu)}} \lambda^{\nu-k-3} M(\lambda)$, then the equation $[ab_1^\kappa] = [\mathcal{M}]$ can be satisfied if $s + \delta(k+3-\nu) = 0$ and if $\kappa = \nu - k - 3$; in this case, the integral (3.7) yields $\lambda^{\nu-k-3} R(V-\delta) = C$. If $C = 0$, then $V = \delta$ and, therefore, $v = (r/t)\,\delta$. We will study this particular solution in §15.

6*

in the spherical case, and to the energy calculated per unit length or area in the cylindrical or plane cases, respectively, can be formed from the characteristic constants a and b_1.

We first consider the case when gravitational forces are absent, and ν takes values 1, 2, 3.

The total energy between the moving surfaces $r'(t)$ and $r''(t)$ is given by

$$\mathscr{E} = \int\limits_{r'}^{r''} \left(\frac{v^2}{2}+\epsilon\right) \sigma_\nu \rho r^{\nu-1} dr,$$

where ϵ is the internal energy per unit mass. The change in the energy of particles contained between the surfaces $r'' = $ const and $r' = $ const at a given instant equals the work of the pressure forces on these surfaces; consequently

$$\frac{d\mathscr{E}}{dt} = -\sigma_\nu(p'' v'' r''^{\nu-1} - p' v' r'^{\nu-1}).$$

Furthermore, it follows from dimensional considerations applied to any self-similar motions ($m \neq 0$) that the quantity $\mathscr{E}$, which has dimensions $ML^{\nu-1}T^{-2}$, is given by a relation

$$\mathscr{E} = ab^{\nu-1-k} t^{\delta(\nu-1-k)-2-s} f(\lambda'', \lambda', \alpha_1, \alpha_2, ...),$$

where $f(\lambda', \lambda'', \alpha_1, ...)$ is an arbitrary function. Now, let us assume that $r'(t)$ and $r''(t)$ are determined from the conditions $\lambda' = $ const and $\lambda'' = $ const. Then in the general case:

$$\frac{d\mathscr{E}}{dt} = [\delta(\nu-1-k)-2-s]\frac{\mathscr{E}}{t}.$$

Now using these formulas, formulas (3.5) and (3.6), and replacing $F(r, t)$ by $v^2/2 + \gamma p/(\gamma-1)\rho$ (to be definite we assume that $\epsilon = \gamma p/(\gamma-1)\rho$), we easily derive the following relation which holds for any self-similar gas motion,

$$[s+2-\delta(\nu-1-k)]f(\lambda'', \lambda', \alpha_1, \alpha_2, ...)$$
$$= \sigma_\nu\left\{\lambda^{\nu+2}\left[PV + (V-\delta)\left(\frac{RV^2}{2}+\frac{P}{\gamma-1}\right)\right]\right\}_{,}^{''}$$

The relation contains the unknown function $f(\lambda'', \lambda', \alpha_1, \alpha_2)$ which is eliminated if

$$s-\delta(\nu-1-k) = -2. \tag{3.10}$$

Therefore, if (3.10) holds, we obtain one essential integral

$$\lambda^{\nu+2}\left[PV + (V-\delta)\left(\frac{PV^2}{2}+\frac{P}{\gamma-1}\right)\right] = \text{const}, \tag{3.11}$$

from the law of conservation of energy. Noting that we took $m = 1$ in the previous calculations, it is easy to see that (3.10) is equivalent to the relation

$$[\mathscr{E}] = [ab_1^{-1-k}].$$

Therefore, the existence of the energy integral is equivalent to the condition that the constant ab_1^{v-1-k} has the dimensions of the energy $\mathscr{E}$; the appropriate self-similar motions can be determined by the constant $\mathscr{E}$ and the constant b_1:

$$[b_1] = LT^{-\delta},$$

where the exponent δ can be arbitrary.

This result is not dependent on the use of the relation $\epsilon = \gamma p/[(\gamma - 1)\rho]$; an integral similar to the integral (3.11) can be written for other forms of the function $\epsilon(p, \rho)$ if we have self-similarity of the type described.

Let us consider the case of gas motion with spherical symmetry taking Newtonian gravitational forces into account.

In motion with spherical symmetry the total energy of gas contained in the volume 0 between two spheres of radii r'' and r' is

$$\mathscr{E} = \int_{r'}^{r''} \left[\frac{\rho v^2}{2} + \rho\epsilon - \frac{\rho f(\mathscr{M} - \mathscr{M}')}{r} \right] 4\pi r^2 \, dr. \tag{3.12}$$

The first term in the integrand defines the kinetic energy, the second is the thermal, internal energy and the third is the part of the internal energy due to the internal gravitational forces. The mass of gas within the sphere of radius r' is denoted by $\mathscr{M}'$. We can simplify the third term, which gives the internal energy of gravitational interaction within the gas. In deriving this formula the interaction energy is put equal to zero for particles an infinite distance apart. Evidently, the potential energy due to interaction between two masses m_1 and m_2 equals

$$-\frac{fm_1 m_2}{r_{12}}.$$

As is known, a spherical layer of material of total mass $\mathscr{M}_1$, with a density dependent only on the radius (Fig. 29), has the same effect on an external point A with mass m_2 as a point mass $\mathscr{M}_1$ concentrated at the centre of the spherical layer.

The total force of attraction of the spherical layer on the interior point B is exactly equal to zero. Consequently, the potential energy due to attraction of a spherical layer of mass $\mathscr{M}_1$ on a point mass m_2 located at a distance r from the centre of the spherical layer is

$$-\frac{f\mathcal{M}_1 m_2}{r}.$$

The potential energy due to attraction of a spherical layer on the point B is zero.

The mass of the spherical layer between the spheres of radius r and r' equals $\mathcal{M} - \mathcal{M}'$; consequently, the potential interaction energy of the mass within the space 0 between the spheres r' and r'' is

$$-\int_{r'}^{r''}\frac{f(\mathcal{M}-\mathcal{M}')\,d\mathcal{M}}{r} = -\int_{r'}^{r''}\frac{f(\mathcal{M}-\mathcal{M}')\,\rho 4\pi r^2\,dr}{r}.$$

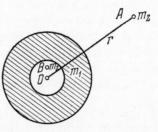

FIG. 29. A spherical layer exerts the same attraction on the external point A as a point mass placed at the centre O. The resultant attraction of the layer is zero for the interior point B.

The law of conservation of energy applied to the mass within the space 0 gives

$$\frac{d\mathscr{E}}{dt} = \frac{dA^{(e)}_{\text{surf}}}{dt} + \frac{dA^{(e)}_{\text{mass}}}{dt} + \frac{dQ^{(e)}}{dt},$$

where $dA^{(e)}_{\text{surf}}/dt$ is the work done per unit time by the external surface forces given by

$$\frac{dA^{(e)}_{\text{surf}}}{dt} = -p''\,v''\,4\pi r''^2 + p'\,v'\,4\pi r'^2,$$

$dA^{(e)}_{\text{mass}}/dt$ is the work done per unit time by the external gravitational forces which are equivalent to the attractive force of a point mass $\mathcal{M}'$, at the centre of symmetry. The work of this force on the element $d\mathcal{M}$ in unit time equals:

$$-\frac{f\mathcal{M}'\,d\mathcal{M}}{r^2}\frac{dr}{dt} = f\mathcal{M}'\frac{d}{dt}\left(\frac{d\mathcal{M}}{r}\right).$$

Hence, it follows that

$$\frac{dA_{\text{mass}}^{(e)}}{dt} = f\mathcal{M}' \frac{d}{dt} \int\limits_{r'}^{r''} \frac{4\pi r^2 \rho \, dr}{r},$$

$dQ^{(e)}/dt$ is the external heat flow per unit time, which equals zero since the process is adiabatic by hypothesis.

The law of conservation of energy applied to the gas inside the volume 0 at a given instant then leads to the equation

$$\frac{d}{dt} \int\limits_{r'}^{r''} \left[\frac{\rho v^2}{2} + \rho\epsilon - \frac{\rho f \mathcal{M}}{r} \right] 4\pi r^2 \, dr = 4\pi (r'^2 p' \, v' - r''^2 p'' \, v''). \qquad (3.13)$$

For slow adiabatic compression of the gas within the spheres, $(v = 0)$, with no work done by external forces, conditions which apply to stars, Equation (3.13) yields

$$\int\limits_{0}^{R} \epsilon \, d\mathcal{M} = \int\limits_{0}^{R} \frac{f \mathcal{M}}{r} \, d\mathcal{M} + \text{const.}$$

Since, the right side increases with compression, then evidently the gas temperature must be raised. The thermal radiation energy is consumed because of the gravitational energy.

If the gas motion is self-similar with the characteristic constants $[a] = [1/f] = ML^{-3}T^2$ ($k = -3$, $s = 2$) and $[b_1] = LT^{-\delta}$, then we have

$$\mathcal{E}^* = \int\limits_{r'}^{r''} \left(\frac{v^2}{2} + \epsilon - \frac{f\mathcal{M}}{r} \right) d\mathcal{M} = f^{-1} b_1^5 t^{5\delta-4} f(\lambda', \lambda''). \qquad (3.14)$$

Let us now use (3.6) putting $F = v^2/2 + \epsilon - f\mathcal{M}/r$. Taking (3.13) and (3.14) into account, we find

$$\frac{5\delta - 4}{t} \mathcal{E}^* = \left\{ 4\pi r^2 \left[pv + \left(\frac{v^2}{2} + \epsilon - \frac{f\mathcal{M}}{r} \right) \rho \left(v - \frac{dr}{dt} \right) \right] \right\}_{''}',$$

hence, we obtain from (3.5) with $\epsilon = \gamma p/(\gamma - 1)\rho$

$$(5\delta - 4) f(\lambda', \lambda'') = \left\{ 4\pi\lambda^5 \left[PV + \left(\frac{V^2}{2} + \frac{\gamma P}{(\gamma-1) R} - M \right) R(V - \delta) \right] \right\}_{''}'.$$

The unknown function $f(\lambda', \lambda'')$ drops out of this relation if $\delta = 4/5$. Hence, for $\delta = 4/5$ we get one integral

$$\lambda^5 \left[PV + \left(\frac{RV^2}{2} + \frac{\gamma P}{\gamma - 1} - RM \right) \left(V - \frac{4}{5} \right) \right] = \text{const.} \qquad (3.15)$$

174 SIMILARITY AND DIMENSIONAL METHODS IN MECHANICS

If $\delta = 4/5$, then

$$\left[\frac{1}{f}\, b_1^{+5}\right] = ML^2 T^{-2} = [\mathscr{E}*].$$

Therefore, the gravitational constant f and a particular value of the energy $\mathscr{E}*$ can be taken as the independent dimensional constants in this case.

4. MOMENTUM INTEGRAL

We consider the case of one-dimensional unsteady wave motion when the constant $c = ab^\chi$ can be formed from the characteristic constants a and b, for a certain χ, with the dimensions of momentum per unit area $[c] = ML^{-1}T^{-1}$. In this case, the integral

$$P - (V - \delta)\, RV = \text{const}, \tag{3.16}$$

can be established by arguments similar to those given above. The previous methods of obtaining integrals are applied to establish integrals for linearized solutions of almost self-similar motions (Lidov, 1955a) and for any approximation (Korobeinikov, 1955) in the expansion of non-self-similar solutions in series of self-similar functions.

We have established closed integrals of the ordinary differential equations for self-similar gas motions. Considerations of dimensional analysis and general theorems of mechanics consistent with the system of equations of motion were used in the derivation. It is clear from the general reasoning that the integrals can also be obtained from the system of ordinary equations by formal calculations.

The closed integrals are derived from a system of ordinary equations which retain the same form for any formulation of the problem; in particular, for polytropic motion of a perfect gas when $\gamma \neq c_p/c_v$ and, therefore, when the entropy per particle is variable and external heat flow results.

§4. MOTIONS WHICH ARE SELF-SIMILAR IN THE LIMIT

Starting from a given family of self-similar motions, which depend on several parameters, other families of exact solutions can be constructed by applying certain limit processes to the governing system of partial differential equations.

Let us clarify this by an example.

We take a solution of the type (3.5) and write it as follows:

$$\left.\begin{array}{l} v = \dfrac{r}{t+t_0}\,\delta\, \tilde{V}(\lambda) \sim \rho = \dfrac{a\delta^s \tilde{R}(\lambda)}{r^{k+3}(t+t_0)^s} \\[4mm] p = \dfrac{a\delta^{s+2}}{r^{k+1}(t+t_0)^{s+2}}\,\tilde{P}(\lambda); \quad \mathscr{M} = \dfrac{a\delta^s \tilde{M}(\lambda)}{r^{k+3-\nu}(t+t_0)^s} \quad \text{and} \quad \lambda = \dfrac{r}{b(t+t_0)^\delta} \end{array}\right\} \tag{4.1}$$

Clearly if the time t is replaced by $t + t_0$ in the solution (3.5), we again obtain a solution containing one constant parameter t_0. We now introduce the new notation

$$V = \delta \tilde{V}; \quad R = \delta^s \tilde{R}; \quad P = \delta^{s+2} \tilde{P}; \quad M = \delta^s \tilde{M}; \quad z = \delta^2 \tilde{z}. \quad (4.2)$$

Equations (3.1), (3.2), (3.3) and (3.4) have a solution of the type (4.1) for $k = -3$, $s = 2$ and any values of δ, t_0 and b. In Equation (1.3), $f = 0$ and the constants k and s can be arbitrary. Now, let us put

$$t_0 = \delta \tau \quad \text{and} \quad b = r_0 (\delta \tau)^{-\delta}.$$

Evidently, $|\tau| = |t|$ and $|r_0| = |r|$, where τ, r_0 and δ are arbitrary constants.

The limiting form of Equation (4.1) when $\delta \to \infty$ for fixed τ, r_0 and finite $\tilde{V}$, $\tilde{R}$, $\tilde{P}$ and $\tilde{M}$, is

$$\left.\begin{array}{l} v = \dfrac{r}{\tau} \tilde{V}(\lambda); \quad \rho = \dfrac{a}{r^{k+3} \tau^s} \tilde{R}(\lambda); \\[4mm] p = \dfrac{a}{r^{k+1} \tau^{s+2}} \tilde{P}(\lambda); \quad \mathcal{M} = \dfrac{a}{r^{k+3-\nu} \tau^s} \tilde{M}(\lambda) \quad \text{and} \quad \lambda = \dfrac{r}{r_0} e^{-t/\tau}. \end{array}\right\} \quad (4.3)$$

It follows that the equations of one-dimensional unsteady gas motion have solutions of the type (4.3), which can be considered to define limiting forms of self-similar motion. Equations for $\tilde{z}$, $\tilde{R}$, $\tilde{P}$ and λ can easily be obtained from (2.1), (2.2) and (2.3); we find for finite k and s

$$\left.\begin{array}{c} \dfrac{d\tilde{z}}{d\tilde{V}} = \\[2mm] \dfrac{\tilde{z}\left\{[2+\nu(\gamma-1)]\tilde{V}(\tilde{V}-1)^2 - (\gamma-1)\tilde{V}^2(\tilde{V}-1) - \left[2\tilde{V}+\dfrac{k+1}{\gamma}(\gamma-1)\right]\tilde{z}\right\}}{(\tilde{V}-1)\left[\tilde{V}^2(\tilde{V}-1) + \left(\dfrac{k+1}{\gamma}-\nu\tilde{V}\right)\tilde{z}\right]} \\[6mm] \dfrac{d\ln\lambda}{d\tilde{V}} = \dfrac{\tilde{z} - (\tilde{V}-1)^2}{\tilde{V}^2(\tilde{V}-1) + \left(\dfrac{k+1}{\gamma}-\nu\tilde{V}\right)\tilde{z}} \\[6mm] (\tilde{V}-1)\dfrac{d\ln\tilde{R}}{d\ln\lambda} = -\dfrac{\tilde{V}^2(\tilde{V}-1) + \left(\dfrac{k+1}{\gamma}-\nu\tilde{V}\right)\tilde{z}}{\tilde{z} - (\tilde{V}-1)^2} + (k+3-\nu)\tilde{V}. \end{array}\right\} \quad (4.4)$$

The mass and adiabatic integrals for these limiting motions become

$$\left. \begin{array}{l} \lambda^{\nu-3-k}\{\tilde{M}(k+3-\nu)-[2(\nu-1)\,\pi+\delta_{\nu1}]\tilde{R}(\tilde{V}-1)\} \,=\, \text{const} \\[3mm] \dfrac{\tilde{z}}{\tilde{R}\gamma-1} \,=\, \tilde{M}^{-[(k+3)\,(\gamma-1)+2]/[k+3-\nu]}.\,\text{const.} \end{array} \right\} \quad (4.5)$$

The energy integral is

$$\tilde{P}\tilde{V}+(\tilde{V}-1)\left(\frac{\tilde{R}\tilde{V}^2}{2}+\frac{\tilde{P}}{\gamma-1}\right) \,=\, \text{const} \qquad (4.6)$$

for values $k=\nu-1$ and $s=-2$.

It can be shown that the case $s=s_0\delta+s_1$, i.e. $s\to+\infty$ as $\delta\to\infty$, reduces to the case considered if the constant a/τ^s is replaced by a_1 in Equation (4.3) and k by $k+s_0$ in Equations (4.4) and (4.5).

Similar considerations apply to the coordinate r which can be replaced by $x+x_0$ for plane wave motions of a gas.

Formulas (3.5) can be written as

$$\left. \begin{array}{ll} v = \dfrac{(x+x_0)\,\delta}{t}\,\hat{V}(\hat{\lambda}); & \rho = \dfrac{a\hat{R}(\hat{\lambda})}{[(x+x_0)\,\delta]^{k+3}t^s}; \\[4mm] p = \dfrac{a\hat{P}(\hat{\lambda})}{[(x+x_0)\,\delta]^{k+1}t^{s+2}}; & \mathscr{M} = \dfrac{a\hat{M}(\hat{\lambda})}{[(x+x_0)\,\delta]^{k+2}t^s} \\[4mm] \multicolumn{2}{c}{\hat{\lambda} = \lambda^{1/\delta} = \left(\dfrac{x+x_0}{b}\right)^{1/\delta}\dfrac{1}{t}} \end{array} \right\} \quad (4.7)$$

Here we used the notation

$$V = \delta\hat{V}; \quad R = \frac{1}{\delta^{k+3}}\hat{R}; \quad P = \frac{1}{\delta^{k+1}}\hat{P}; \quad M = \frac{\hat{M}}{\delta^{k+2}}; \quad (z = \hat{z}\delta^2). \quad (4.8)$$

Let us put

$$x_0 = \frac{r_0}{\delta}; \quad (b)^{1/\delta} = \frac{1}{\tau}\left(\frac{r_0}{\delta}\right)^{1/\delta}$$

We have the following dimensional relations

$$[x] = [r_0] \quad \text{and} \quad [t] = [\tau],$$

where τ and r_0 can be arbitrary. If we fix k, s, r_0 and τ and let δ tend to zero, we obtain

$$v = \frac{r_0}{t}\,\hat{V}(\hat{\lambda}); \quad \rho = \frac{a}{r_0^{k+3}t^s}\,\hat{R}(\hat{\lambda}); \quad p = \frac{a\hat{P}(\hat{\lambda})}{r_0^{k+1}t^{s+2}}; \quad \hat{\lambda} = \frac{\tau}{t}e^{x/r_0} \quad (4.9)$$

It is easy to obtain equations for $\hat{z}$, $\hat{P}$, $\hat{R}$ and $\hat{\lambda}$ from (2.1), (2.2) and

(2.3); after substituting (4.8) into (2.1), (2.2) and (2.3) and proceeding to the limit as $\delta \to 0$, we obtain

$$\frac{d\hat{z}}{d\hat{V}} = \frac{\hat{z}\left[2(\hat{V}-1)^2 - (\gamma-1)\,\hat{V}(\hat{V}-1) + \left(s-\frac{s+2}{\gamma}\right)\hat{z}\right]}{(\hat{V}-1)\left[\hat{V}(\hat{V}-1) - \frac{s+2}{\gamma}\hat{z}\right]}.$$

$$\left.\begin{array}{l}
\dfrac{d\ln\hat{\lambda}}{d\hat{V}} = \dfrac{\hat{z}-(\hat{V}-1)^2}{\dfrac{s+2}{\gamma}\hat{z} - \hat{V}(\hat{V}-1)}. \\[3em]
(\hat{V}-1)\dfrac{d\ln\hat{R}}{d\ln\hat{\lambda}} = s - \dfrac{\dfrac{s+2}{\gamma}\hat{z} - \hat{V}(\hat{V}-1)}{\hat{z}-(\hat{V}-1)^2}.
\end{array}\right\} \quad (4.10)$$

In this case the mass and adiabatic integrals can be written

$$\left.\begin{array}{l}
s\hat{M} - \hat{R}(\hat{V}-1) = \text{const}, \\[1.5em]
\dfrac{\hat{z}}{\hat{R}^{\gamma-1}} = \hat{M}^{(1-\gamma)+(2/s)}\,\text{const}.
\end{array}\right\} \quad (4.11)$$

The following energy integral† holds for $k = 0$ and $s = -2$:

$$\hat{P}\hat{V} + (\hat{V}-1)\left(\frac{\hat{R}\hat{V}^2}{2} + \frac{\hat{P}}{\gamma-1}\right) = \text{const}. \qquad (4.12)$$

Proceeding to the limit once more as $\mu \to +\infty$ in (4.3) and (4.9), we obtain general formulas for the steady gas motion after replacing t by $t+\mu t_1$ and τ by μt_1. If we let $\mu \to +\infty$ in (4.9) after replacing x by $x+\mu l$ and r_0 by μl we obtain motion in which the pressure varies with time.

K. P. Staniukovich (1959) considered solutions of the type (4.3) and (4.9) and G. I. Barenblatt (1954) showed a method of constructing such solutions as limits of self-similar motions by formal substitution.

§5. INVESTIGATION OF THE FIELDS OF THE INTEGRAL CURVES IN THE z, V PLANE

To analyse the existence, uniqueness and construction of solutions of different boundary and initial value problems, it is necessary to investigate the family of integral curves in the z, V plane. When gravitational forces are absent, we consider the ordinary differential equation (2.1), which contains two characteristic parameters κ and δ in addition to

† The equations of motion can be integrated by quadrature if use is made of the integrals (4.5) and (4.6) in addition to (4.11) and (4.12). N. N. Kochina performed this integration.

the parameter $\nu = 1, 2, 3$. The following two families of solutions, the characteristic parameters of which have the dimensions described, are especially important in the problems discussed above in §1 and in later applications.

1. One of the constants u has the dimensions of velocity; the arbitrary dimensions of the second constant A can be taken equal to $ML^{\omega-3}(k = \omega - 3, s = 0)$.

2. One of the constants E has the dimensions of energy $ML^{\nu-1}T^{-2}$; the dimensions of the second constant A can either be arbitrary or taken equal to $ML^{\omega-3}$.

The first case includes problems in which the phase velocities are constant; the motion of a piston with constant speed in a medium with constant initial pressure and density ($\omega = 0$); the detonation and combustion in a medium with constant density or density varying like $\rho_1 = A/r^\omega$; the decay of an arbitrary discontinuity in a combustible mixture with uniform conditions in the gas ahead of and behind the front.

Case 2 includes the problems of a strong explosion ($p_1 = 0$) with constant initial density, when $\omega = 0$, or with variable density $\rho_1 = A/r^\omega$ when $\omega \neq 0$.

We have, in case 1

$$\delta = 1; \quad \kappa = \frac{\omega}{\gamma}; \quad \lambda = \beta\,\frac{r}{ut}, \tag{5.1}$$

and, in case 2

$$b = \left(\frac{E}{A}\right)^{1/(2+\nu-\omega)}; \quad |b| = LT^{-2/(2+\nu-\omega)};$$

$$\delta = \frac{2}{2+\nu-\omega}; \quad \kappa = \frac{\nu\delta}{\gamma}; \quad \lambda = \beta\,\frac{r}{bt^\delta} \tag{5.2}$$

where β is an arbitrary constant which must be chosen suitably in each specific solution. In both cases, families of solutions depending only on the single characteristic parameter ω are obtained in the z, V plane.

We now consider the differential equations (2.1) and (2.2) in detail.

We have in case 1:

$$\frac{dz}{dV} = \frac{z\left\{2(V-1)^3 + (\nu-1)(\gamma-1)\,V(V-1)^2 - \left[2(V-1) + \omega\dfrac{\gamma-1}{\gamma}\right]z\right\}}{(V-1)\left[V(V-1)^2 + \left(\dfrac{\omega}{\gamma} - \nu V\right)z\right]} \tag{5.3}$$

$$\frac{d\ln\lambda}{dV} = \frac{z - (V-1)^2}{V(V-1)^2 + \left(\dfrac{\omega}{\gamma} - \nu V\right)z}. \tag{5.4}$$

Equation (2.3) can be replaced by the adiabatic integral. From (3.9), for $\delta = 1$, $s = 0$, $k = \omega - 3$ and using (3.7), we obtain

$$\frac{z}{R^{\gamma-1}} = C_1 \left[R(V-1) + \frac{C_2}{\lambda^{\nu-\omega}} \right]^{[\omega(\gamma-1)]/[\nu-\omega]} \frac{1}{\lambda^2}. \tag{5.5}$$

When (5.3) is integrated and $\lambda(V)$ is calculated from (5.4) by quadrature, (5.5) determines the function $R(V)$; C_1 and C_2 are arbitrary constants. If the mass $\mathscr{M}$ equals zero for $\lambda = r/r_2 = 0$ and there is no mass source, then

$$C_2 = 0.$$

The ordinary differential equation (5.3) in the half-plane $z \geqslant 0$ has the following singular points in the spherical case for $\nu = 3$ (see Figs. 30–33).

The point $O(z = 0,\ V = 0)$ is a node; in general integral curves touch the V axis at O, but there is a curve which approaches the point O with the slope γ/ω. The following asymptotic formulas hold for the integral curves near the point O:

$$z = CV^2; \quad \lambda = \frac{C_1}{V} \quad \text{and} \quad z = \frac{\gamma}{\omega}V; \quad \lambda = \frac{C_1}{\sqrt{V}}. \tag{5.6}$$

The point O in the z, V plane corresponds to points at infinity in the gas. The point $C(z = 0,\ V = 1)$ is a multiple node; one set of integral curves touch the V axis at C. When $\omega > 0$ a second set of integral curves touches the line $V = 1$ at C.

The asymptotic formulas for the integral curves touching the V axis for $\gamma < 2$ are

$$\left.\begin{array}{l} \lambda - \lambda^* = \dfrac{\gamma+1}{3(\gamma-1)} \lambda^*(1-V), \\[4mm] z = \dfrac{2(2-\gamma)}{6 - \dfrac{\gamma+1}{\gamma}\omega}(1-V)^2, \\[4mm] R = C(1-V)^{[6-\omega-\gamma\omega]/[3(\gamma-1)]}, \\[4mm] P = C\dfrac{2(2-\gamma)}{6\gamma - (\gamma+1)\omega}(1-V)^{[6\gamma-\omega(\gamma+1)]/[3(\gamma-1)]} \end{array}\right\} \tag{5.7}$$

where C and λ^* are certain constants; the quantity $P \to 0$ as $\lambda \to \lambda^*$ for $\omega < 6\gamma/(\gamma+1)$.

The following asymptotic formulas are valid for the curves perpendicular to the V axis:

$$\left.\begin{array}{c} \lambda - \lambda^* = \dfrac{\gamma}{\nu\gamma - \omega}(1 - V) \quad \text{where} \quad \omega < \dfrac{6\gamma}{\gamma + 1}, \\[3mm] z = A(1 - V)^{[(\gamma-1)\,\omega]/[3\gamma-\omega]}, \\[3mm] R = B(1 - V)^{-[(\gamma-1)\,\omega]/[3\gamma-\omega]}, \\[3mm] P = \dfrac{AB}{\gamma}, \end{array}\right\} \qquad (5.8)$$

where λ^*, A, B are certain constants. As the point C is approached in the gas, we approach a surface at a finite distance from the centre, a piston or the boundary of a vacuum, at which the phase velocity equals the gas particle velocity.

The point

$$B\!\left(z = \frac{18\gamma(\gamma-1)^2}{(3\gamma-1)^2[\omega + 3\gamma(2-\omega)]}, \quad V = \frac{2}{3\gamma-1}\right)$$

is a node for $\omega > 4\gamma/(3\gamma-1)$ and a saddle-point for $\omega < 4\gamma/(3\gamma-1)$; the point B for $\omega = 6\gamma/(3\gamma-1)$ coincides with the singular point at infinity D. The point B transforms into the lower half-plane for large ω. If the point B does not lie on the parabola $z = (1 - V)^2$ then the variable λ on the integral curves entering at this point approaches zero or infinity, i.e. the centre of symmetry or the point at infinity. If the point B lies on the parabola $z = (1 - V)^2$, which is possible only for $\omega = 4\gamma/(3\gamma-1)$, then the value of λ in this case remains finite as the point B is approached. The point B can correspond to a moving point in the gas in this case.

If $\omega = [3(\gamma+1)]/[3\gamma-1]$, then the point B lies on the parabola $z_2 = \gamma V_2(1 - V_2)$; for smaller ω, it lies below and for larger ω it lies above this parabola. This result is of importance in solving problems of propagation of detonation in a medium with density varying like $\rho_1 = A/r^\omega$.

The point $D(z = \infty,\ V = \omega/3\gamma)$ is a saddle-point for $\omega < 6\gamma/(3\gamma-1)$ and a node for $6\gamma/(3\gamma-1) < \omega < 3\gamma$. In the first case there is one integral curve which has the z axis as an asymptote; $\lambda \to 0$ along this curve as infinity is approached and, therefore, as we approach the centre of symmetry. In the second case, we withdraw toward infinity in the gas as the node D is approached along the integral curves since

$\lambda \to +\infty$. The following asymptotic formulas hold near the point D for $\omega < 6\gamma/(3\gamma - 1)$:

$$\left. \begin{aligned}
&\lambda = C\left(V - \frac{\omega}{3\gamma}\right)^{[3\gamma - \omega]/[6\gamma + \omega - 3\gamma\omega]}, \\[2mm]
&z = \frac{\omega(3\gamma - \omega)^3}{27\gamma^3(15\gamma - 3\gamma\omega - 2\omega)}\frac{1}{V - \dfrac{\omega}{3\gamma}}, \\[2mm]
&R = B\left(V - \frac{\omega}{3\gamma}\right)^{[\omega(3-\omega)]/[6\gamma + \omega - 3\gamma\omega]} = B_1 \lambda^{[\omega(3-\omega)]/[3\gamma - \omega]}, \\[2mm]
&P = \\
&\quad B_1 C^{(6\gamma + \omega - 3\gamma\omega)/(3\gamma - \omega)}\frac{\omega(3\gamma - \omega)^3}{27\gamma^4(15\gamma - 3\gamma\omega - 2\omega)}\lambda^{[3\gamma\omega + 2\omega - \omega^2 - 6\gamma]/[3\gamma - \omega]},
\end{aligned}\right\} \quad (5.9)$$

where B_1 and C are arbitrary constants; B is expressed in terms of B_1 and C.

Asymptotic formulas for the velocity, pressure and density are easily obtained near the centre of symmetry from (5.9) and the basic formulas (1.3).

The point $A\{z = [1 - (\omega/2\gamma)]^2; \ V = \omega/2\gamma\}$ is a focus for $\omega < 0$, a node for $0 \leqslant \omega < 4\gamma/(3\gamma - 1)$, a saddle-point for $4\gamma/(\gamma - 1) < \omega < 2\gamma$, a node for $2\gamma < \omega < [4\gamma(\gamma + 1)]/[6\gamma - 2 - \gamma^2]$ and a centre for $[4\gamma(\gamma + 1)]/[6\gamma - 2 - \gamma^2] < \omega$.

The integral curves approach the node tangentially to the z axis for $\omega = 0$.

There are two directions in which the integral curves may approach the point A for $0 < \omega < [4\gamma(\gamma + 1)]/[6\gamma - 2 - \gamma^2]$; the corresponding slopes along these directions are given by

$$k_{1,2} = \gamma\left(\frac{\omega}{2\gamma} - 1\right)\left[1 \pm \sqrt{\left(\frac{\omega(\gamma^2 - 6\gamma + 2) + 4\gamma(\gamma + 1)}{\omega\gamma^2}\right)}\right].$$

The point A is always on the parabola $z = (1 - V)^2$, consequently, the variable λ at the point A must be finite. Passing through the point A where the phase and particle velocities are different corresponds to crossing a characteristic and, consequently, this point corresponds to a weak discontinuity.

The point $\mathscr{E}(z = \infty, V = 1)$ is a node for $\omega < 0$ and a saddle-point for $0 < \omega < 3\gamma$.

The point $G(z = \infty, V = \infty)$ is a saddle-point for any ω.

A simple exact solution of the equations of gas dynamics for which $z = $ const and $V = $ const corresponds to each singular point; the variable

λ remains free. The function $R(\lambda)$ is determined from (5.5); it is found that R is a power of λ for $C_2 = 0$ and, therefore, P is the same power function. Hence, this is a particular solution for which v, ρ, p are simple powers of r and t.

In particular, if $z = 0$ and $V = 0$, then we have a state of rest with zero pressure, which is the initial state in certain self-similar motions.

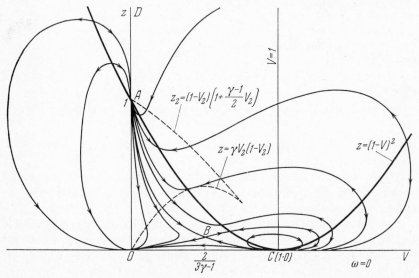

Fig. 30. The family of integral curves for $\nu = 3$, $\delta = 1$, $\omega = 0$. Points corresponding to a state of rest on the line OA transform after passing through the shock wave into points of the parabola $z_2 = (1 - V_2)\{1 + [(\gamma - 1)/2] V_2\}$. Arrows indicate the direction of increasing $\lambda = \beta(r/ut)$.

Qualitative pictures of the fields of the integral curves are given in Figs. 30–33 for $\gamma = 5/3$ in the cases

$$\omega = 0; \quad 0 < \omega < \frac{2\gamma}{\gamma + 1}; \quad \omega = \frac{3(\gamma + 1)}{3\gamma - 1} \quad \text{and} \quad \frac{3(\gamma + 1)}{3\gamma - 1} < \omega < \frac{6\gamma}{3\gamma - 1}.$$

The direction of increasing λ is shown by arrows. The parameter λ attains a maximum or minimum on the parabola $z = (1 - V)^2$, consequently, a continuous passage along the integral curve across this parabola is impossible since it leads to a double-sheet in the space of gas motion, and the solution is not unique. However, transition across the parabola $z = (1 - V)^2$ along the integral curve is possible if this passes through the singular point A on the parabola. A change from a maximum to a minimum for λ occurs at A on the parabola. When moving along the

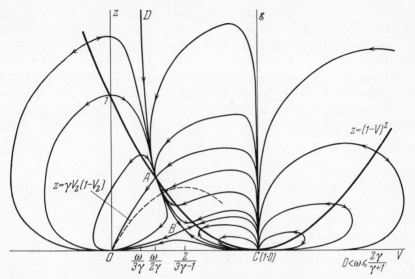

FIG. 31. The family of integral curves for $\nu = 3$, $\delta = 1$ and $0 < \omega < 2\gamma/(\gamma+1)$. Note that the singular point A is shifted along the parabola $z = (1 - V)^2$; the point D is also displaced and the property of the singular point C varies.

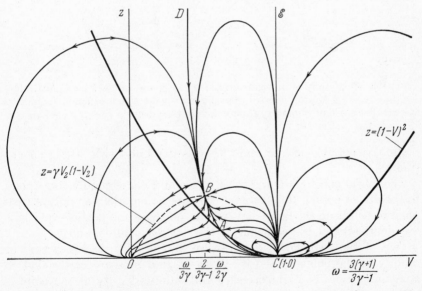

FIG. 32. The family of integral curves for $\delta = 1$ and $\omega = 3(\gamma+1)/(3\gamma-1)$. The singular point B lies on the parabola $z = \gamma V_2(1 - V_2)$.

intersecting integral curve the parameter λ has a finite value at A and varies monotonically. We shall see below that this fact isolates the intersecting integral curve as a solution of the appropriate detonation problem.

We can have $p_1 \neq 0$ for $\omega = 0$ in self-similar solutions: the whole line $V = 0$ then corresponds to a state of rest.

The parameter λ must vary from 0 to ∞ along the integral curves in the solutions being investigated, if the gas is unbounded. An appropriate

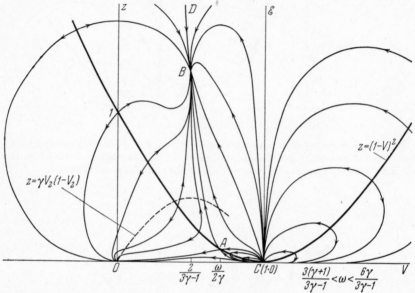

FIG. 33. The family of integral curves for $\nu = 3$, $\delta = 1$ and $[3(\gamma + 1)]/[3\gamma - 1] < \omega < (6\gamma)/(3\gamma - 1)$. Motion toward the singular point C corresponds to motion toward the centre of symmetry from points of the parabola $z = \gamma V_2(1 - V_2)$.

motion is possible in the majority of cases only with strong shocks (jumps).

The initial point $\lambda = 0$ and the final point $\lambda = \infty$ can only correspond to the above points, the piston or vacuum boundaries can only correspond to points on the line $V = 1$ (where the phase and particle velocities are identical) and, in particular, to the singular points C and $\mathscr{E}$.

One of the two variables, pressure and density, is either zero or infinity at the points C and $\mathscr{E}$. Cases of zero pressure correspond to a vacuum, cases of finite pressure and infinite density at the point C and zero density at the point $\mathscr{E}$ correspond to motion of a spherical piston.

Let us consider the field of integral curves of the differential equations (2.1) for case 2. We have

$$\frac{dz}{dV} = \frac{z[2(V-1)+\nu(\gamma-1)\,V](V-\delta)}{\left[V(V-1)(V-\delta)+\nu\left(\frac{\delta}{\gamma}-V\right)z\right]}$$

$$-\frac{z\left\{(\gamma-1)\,V(V-1)(V-\delta)+\left[2(V-1)+\frac{\nu\delta}{\gamma}(\gamma-1)\right]z\right\}}{(V-\delta)\left[V(V-1)(V-\delta)+\nu\left(\frac{\delta}{\gamma}-V\right)z\right]}, \qquad (5.10)$$

$$\frac{d\ln\lambda}{dV} = \frac{z-(V-\delta)^2}{V(V-1)(V-\delta)+\nu\left(\frac{\delta}{\gamma}-V\right)z}. \qquad (5.11)$$

where $\delta = 2/(2+\nu-\omega)$. The third equation defining R can be replaced by the adiabatic integral. We obtain from (3.9) for $\delta = 2/(2+\nu-\omega)$, $s = 0$, $k = \omega-3$ and from (3.7) for $\omega < \nu$:

$$z = R^{(2\nu-\gamma\nu-\omega)/(\omega-\nu)}\left(V-\frac{2}{2+\nu-\omega}\right)^{(\nu-\gamma\omega)/(\omega-\nu)}\lambda^{-(2+\nu-\omega)}C, \qquad (5.12)$$

where C is an integration constant and the condition $\omega < \nu$ can be replaced by the inequalities $\omega \neq \nu$ and $\omega \neq \nu+2$. The integral, defining the mass

$$\mathcal{M} = \sigma_\nu \int\limits_0^r \frac{A}{r^\omega} r^{\nu-1}\, dr,$$

diverges for $\omega \geqslant \nu$. The constant on the right side of (3.7) is taken as zero on the basis of the remarks made on page 179.

Relation (5.12) defines R in terms of z, V and λ.

The constant E with the dimensions $ML^{\nu-1}T^{-2}$ is characteristic; consequently, an energy integral holds in case 2. We obtain on the basis of (3.11)

$$\lambda^{\nu+2}\left[zV+\left(V-\frac{2}{2+\nu-\omega}\right)\left(\frac{\gamma V^2}{2}+\frac{z}{\gamma-1}\right)\right]R = C_1. \qquad (5.13)$$

The integral (5.13) can replace Equation (5.10). It is possible to determine z from (5.13) and to substitute it into (5.11) and we then obtain an ordinary equation containing only λ and ν. If $C_1 = 0$, then (5.13) yields an integral of (5.10) and (5.11) is integrated by quadrature. If $C_1 \neq 0$, then (5.10) can be integrated to determine $z(V)$ and afterwards the function of λ can be calculated by using (5.13) without integrating (5.11).

A qualitative picture of the family of integral curves in the z, V plane is shown in Fig. 34 for (5.10) with $\nu = 3$, $\omega = 0$ and $\gamma < 2$. The direction of increasing λ is shown by arrows. The character of the singular points is seen in the sketch.

We note that the singular point $C(z = \infty$, $V = 2/[2 + \nu - \omega])$ is a saddle-point. Along a single integral curve terminating at a point, the variable

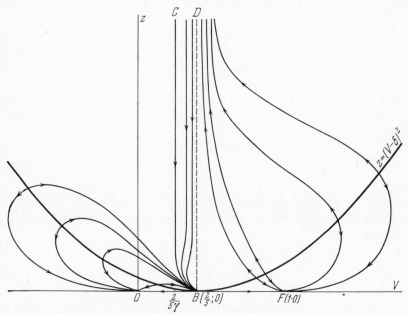

FIG. 34. The family of integral curves for $\nu = 3$, $\omega = 0$, $\delta = 2/5$, $\gamma < 2$.

λ approaches zero as the point C is left behind. It should be noted that we obtain the equation of this curve from (5.13) for $C_1 = 0$.

When analysing the family of integral curves in the z, V plane as functions of ω and γ, it is necessary to take into account the fact that (5.10) has the singular point

$$\mathcal{E}\left(V^* = \frac{2}{[2 + \nu(\gamma - 1)]}, \quad z^* = -\frac{2(\gamma - 1)\gamma[(2 - \gamma)\nu - \omega]}{[2 + \nu(\gamma - 1)]^2[(2 - \omega)\gamma + \nu - 2]}\right), \quad (5.14)$$

which can pass through the point B from the lower half-plane $z < 0$ into the upper half-plane $z > 0$. It is easy to verify that the singular point lies on the integral curve passing through the singular point C, i.e. z^* and V^* satisfy (5.13) for $C_1 = 0$.

The condition that the coordinate z^* be positive is

$$\nu(2-\gamma) \leqslant \omega \leqslant \frac{2\gamma+\nu-2}{\gamma}. \tag{5.15}$$

The point $\mathscr{E}$ coincides with the point B at the lower bound in (5.15) and with the point C at the upper bound.

In Fig. 35 the family of integral curves is shown for the case when $\mathscr{E}$ is in the upper half-plane.

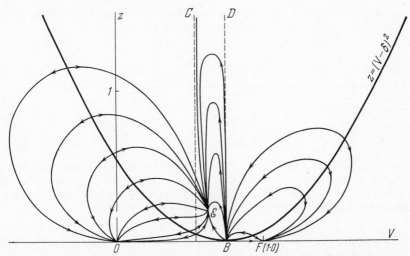

FIG. 35. The family of integral curves for $\nu = 3$, $6 - 3\gamma < \omega < (2\gamma + 1)/(5\gamma\omega)$, $\delta = 2/(5 - \omega)$. The singular nodal point $\mathscr{E}$ appears in the upper half-plane.

Only the point C and the singular points $z = 0$, $V = \pm \infty$ can correspond to the centre of symmetry $\lambda = 0$.

The singular points O, F and $\mathscr{E}$ correspond to the infinitely distant point $\lambda = \infty$.

The parameter λ has a finite value at the singular points B and D which correspond to the boundaries of an expanding spherical piston or the boundary of a vacuum.

§6. The Piston Problem

We now analyse the disturbance at a certain instant t in a gas forced out by a spherical piston expanding with constant velocity U. The initial pressure p_1 and the initial density ρ_1 are constant and not zero.†

Since $v = U$ and $r = Ut$ for fluid particles adjacent to the piston, then $V = v/(r/t) = 1$ on the piston. Therefore, the image point in the V, z plane corresponding to the piston (see Fig. 36) must be on the line $V = 1$. Motion along the integral curve in the directions of increasing λ corresponds to motion along the radius from the piston to infinity. The integral curve intersects the parabola $z = (V-1)^2$. Since continuous passage through this curve is impossible, extension of the motion to the point O, which corresponds to the point at infinity, can only be achieved by means of a jump.

On the other hand, the gas is at rest in the undisturbed region, i.e. a point on the z axis corresponds to the far side of the jump.

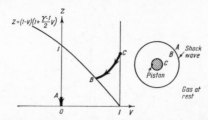

FIG. 36. Integral curve in the x, V plane for the solution of the spherical piston problem. Passage from the point A to the point B occurs by a jump across the shock wave. The point C corresponds to the piston. The curve BC corresponds to adiabatic compression between the piston and the shock wave.

As was shown above, points of the z axis transform discontinuously into points of the parabola

$$z = (1-V)\left(1 + \frac{\gamma-1}{2}V\right). \qquad (6.1)$$

Thus the image point for motion between the piston and infinity in physical space moves along an integral curve from a certain point on the line $V = 1$ to the intersection with the parabola (6.1), and then jumps to the line $V = 0$ discontinuously (Fig. 36).

The physical interpretation of this motion is that a shock wave is propagated into gas at rest, followed by adiabatic compression of the gas between the shock wave and the piston.

† The solution of the motion of a gas forced out by a sphere expanding with constant velocity was first published with numerical computations in Sedov (1945a).

The solution of the problem of the motion of a gas forced out by a sphere expanding with a velocity $U = ct^n$ with a comparative analysis for various n (which takes viscosity and heat conduction into account for $n = -\frac{1}{2}$) is given in the thesis by N. L. Krasheninnikova, presented at Moscow University in Spring, 1954. (See Krasheninnikova, 1955).

The case of a cylindrical piston does not differ qualitatively from that of the spherical piston. It is evident that a region of gas motion with constant velocity, density and pressure exists between the shock wave and the piston for a plane piston moving in a cylindrical pipe, in contrast to the spherical case.

The curves in Figs. 37 and 38 give the ratio of the gas pressure and density respectively adjacent to the piston to corresponding values in

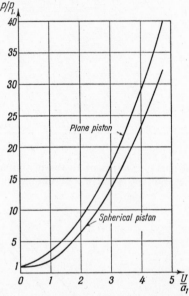

FIG. 37. Pressure on a piston p as a function of piston velocity U; p_1 is the pressure and a_1 is the speed of sound in the undisturbed gas.

the gas at rest, as a function of the ratio of the piston velocity to the initial speed of sound.

Figure 39 shows the ratio of shock wave velocity to speed of sound as a function of the ratio of piston velocity to speed of sound.

It appears from the graphs that for equal piston velocities the gas compression is larger in the plane case than in the spherical case. The shock wave is stronger in the plane than in the spherical case (especially for low piston velocities).

In the plane case, in contrast to the spherical and cylindrical cases, problems of a piston being drawn away from the gas (backstroke) can arise, in addition to the problem of gas compression by the piston. The solution of this problem is easily constructed by using the progressive

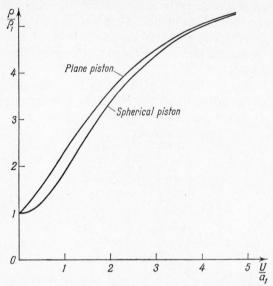

FIG. 38. Density ρ of particles adjacent to the piston as a function of piston speed U; ρ_1 is the density, a_1 is the speed of sound in undisturbed gas.

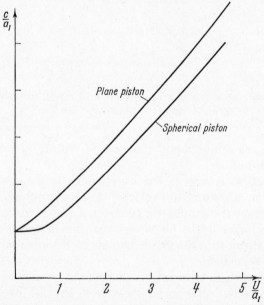

FIG. 39. Shock wave velocity c ahead of the piston as a function of piston velocity U; a_1 is the speed of sound in the disturbed gas.

motions and the singular solution of the system of Equations (1.3) for $\nu = 1$, which is, in the z, V, variables:

$$z = (1 - V)^2, \quad V - \frac{2}{\gamma - 1}(1 - V) = C\lambda^{-1}, \tag{6.2}$$

where C is an integration constant.

After transforming to dimensional variables, we obtain the nonlinear solution for a progressive rarefaction wave in a gas (the most simple case of the Riemann solution):

$$x = (v \pm a)t, \quad v = \pm \frac{2}{\gamma + 1}a + \text{const}, \tag{6.3}$$

where a is the speed of sound.

§7. PROBLEM OF IMPLOSION AND EXPLOSION AT A POINT

We consider the implosion and explosion problem in the case when the initial velocity, density and pressure is everywhere uniform, i.e. $\omega = 0$, $\delta = 1$. The appropriate field of the integral curves in the z, V plane is given in Fig. 30.

The point O corresponds to the point at infinity. From (5.6), the asymptotic formulas near the point O are:

$$z = CV^2, \quad \lambda = \frac{C_1}{V}.$$

We obtain from the condition at infinity or the initial condition:

$$z = \frac{\gamma p_1}{\rho_1}; \quad \frac{r^2}{t^2} = C\frac{t^2}{r^2}v_1^2; \quad \text{where } C = \frac{\gamma p_1}{\rho_1 v_1^2}. \tag{7.1}$$

When $v_1 < 0$ we follow the left branch of the parabola $z = CV^2$ with $V < 0$; when $v_1 > 0$ we follow the right branch with $V > 0$.

Let us put $\lambda = r/(a_1 t)$, $a_1 = (\gamma p_1)/\rho_1$; according to (5.2), we then fix the constant β.

Using this, the asymptotic formula for λ yields:

$$\frac{r}{a_1 t} = \frac{C_1 r}{v_1 t}; \quad C_1 = \frac{v_1}{a_1}. \tag{7.2}$$

If the initial velocity is directed toward the centre (implosion), i.e. $v_1 < 0$, then the motion along an integral curve starting at 0 for negative V corresponds to motion from infinity towards the centre for fixed t.

Since the gas velocity at the centre is zero for $t \neq 0$, then the centre can be reached either by moving along the integral curve passing through the singular point B or by moving along the integral line $V = 0$ passing through the singular point $D(z = \infty,\ V = 0)$. But it is generally impossible to pass from the region $V < 0$ where the image point is found, to

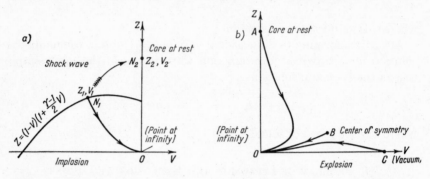

Fig. 40. Integral curves in the z, V plane corresponding to: a, implosion at a point; and b, explosion from a point.

the integral curve passing through the point B and it is only possible to cross to the z axis discontinuously.

Therefore, the image point follows the integral curve as far as its intersection with the parabola (2.11) $z = (1 - V)\{1 + [(\gamma - 1)/2]V\}$, on to which points of the z axis transform discontinuously. After intersecting

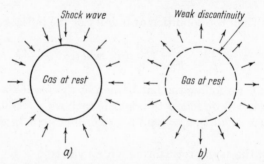

Fig. 41. Motion diagrams: a, implosion, and b, explosion. Adiabatic compression or rarefaction arises in front of the core.

the parabola, the image point transforms into a certain point of the z axis by a jump. The shape of the integral curve is shown on Fig. 40a.

Hence, in physical space, the gas is first compressed adiabatically in motion from infinity towards the centre and is afterwards brought to rest by a shock (Fig. 41a).

The image point in the explosion case ($v_1 > 0$ is the gas particle velocity directed from the centre) moves along the integral curve defined by (7.1) from the point O to the point A, for low values of the initial velocity, and then along the integral line $V = 0$ (Fig. 40b). In physical space this corresponds to motion from infinity towards the centre up to a fixed instant, accompanied by a drop in the gas density and pressure to certain definite values; afterwards, the gas is brought to rest across a weak shock (Fig. 41b). The singular point A (Fig. 40b) corresponds to the boundary of the spherical core of the gas at rest.

The integral curve OB corresponds to the solution in the (z, V) plane for a certain initial velocity $v = v_1^*$. The point B corresponds to the centre of symmetry. In this case, the velocity equals zero only at the centre of symmetry in physical space.

If $v_1 > v_1^*$, then the integral curve can only be followed up to the singular point C where the parameter λ has a finite, non-zero, constant value λ^*, and the density and pressure equal zero.

Therefore, a vacuum, propagating with constant velocity defined by the value λ^*, is formed in the gas.

§8. Spherical Detonation

Let us consider the disturbance in a gas due to a detonation at the centre of symmetry at time $t = 0$; a spherical detonation wave is propagated through the initially undisturbed gas when $t > 0$. We assume that heat changes occur only at the shock front; the gas motion is adiabatic beyond the shock, which is a detonation wave.

Let us first study the case when the density ρ_1 and the pressure p_1 are constant and non-zero in the initially undisturbed gas (Zel'dovich, 1942). The disturbed motion of a perfect gas is determined by the parameters

$$r, t, \gamma, \gamma_1, p_1, \rho_1, Q,$$

where Q is the heat liberated at the front by unit mass of gas; γ_1, γ are the appropriate values of the specific heat ratio; γ_1, ahead of the front and γ behind the front. The motion is self-similar and belongs to type 1 defined in §5.

The family of integral curves in the z, V plane for the differential equation (5.3) ($\omega = 0$) is shown in Fig. 30; $\lambda = [\beta r / \sqrt{(Q)} t]$.

The gas is at rest in the region ahead of the detonation wave; consequently, the point H_1, on the z axis corresponds to the far side of the detonation wave in the z, V plane. At this point $\lambda_2 = \beta(c / \sqrt{Q})$; we determine β from the condition $\lambda_2 = 1$. It follows from the Hugoniot condition

7

(2.27) that the inner side of the detonation wave in the z, V plane corresponds to the parabola

$$z_2 = (1 - V_2)^2 \frac{1 + \gamma \Lambda}{1 - \Lambda}. \tag{8.1}$$

The Chapman-Jouguet condition is satisfied for $\Lambda = 0$ and the parabola (8.1) coincides with the parabola

$$z = (1 - V)^2. \tag{8.2}$$

If $\Lambda > 0$, the parabola (8.1) lies above the parabola (8.2) (see Fig. 30 and the diagram on Fig. 42). Since the variable λ has a stationary value

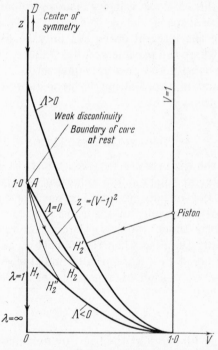

FIG. 42. Integral curves in the z, V plane corresponding to a spherical detonation. Case $p_1 = \text{const} \neq 0$, $\rho_1 = \text{const}$ ($\omega = 0$).

on the parabola (8.2), it is impossible to extend the solution continuously between points of parabola (8.1) and the centre of symmetry. It is also easy to see that a solution with an additional compression shock is impossible.

However, the solution can be continued as far as the line $V = 1$; the points of this line can be considered to represent a spherical piston.

Hence, the solution of the problem of piston expansion in a detonated mass of gas can be obtained. The choice of the appropriate integral curve and of the parameter Λ is fixed by the values of the pressure or velocity (λ^*) on the piston since the point H' on the parabola (8.1) (see Fig. 42) is determined for each Λ by the value of the parameter $p_1/Q\rho_1$.

If $\Lambda < 0$, then the parabola (8.1) is situated below the parabola (8.2). In this case, the solution of the problem is not unique.

There is a whole pencil of curves issuing from the node A which give solutions satisfying all the boundary conditions. The position of a point behind the shock front H_2'' and the parameter $\Lambda < 0$ are determined by finding the detonation front velocity. A core of gas at rest is obtained at the centre, whose boundary corresponds to the point A and which is a weak discontinuity. The line AD corresponds to points of the core. The point D corresponds to the core centre $\lambda = 0$ (Fig. 43). If the point H_2

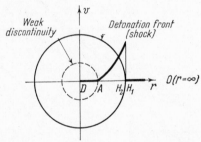

FIG. 43. Flow picture for a spherical detonation.

lies on the parabola (8.2), then the Chapman-Jouguet condition is satisfied and the detonation velocity will be a minimum.

If $\Lambda \leqslant 0$, then a rarefaction wave extends from the rear of the detonation front to the gas at rest.

Figures 44, 45 and 46 show the computed distributions of pressure, velocity and temperature in an example in which $\gamma = \gamma_1 = 5/3$, $p_1 = 0$ and the Chapman-Jouguet condition is satisfied (curves corresponding to $\omega = 0$).

The solution of the detonation problem in the cylindrical and plane wave cases can be obtained in a similar manner.

We now consider the problem of detonation in a medium with a variable initial density (Sedov, 1956; Iavorskaia, 1956).

If

$$0 < \omega \leqslant \frac{2\gamma}{\gamma+1},$$

then the field of integral curves is shown in Fig. 31.

Figure 47 shows the integral curves in the z, V plane relevant to this problem.

In order to continue the solution inwards behind the detonation wave front, it is necessary to use the integral curve starting from the point D at which $\lambda = 0$ and which passes through the singular point A. A weak discontinuity can arise at the point A on the parabola $z = (1 - V)^2$; the

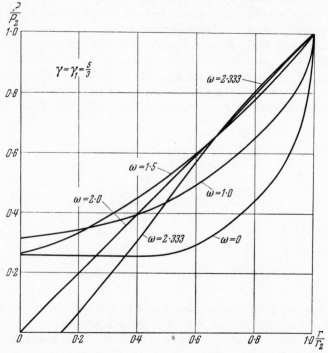

FIG. 44. Distribution of pressure behind the detonation wave front: initial pressure $p_1 = 0$, initial density $\rho_1 = A/r^\omega$; a vacuum is formed near the centre for $\omega > [3(\gamma + 1)]/(3\gamma - 1)$ within which the pressure is zero.

variable λ has a finite value and increases as we move downwards along any integral curve; the only integral curves which yield a solution are those intersecting the parabola given by Equation (2.29), namely,

$$z_2 = \gamma V_2(1 - V_2). \tag{8.3}$$

Just as in the $\omega = 0$ case, the solution is not unique when $\varLambda \leqslant 0$ on the detonation front. The solution satisfying the Chapman-Jouguet condition $\varLambda = 0$ corresponds to the points of intersection of the parabolas (8.2) and (8.3).

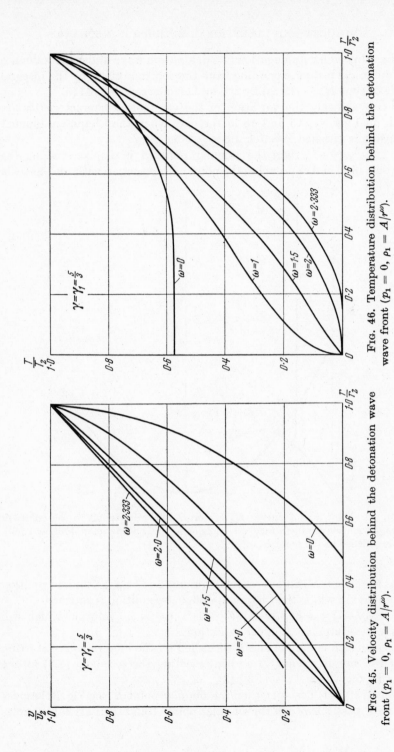

FIG. 45. Velocity distribution behind the detonation wave front ($p_1 = 0$, $\rho_1 = A/r^\omega$).

FIG. 46. Temperature distribution behind the detonation wave front ($p_1 = 0$, $\rho_1 = A/r^\omega$).

The solution for propagation of a detonation wave under the influence of a spherical piston expanding from the centre is given by the integral curve of type $H_2'' C$, shown by the broken curve on Fig. 47.

Curves showing the variation of the gas motion characteristics are given on Figs. 44, 45 and 46 in the case when the Chapman-Jouguet condition is satisfied, $\gamma = 5/3$ and $\omega = 1\cdot 5$.

If $\omega \to 2\gamma/(\gamma+1)$ then the singular point A is displaced along the parabola (8.2) and approaches the point of intersection with the parabola

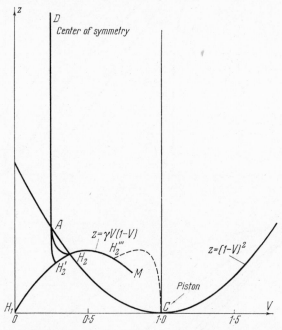

Fig. 47. The integral curves $H_2 AD$ or $H_2' AD$ correspond to the solutions of the detonation problem for $0 < \omega < [2\gamma/(\gamma+1)]$. The Chapman-Jouguet condition is satisfied at the point H_2.

(8.3); the points H_2 and A coincide for $\omega = 2\gamma/(\gamma+1)$; a unique solution is obtained in which the Chapman-Jouguet condition is satisfied.

If $2\gamma/(\gamma+1) < \omega < [3(\gamma+1)]/(3\gamma-1)$ there is no solution which will satisfy the Chapman-Jouguet condition.

The solution is unique and is furnished by the integral curve starting from the singular point D and intersecting the parabola (8.3) at the point H_2 for which $\Lambda > 0$ (Fig. 48).

If $\omega \to [3(\gamma+1)]/(3\gamma-1)$, then the singular point B (see Fig. 31) moves upwards, passes through the singular point A and then exchanges roles with this point. When $\omega = [3(\gamma + 1)]/(3\gamma - 1)$, the point B lies on the parabola (8.3) (see Fig. 32). In this case, it is necessary to make a jump from the point O to the point B in order to obtain the solution. All the gas motions behind the wave front in the z, V plane correspond to the single point B. In this case, $z = z_2 = \text{const}$, $V = V_2 = \text{const}$ behind the wave front, and the appropriate solution is given by the simple formulas

$$\frac{v}{v_2} = \frac{r}{r_2}, \quad \frac{\rho}{\rho_2} = \frac{r_2}{r}, \quad \frac{p}{p_2} = \frac{r}{r_2}. \tag{8.4}$$

The velocity behind the wave front is directly proportional to the coordinate r.

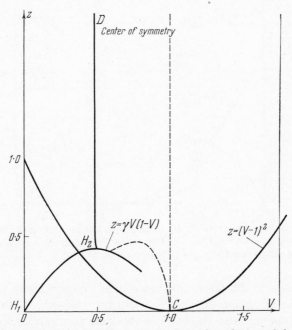

FIG. 48. The integral curve DH_2 corresponds to the solution of the detonation problem for $2\gamma/(\gamma + 1) < \omega < [3(\gamma + 1)]/(3\gamma - 1)$. The Chapman-Jouguet condition is not satisfied.

If $[3(\gamma + 1)]/(3\gamma - 1) < \omega$, then it is evident from Fig. 33 that the integral curve passing through the point A and subsequently through the point C corresponds to the solution in the z, V plane. An expanding sphere within which the pressure is zero corresponds to the singular point C. The value of Λ and, therefore, of the detonation velocity, is determined by the point of intersection of the integral curve with the parabola (8.3) (see Fig. 49).

The variation of the gas characteristics is given for $\gamma = 5/3$ and $\omega = 7/3 = 2 \cdot 33\ldots$ in Figs. 44, 45 and 46.

Clearly all conclusions about increasing the detonation velocity and on the formation of a vacuum at the centre of symmetry based on the self-similarity property, are independent of the magnitude of the heat liberated Q and only depend on the law under which the initial density falls off, and this is determined by the value of the exponent ω.

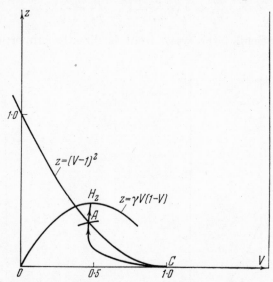

Fig. 49. The integral curve H_2AC corresponds to the solution of the detonation problem for $\omega > [3(\gamma+1)]/(3\gamma-1)$. The point C corresponds to an expanding vacuum. The Chapman-Jouguet condition is not satisfied.

The increase in the detonation velocity as compared with the velocity given by the Chapman-Jouguet rule can be obtained also for gas detonations in tapered tubes. If the cross-sectional area varies according to a power law, then we find, using the hydraulic approximation, that the gas motion is self-similar and is determined by (5.3), (5.4) and (5.5), but for $\nu < 1$. The magnitude of ν is determined by the law of stream tube contraction.

§9. FLAME PROPAGATION

We consider the disturbed motion of a combustible mixture resulting from combustion in a very thin moving layer. Analysis† shows that the

† The problem of spherical flame propagation when $\rho_1 = \mathrm{const}$ and $p_1 = \mathrm{const}$ was studied by G. M. Bam-Zelikovich.

thickness of the layer in which the chemical reaction occurs can be neglected in a number of cases and we arrive at the problem of gas motion in which the chemical reaction and the heat liberation are performed instantaneously at a certain surface, across which variables characterizing the state and motion of the gas vary discontinuously; the surface is called a flame front. In contrast to the detonation front, the flame front is a rarefaction jump, the velocity of propagation the flame front, u, through the combustible mixture is a known chemical constant. The flame propagation velocity is small in comparison with the speed of sound and, therefore, in comparison with the detonation velocity.

Conditions (2.12), (2.13) and (2.14) are satisfied on the flame front, just as on the detonation front; the difference between the flame and detonation fronts is only that the flame front velocity is small and known in advance. Disturbances caused by combustion are propagated into the gas in front of and behind the flame front. It is necessary to take the smaller root $\rho_1/\rho_2 > 1$ as the solution of (2.14). This corresponds to states which are obtained from the initial state with continuous heat liberation (without a heat absorption zone). The reaction thus occurs in a thin, finite layer.

We consider the problem of propagation of a plane flame front through a gas at rest with density ρ_1 and pressure p_1 in a cylindrical tube. Burning progresses towards a closed end of the tube. This solution of the problem is very simple and is as follows: a shock wave moves from the closed end through the undisturbed gas; behind the shock wave front the gas motion is directed forwards toward the shock wave. A plane flame front is propagated through the moving gas leaving the gas behind it at rest, a consequence of the boundary conditions at the closed end. In order to solve the problem completely, it is sufficient to write and solve six equations simultaneously; three on the flame front and three at the shock. The six unknowns to be determined from the six equations are: the density and pressure behind the flame front and behind the shock, the gas velocity behind the shock wave and the velocity of shock wave propagation.

Consider the problem of a spherical flame front under the assumption that combustion starts at $t = 0$ at a point, is then propagated by means of a spherical wave through the undisturbed gas with constant density ρ_1 and constant pressure p_1. Evidently, the disturbed gas motion is self-similar and is determined by the same constants as in the detonation phenomenon. The integral curves in the z, V plane for a spherical flame are given in Fig. 30 just as for the spherical detonation case.

The gas particles sufficiently far removed from the ignition centre at any instant $t > 0$ will be at rest. The stationary region corresponds to

7*

the integral line $V = 0$. It is impossible to effect the transition from rest, $V = 0$, to motion on another integral curve in the left part of the $z > 0$, $V < 0$ half-plane by means of a rarefaction jump, a flame front through the singular point A with a weak discontinuity, since the subsequent motion cannot be continued to the centre of symmetry. In these cases, continuous motion or motion in the presence of a shock follows an integral curve intersecting the parabola $z = (1 - V)^2$.

Therefore, the transition from rest, $V = 0$, to motion on another integral curve is only possible by means of a transition through a simple compression shock originating at $z_1 < 1$. According to (2.10), compression shocks transform the $V = 0$ axis into points of the parabola.

$$z_2 = (1 - V_2)\left(1 + \frac{\gamma - 1}{2} V_2\right). \tag{9.1}$$

To continue the solution up to the centre of symmetry where $V = 0$, the flame front must be so determined that a point behind it would be located either on the integral line $V = 0$ leaving the singular point $D(z = \infty, V = 0)$ or on the integral curve L entering the singular point $B(z = [3(\gamma - 1)^2]/(3\gamma - 1)^2, V = 2/[3\gamma - 1])$† (see Fig. 50). It follows from condition (2.25) that transition through the flame front is possible on the $V = 0$ axis only from points of the curve

$$z_3 = \frac{V_3(1 - V_3)\left(1 + \frac{\gamma' - 1}{2} V_3\right) + (\gamma' - 1)\dfrac{Q}{u^2}(1 - V_3)^3}{\dfrac{\gamma'}{\gamma} - \dfrac{\gamma' - 1}{\gamma - 1}(1 - V_3)}. \tag{9.2}$$

Equation (9.2) has been obtained from (2.25) after replacing the subscripts 1, 2 by 4, 3, putting $V_4 = 0$ and changing the sign in $Q/c^2 = [Q(1 - V_3)^2]/u^2$ since combustion occurs as the transition is made from state 3 to state 4; u is the flame propagation velocity:

$$u = c - v = \frac{r}{t}(1 - V_3) = c(1 - V_3).$$

In (9.2), γ refers to the moving combustible mixture; γ' to the combustion products at rest.

Figure 50 shows the method of constructing the solution.

The image point appears to be at a certain point z_2, V_2 on the parabola (9.1) following the jump; movement along the integral curve passing through this point in the direction of decreasing λ corresponds to further

† If the initial density varies according to the law $\rho_1 = A/r^\omega$, then the coordinates of the point B depend on ω, see §5.

motion toward the centre, i.e. motion in a region above the parabola
(9.1). The point of intersection of the integral curve under consideration

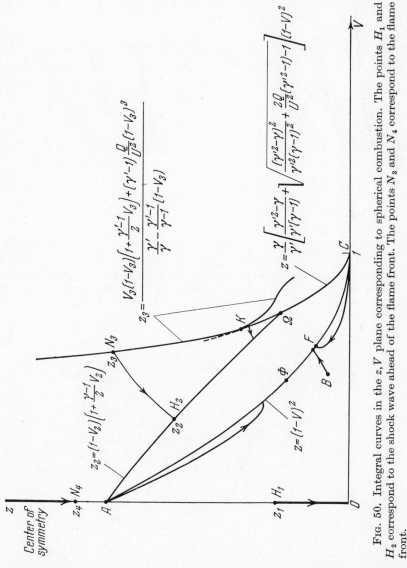

$$z_3 = \frac{V_3(1-V_3)\left(1+\frac{\gamma-1}{2}V_3\right)+(\gamma-1)\frac{Q}{U^2}(1-V_3)^3}{\frac{\gamma'-1}{\gamma'}-\frac{\gamma'-1}{\gamma-1}(1-V_3)}$$

$$z = \frac{\gamma}{\gamma'}\left[\frac{\gamma'^2-\gamma}{\gamma'\gamma'(\gamma-1)} + \sqrt{\left[\frac{(\gamma'^2-\gamma)^2}{\gamma'^2(\gamma-1)^2} + \frac{2Q}{U^2}(\gamma'^2-1)-1\right](1-V)^2}\right]$$

$$z_2 = (1-V_2)\left(1+\frac{\gamma-1}{2}V_2\right)$$

$$z = (1-V)^2$$

Fig. 50. Integral curves in the z, V plane corresponding to spherical combustion. The points H_1 and H_2 correspond to the shock wave ahead of the flame front. The points N_3 and N_4 correspond to the flame front.

with the curve (9.2) corresponds to the leading edge of the flame front.
Behind this there is a stationary core of gas, corresponding to points of
the $V = 0$ axis.

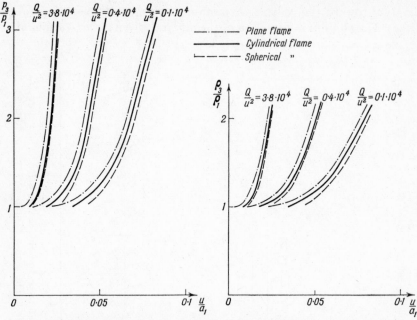

FIG. 51. Density and pressure ahead of the flame front for various amounts of heat released Q/u^2 (Q is the energy liberated per unit mass; u is the flame front velocity through the gas; p_1 is the pressure; ρ_1 is the density and a_1 is the speed of sound in the combustible mixture.

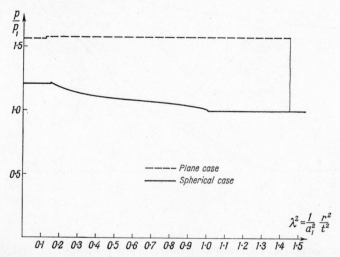

FIG. 52. Pressure distribution for propagation of combustion from a plane wall (plane flame front) and from a point (spherical flame front): $Q/a_1^2 = 60$ and $u/a_1 = 0\cdot016$.

The construction described is always possible since any integral curve starting from the parabola (9.1) intersects the curve (9.2) for $V \leqslant 2/(\gamma+1)$ and a transition is possible from any point of the curve (9.2) to points of the $V = 0$ axis through the flame front.

However, we assume in this construction that $z_4 \geqslant 1$. Points of the curve (9.2) for which $z_4 < 1$ is obtained cannot correspond to the leading edge of the flame front since this leads to a supersonic flame front velocity relative to the particles behind the front. In this case, the flame front can be constructed by using a jump on to the parabola $z = (1 - V)^2$ from

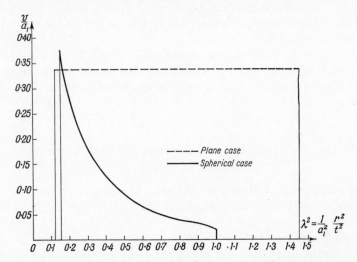

FIG. 53. Velocity distribution for propagation of combustion from a plane wall (plane flame front) or from a point (spherical flame front): $Q/a_1^2 = 60$; and $u/a_1 = 0\cdot016$.

the point of intersection of the integral curve for the gas motion behind the shock wave, with the parabola†

$$z = \frac{\gamma}{\gamma'} \left\{ \frac{\gamma'^2 - \gamma}{\gamma'(\gamma - 1)} + \sqrt{\left[\frac{(\gamma'^2 - \gamma)^2}{\gamma'^2(\gamma - 1)^2} + \frac{2Q}{u^2}(\gamma'^2 - 1) - 1 \right]} \right\} (1 - V)^2 \quad (9.3)$$

(Fig. 50, curve $K\Omega C$). The propagation velocity of such a jump through the gas behind the front exactly equals the speed of sound, while an additional rarefaction wave is formed behind the front. This rarefaction wave corresponds to an integral curve proceeding from a point on the parabola $z = (1 - V)^2$ to either the point A, which corresponds to the

† The parabola (9.3) transforms into the parabola $z = (1 - V)^2$ for a jump of the flame front.

boundary of the core at rest, or from the point F to the point B, in which case a core at rest is not formed, and the motion can be continued up to the centre of symmetry, or to the singular point $C(z = 0, V = 1)$; a vacuum is formed near the centre in the latter case.

A solution of the problem exists for all points z_2, V_2 on the parabola (9.1) located above the point Ω at which the parabola (9.1) intersects the parabola (9.3). The point Φ corresponds to the point Ω on the parabola $z = (1 - V)^2$. If the point Φ lies above the point F, then a stationary core is always formed near the centre of symmetry. If p_1 and ρ_1 are constant then a stationary gas core occurs near the centre of symmetry. If $\rho_1 = A/r^\omega$, a vacuum can form near the centre of symmetry for a certain value of ω.

The solution of problems of propagation of cylindrical flames can be constructed by using analogous methods. The results of numerical computations and a comparison of various cases are given in Figs. 51, 52 and 53.

§10. Collapse of an Arbitrary Discontinuity in a Combustible Mixture

We now discuss, in outline only, the general character of the problem of the collapse of an arbitrary discontinuity (formulated in §1). Detailed analysis will be omitted.†

We first consider two inert gases separated by a surface of discontinuity and assume that the pressure in the second gas to the left is larger than that in the first gas to the right of this surface (the converse case is exactly similar). Then, if the x axis runs from left to right and if the difference $v_1 - v_2$ of the initial gas velocities is negative and large in absolute value (this case will occur, for example, if both initial gas velocities are directed toward the surface of discontinuity), then shock waves will develop on both sides of the discontinuity. At the gas interface there is a stationary discontinuity at which the pressure and the normal velocity are continuous but the density changes discontinuously. A graph of the pressure in this case is shown in Fig. 54.

As the difference between the initial velocities increases, the shock wave in the second gas changes to a rarefaction wave (Fig. 54b) and then a rarefaction wave in the first replaces the shock wave (Fig. 54c). When the initial velocity difference becomes a very large positive quantity, a vacuum forms between the rarefaction waves on both sides (Fig. 54d).

† These problems are analysed in detail in Landau and Lifshitz (1959).

A more complex case arises when a combustible mixture is on the right of the surface of discontinuity so that a flame front can develop when the surface collapses. The general features of the motion which occurs in this case will be analogous to that considered above.†

When the initial velocities differ by a small amount a shock wave is propagated through the inert gas, a shock wave also develops in the combustible mixture followed by a flame front. A stationary discontinuity can exist between the inert gas and the combustion products.

A graph of the pressure in this case is shown in Fig. 55a (the flame front is denoted by the wide vertical band, the stationary discontinuity by dashes). As the difference in the initial velocities increases, the shock

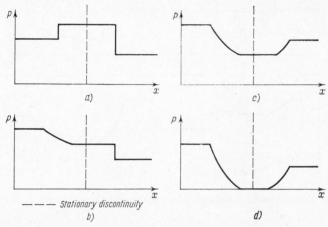

FIG. 54. Various forms of decay of an arbitrary discontinuity in an inert gas.

wave in the inert gas first changes into a rarefaction wave (Fig. 55b) and then a rarefaction wave arises ahead of the flame front in place of the shock wave (Fig. 55c). Here, the velocity of the combustion products relative to the flame front increases just until sonic speed is attained.

As the difference in the initial velocities increases further, the flow ahead of the flame front does not change but still another rarefaction wave appears directly behind the front (Fig. 55d). A vacuum can form between the inert gas and the combustion products if the difference in the initial velocities is very large.

If the pressure in the inert gas is less than in the combustible mixture, a rarefaction wave may develop in the combustible mixture ahead of the flame front but there is a shock wave in the inert gas.

† This question was worked out quantitatively in detail by Bam-Zelikovich (1949).

Similarly, if there is a detonation wave in the combustible mixture, then by increasing the difference in the initial velocities from $-\infty$ to $+\infty$, we find that there is first a shock wave in the inert gas and a detonation wave with velocity as large as desired in the combustible mixture; then the shock wave in the inert gas is replaced by a rarefaction wave and the velocity of the detonation wave decreases to a certain definite value. Here the velocity of the detonation products relative to the front increases until sonic speed is attained. Later, the detonation wave velocity varies and a rarefaction wave forms behind it.

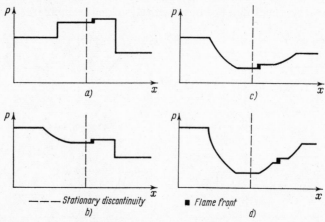

FIG. 55. Various cases of collapse of a surface of discontinuity in a combustible mixture.

We now record the various types of formation and collapse of an arbitrary discontinuity.

1. A shock wave is propagated into the gas and a second shock wave overtakes it from behind. At the instant of overtaking, a surface of discontinuity is formed across which the conditions of conservation of mass, momentum and energy are not satisfied, i.e. an arbitrary discontinuity is formed.

Computations show that shock waves will develop on both sides in this case after the discontinuity has collapsed.

2. A shock wave approaches the interface of two media of different densities. When the shock wave crosses from one medium into the other an arbitrary discontinuity is formed. Two types of motion are possible when this discontinuity collapses.

Shock waves will occur on both sides when the wave crosses from the less dense to the more dense medium (for example, from air to water).

If the wave crosses from the more dense to the less dense medium (from water to air, say) then a shock wave occurs in the front (in the air) and a rarefaction wave in the rear (in the water).

3. A low intensity shock wave overtakes a flame front. (This case is encountered in pulsating combustion in closed vessels.) After the wave has overtaken the flame, shock waves will occur on both sides of the

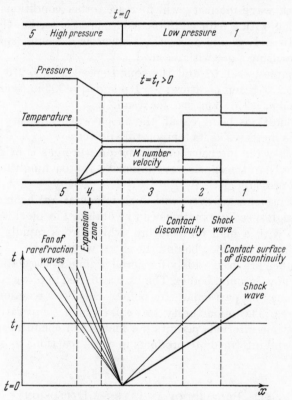

FIG. 56. Diagram of the motion in a shock tube.

flame front. If a low intensity shock wave encounters a flame front, then a rarefaction wave will occur ahead of the flame front after the arbitrary discontinuity has collapsed and a shock wave will develop in the combustion products.

The problem of the collapse of a given discontinuity is important in the study of the initial stages of gas motion in shock tubes. A diagram of the motion in a shock tube is pictured on Fig. 56. Two gases of high and low pressure are separated by a membrane. An arbitrary discontinuity forms after the membrane is suddenly destroyed; consequently, a shock wave occurs in the low pressure gas. The high pressure gas is either at rest or is moving at the moment the membrane bursts if a shock or detonation wave approaches the membrane. The shock wave intensity in the low pressure gas depends on the initial motion and on the pressure drop, on the difference in the temperatures and on the properties of the gases initially separated by the membrane.

The shock wave intensity will increase, other conditions remaining equal, if a gas with a reduced initial sonic speed is used as the low pressure gas. In polyatomic gases the reduction in sound speed can be achieved by using a gas with reduced γ.

For example, $\gamma = 1 \cdot 67$ and the sound speed is $a_1 = 975$ m./sec. for helium at a 273°K temperature; $\gamma = 1 \cdot 4$ and $a_1 = 333$ m./sec. for air and $\gamma = 1 \cdot 15$ and $a_1 = 121 \cdot 5$ m./sec. for freon.

Other conditions being equal, the shock wave intensity in the low pressure gas increases as its temperature decreases.

The shock wave intensity in the low pressure gas is evidently very much larger than the shock wave intensity approaching the membrane in the high pressure gas.

Very intense shock waves, with high temperatures behind the wave front and high speed gas motions can be obtained in shock tubes.

Particles with a high temperature, which drops rapidly within the short time τ, are obtained behind the shock wave front.

Shock tubes are used widely for aerodynamic investigations of very high speed flows around bodies. They are also used in physical chemistry investigations, in particular, to obtain chemical reactions at high temperatures. The opportunity to achieve high temperatures during very short time intervals permits the kinetics of chemical reactions to be studied and intermediate products in chain reactions to be obtained.

§11. PROBLEM OF AN INTENSE EXPLOSION†

1. INTENSE EXPLOSION IN A GAS

The above arguments show that, in an intense explosion the disturbed air region is separated from the undisturbed air by a shock wave.

As already mentioned, the pressure ahead of the shock wave can be neglected in comparison with the pressure behind the shock wave in an intense explosion. Let us first estimate with what accuracy and for which shock waves this statement is valid.

† In this section, we shall explain the exact theoretical formulation and the numerical solution of the problem of an intense explosion for both spherical and cylindrical and plane waves. This was first published in Sedov (1946a) and in Sedov (1946b).

Using the property $v_1 = 0$, we rewrite the shock conditions (2.5) and (2.6) as follows:

$$
\left.
\begin{aligned}
v_2 &= \frac{2}{\gamma+1} c \left[1 - \frac{a_1^2}{c^2} \right] = \frac{2c}{\gamma+1} f_1, \\[2mm]
\rho_2 &= \frac{\gamma+1}{\gamma-1} \rho_1 \left[1 + \frac{2}{\gamma-1} \frac{a_1^2}{c^2} \right]^{-1} = \frac{\gamma+1}{\gamma-1} \rho_1 f_2, \\[2mm]
p_2 &= \frac{2}{\gamma+1} \rho_1 c^2 \left[1 - \frac{\gamma-1}{2\gamma} \frac{a_1^2}{c^2} \right] = \frac{2}{\gamma+1} \rho_1 c^2 f_3
\end{aligned}
\right\} \quad (11.1)
$$

where c is the velocity of shock wave propagation.

As the shock wave increases in strength the ratio a_1/c is reduced.

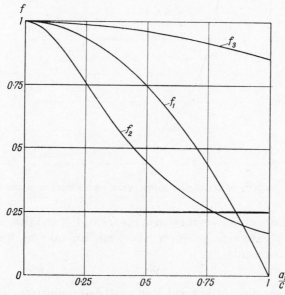

Fig. 57. Relation between the quantities f_1, f_2 and f_3 and the ratio a_1/c, where a_1 is the speed of sound in the undisturbed medium; c is the shock wave velocity.

Figure 57 shows f_1, f_2 and f_3 as a function of a_1/c. Figure 58 shows p_2/p_1 as a function of the ratio a_1/c for $\gamma = 1\cdot 4$. We observe that the quantities f_1, f_2 and f_3 differ from unity by less than 5 per cent when $a_1/c < 0\cdot 1$.

If we put $a_1/c = 0$ and $f_1 = f_2 = f_3 = 1$ into (11.1) (or if we put $p_1 = 0$, which is equivalent), then an error of less than 5 per cent is introduced in the values of v_2, ρ_2 and p_2.

The conditions on the shock wave then become

$$v_2 = \frac{2}{\gamma+1}\,c,$$

$$\rho_2 = \frac{\gamma+1}{\gamma-1}\,\rho_1,$$

$$p_2 = \frac{2}{\gamma+1}\,\rho_1 c^2.$$

(11.2)

The velocity of shock wave propagation c is a characteristic parameter.

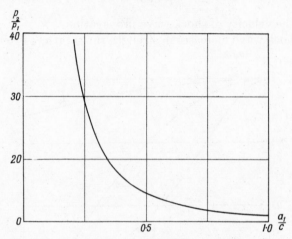

FIG. 58. Pressure drop across the shock wave as a function of the ratio a_1/c.

If we use the equations of motion in the form (1.3) and this formulation of the strong explosion problem, then we can take as fundamental dimensional constants:

$$\rho_1 \quad \text{and} \quad E/\rho_1$$

where E is a certain constant which we shall determine later, and has the same dimensions as the energy E_0 liberated during the explosion; the dimensions of E are

$$[E] = ML^2\,T^{-2} \quad \text{in the spherical case,}$$

$$[E] = MLT^{-2} \quad \text{in the cylindrical case,}$$

$$[E] = MT^{-2} \quad \text{in the plane case,}$$

All three cases can be combined in the one formula

$$[E] = ML^{\nu-1}\,T^{-2}.$$

Evidently, the constant E is directly proportional to E_0:

$$E_0 = \alpha E$$

where α is a constant.

In this case, the single nondimensional variable parameter λ is given by

$$\lambda = \frac{r}{\left(\dfrac{E}{\rho_1}\right)^{1/(2+\nu)} t^{2/(2+\nu)}}.$$

The motion of the shock wave is easily determined without solving the equations of motion.

Different equations of motion can be used provided that these do not contain new essential physical constants with dimensions independent of ρ_1 and E. In particular, it is not necessary to assume that the coefficient $\gamma = c_p/c_v$ in (1.3) is constant.

The shock wave coordinate r_2 is a function of the time t and since it is impossible to form a nondimensional combination of the dimensional quantities t, ρ_1 and E, then

$$r_2 = \left(\frac{E}{\rho_1}\right)^{1/(2+\nu)} t^{2/(2+\nu)} \lambda^*, \tag{11.3}$$

where $\lambda^* = \text{const}$; λ^* can be set equal to any non-zero number and the value of E can be calculated from the magnitude of the charge energy E_0. Later, to be definite, and for the sake of simplicity, we shall set $\lambda^* = 1$. The constant α in the formula $E_0 = \alpha E$ is then determined from the solution of the equations of motion.

Hence, in the spherical symmetry case, the motion of the shock wave is given by

$$r_2 = \left(\frac{E}{\rho_1}\right)^{1/5} t^{2/5}, \quad c = \frac{2}{5}\left(\frac{E}{\rho_1}\right)^{1/5} t^{-3/5} = \frac{2}{5}\sqrt{\left(\frac{E}{\rho_1}\right)} \frac{1}{\sqrt{r_2^3}}; \tag{11.4}$$

and in the cylindrical symmetry case, by

$$r_2 = \left(\frac{E}{\rho_1}\right)^{1/4} \sqrt{t}, \quad c = \frac{1}{2}\left(\frac{E}{\rho_1}\right)^{1/4} \frac{1}{\sqrt{t}} = \frac{1}{2}\sqrt{\left(\frac{E}{\rho_1}\right)} \frac{1}{r_2}; \tag{11.5}$$

while for plane waves

$$r_2 = \left(\frac{E}{\rho_1}\right)^{1/3} t^{2/3}, \quad c = \frac{2}{3}\left(\frac{E}{\rho_1}\right)^{1/3} t^{-1/3} = \frac{2}{3}\sqrt{\left(\frac{E}{\rho_1}\right)} \frac{1}{\sqrt{r_2}}. \tag{11.6}$$

These formulas show that the law of shock wave attenuation depends on the charge shape.

The formula obtained above in the spherical symmetry case is in good agreement with published experimental results in photographs of an atomic bomb explosion in New Mexico in 1945.

Photographs of an atomic bomb explosion published in a paper by G. I. Taylor (1950) are given in Figs. 59, 60 and 61.

The air temperature is very high in the disturbed air region at quite significant distances from the centre of the atom bomb explosion. Consequently, this region is shown as a luminous spot on the photographs. The boundary in the upper part of the spot is spherical, sharply traced and coincides with the shock wave. The shock wave attenuates as time increases and the temperature behind its front decreases. However, the appearance of the wave front is retained in its initial form because of the jump in the density. A relation between the radius r_2 of the expanding spherical shock wave and corresponding time t measured from the instant of initiation is derived from these photographs. The radius ranges from values of 11 to 185 m. at corresponding time intervals between $0 \cdot 1 \times 10^{-3}$ and 62×10^{-3} sec.

The experimental results are shown on Fig. 62 by crosses. The line corresponds to the formula

$$\tfrac{5}{2}\log r_2 \,\mathrm{cm.} - \log t \,\mathrm{sec.} = 11 \cdot 915$$

which agrees with the theoretical formula (11.4) after taking the logarithm of the latter, if it is assumed in addition that

$$\tfrac{1}{2}\log E/\rho_1 = 11 \cdot 915 \text{ from which for } \rho_1 = 0 \cdot 00125 \,\mathrm{gm./cm^3}$$

$$E = 6 \cdot 76 \times 10^{23}\, \rho_1 = 8 \cdot 45 \times 10^{20}\,\mathrm{erg} \tag{11.7}$$

The experimental results are in good agreement with the law of shock wave propagation (11.4) established earlier by using the general reasoning of dimensional analysis and they permit the magnitude of the constant E to be determined according to (11.7).

The formula for the shock wave velocity can be written

$$c = \frac{2}{\nu+2}\frac{r}{t}.$$

Substituting this expression for c into the shock condition (1.2) and transforming to nondimensional variables V, R, P and $z = \gamma P/R$ according to (1.1), we find the values V_2, R_2 and z_2 behind the shock wave

$$V_2 = \frac{4}{(\gamma+1)(\nu+2)}, \quad R_2 = \frac{\gamma+1}{\gamma-1}, \quad z_2 = \frac{8\gamma(\gamma-1)}{(\gamma+1)^2(\nu+2)^2}. \tag{11.8}$$

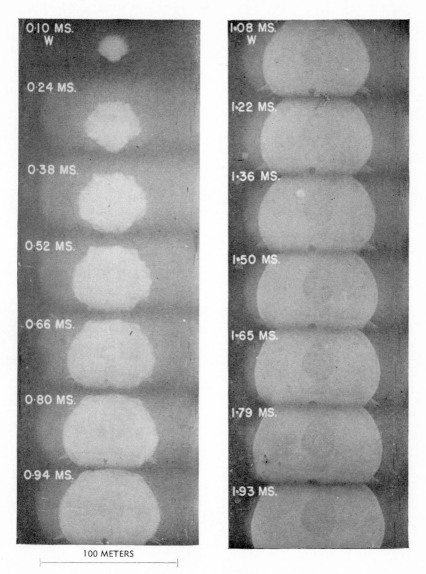

100 METERS

FIG. 59. Successive photographs of the fire ball from $t = 0 \cdot 1 \times 10^{-3}$, $t = 1 \cdot 93 \times 10^{-3}$ sec. in an atomic bomb explosion in New Mexico.

FIG. 60. Fire ball at $t = 15 \times 10^{-3}$.

FIG. 61. Photograph of the explosion at $t = 127 \times 10^{-3}$ sec.

The analysis of the family of integral curves in the z, V plane, given in §5, and the finite integrals established in §3, can be used to determine the field of disturbance due to an intense explosion.

The integral curves are similar for $\nu = 1, 2$ to the fields of the integral curves in the spherical case for $\nu = 3$. This field is mapped in Fig. 34.

The parameter λ can only approach infinity as the singular points O and F are approached during continuous motion along the integral

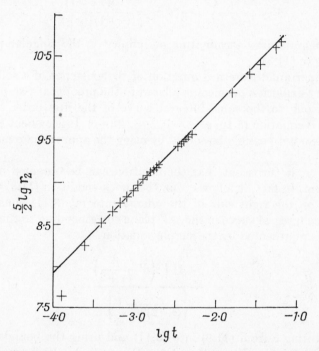

FIG. 62. Experimental results, shown by crosses, lie on a line inclined at 45° to the coordinate axes which is good confirmation of the theoretical formula $r_2 = (E/\rho_1)^{1/5} t^{2/5}$.

curve. Therefore, for continuous motion only the points O and F in the z, V plane can correspond to the points at infinity in the gas. The value $\lambda = 0$ corresponds to the centre of symmetry for $t \neq 0$. The parameter λ approaches O only along the integral curves $z = 0$ and $V = \pm \infty$ and when the singular point C is approached along the particular integral curve which originates at B. Finite values of the parameter $\lambda \neq 0$ correspond to the singular points B and D.

It is not difficult to see that a single integral curve terminating at the singular point C corresponding to the centre of symmetry gives the solution of the problem of an intense explosion.† However, this solution cannot possibly be extended continuously to $r = \infty$, so that continuous gas motion is impossible in a violent explosion. In order that this solution can be continued and joined with the undisturbed solution through the strong shock wave, it is necessary and sufficient that the point M with the coordinates

$$V = \frac{4}{(\gamma+1)(\nu+2)}, \quad z_2 = \frac{8\gamma(\gamma-1)}{(\gamma+1)^2(\nu+2)^2},$$

be an integral curve terminating at infinity at the singular point C, according to (11.8).

The important theoretical question of the existence of a solution of the violent explosion problem is related to the proof that two points M and C belong to the same integral curve of the first order ordinary differential equation (5.10) at $\omega = 0$, $\delta = 2/(2+\nu)$. It is evident that this question cannot possibly be solved by using the approximate numerical solution.

However, it turns out that the solution can be obtained in finite, closed form. In fact, it follows from the shock condition (11.8) that the constant on the right side of the energy integral (3.11) equals zero. Hence, the integral curve in the z, V plane corresponding to the desired solution is represented by the simple equation

$$z = \frac{(\gamma-1)\, V^2\left(V - \dfrac{2}{\nu+2}\right)}{2\left[\dfrac{2}{(\nu+2)\gamma} - V\right]}. \tag{11.9}$$

Substituting z from (11.9) into (5.11) and using the boundary condition $\lambda = 1$ for $V = V_2 = 4/[(\gamma+1)(\nu+2)]$, we find $\lambda(V)$ by means of a simple quadrature.

The function $R(V)$ is determined easily from the adiabatic integral and the functions $z(V)$ and $\lambda(V)$ can then be found. The constant in the adiabatic integral is determined by the shock conditions (11.8).

The integral curve (11.9) is the unique integral curve of (5.10) which terminates at the singular point $C(z = \infty,\ V = 2/[\gamma(\nu+2)])$. The variable λ decreases along this curve from the value unity at the shock wave to zero at the singular point C, where V has the finite value $2/[\gamma(\nu+2)]$. Hence, it follows that the gas velocity equals zero at the centre of

† A detailed analysis of this question is given in Sedov (1945a, b). See also Sedov (1946a).

symmetry $(v = (r/t) V)$, which is a natural mechanical condition for the continuation of the solution to the centre of symmetry.

It has thus been proved that the solution of the self-similar problem of a violent explosion exists and is unique.

This proof is related essentially to the fact that the values of z_2 and V_2 (11.8) behind the shock wave front, belong to the same single integral curve which passes through the singular point C; consequently, the solution can be continued to the centre of symmetry on all the adjacent integral curves.

Standard formulas, tables and graphs are given below which are valid for $p_1 = 0$ for any values of the initial density ρ_1 and explosive energy E_0.

We shall use the following definitions and relations which result from the definition (1.1) and from the conditions on the shock wave (11.2):

$$\left.\begin{array}{l} \dfrac{r}{r_2} = \lambda, \quad \dfrac{v}{v_2} = f = \dfrac{(\nu+2)(\gamma+1)}{4}\lambda V, \\[3mm] \dfrac{\rho}{\rho_2} = g = \dfrac{\gamma-1}{\gamma+1}R, \quad \dfrac{p}{p_2} = h = \dfrac{(\nu+2)^2(\gamma+1)}{8\gamma}\lambda^2 RZ, \end{array}\right\} \quad (11.10)$$

in which

$$\left.\begin{array}{l} v_2 = \dfrac{4}{(\nu+2)(\gamma+1)}\left(\dfrac{E}{\rho_1}\right)^{1/(2+\nu)}\dfrac{1}{t^{\nu/(\nu+2)}} = \dfrac{4}{(\nu+2)(\gamma+1)}\left(\dfrac{E}{\rho_1}\right)^{1/2}\dfrac{1}{r_2^{\nu/2}}, \\[4mm] \rho_2 = \dfrac{\gamma+1}{\gamma-1}\rho_1, \\[4mm] p_2 = \dfrac{8\rho_1}{(\nu+2)^2(\gamma+1)}\left(\dfrac{E}{\rho_1}\right)^{2/(2+\nu)}\dfrac{1}{t^{2\nu/(\nu+2)}} = \dfrac{8E}{(\nu+2)^2(\gamma+1)}\dfrac{1}{r_2^\nu}, \\[4mm] T_2 = \dfrac{\rho_2}{\mathbf{R}\rho_2}. \end{array}\right\} \quad (11.11)$$

Let us denote initial coordinates of the gas particles in a Lagrangian system by r_0. It is evident that the coordinate r_0 for each particle equals the radius of the shock wave r_2 at the moment it passes through the particle.

From the shock conditions and from the adiabatic condition behind the wave front, it follows that

$$\frac{p}{\rho^\gamma} = \frac{p_2}{\rho_2^\gamma}(r_0) = \frac{8(\gamma-1)^\gamma}{(\nu+2)^2(\gamma+1)^{\gamma+1}}\frac{E}{\rho_1^\gamma}\frac{1}{r_0^\nu}. \quad (11.12)$$

Hence, we obtain the formula

$$\left(\frac{r_0}{r_2}\right)^\nu = \frac{8\gamma(\gamma-1)^\gamma}{(\nu+2)(\gamma+1)^{\gamma+1}} \frac{R^{\gamma-1}}{\lambda^2 z}, \qquad (11.13)$$

which can be used to calculate the Lagrangian coordinates. According to (11.11), the following equalities are also true:

$$\frac{v_2}{v_{2_0}} = \left(\frac{r_2}{r_0}\right)^{-\nu/2}, \quad \frac{p_2}{p_{2_0}} = \left(\frac{r_2}{r_0}\right)^{-\nu}$$

where v_{2_0} and p_{2_0} are the velocity and pressure behind the shock wave front at the instant the shock passes through the point with the coordinate r_0.

The solution of the problem of a violent explosion is represented by using final formulas in parametric form with the variable parameter V. The range of variation of V and the character of the motion near the centre of symmetry depend on the relative location of the point M corresponding to the gas motion behind the shock front and on the singular point $\mathscr{E}$, which, as was explained in §5, belongs to the integral (11.9) and which appears in the $z > 0$ region for large γ (see Fig. 35).

According to (11.8) and (5.14), the condition that the values of V_2 at the point M and of V^* at the point $\mathscr{E}$ shall be equal, namely,

$$\frac{2}{2+\nu(\gamma-1)} = \frac{4}{(\gamma+1)(\nu+2)}$$

is attained for $\gamma = 7$ when $\nu = 3$; for $\gamma = \infty$ when $\nu = 2$ and for $\gamma = -1$ when $\nu = 1$.

Therefore, the range of variation of the variable V for $\nu = 1, 2$ and $\gamma > 1$ is determined by the inequality

$$\frac{2}{(\nu+2)\gamma} \leqslant V \leqslant \frac{4}{(\nu+2)(\gamma+1)}. \qquad (11.14a)$$

The value $V = 4/[(\nu+2)(\gamma+1)]$ in (11.14a) corresponds to the shock wave and the value $V = 2/[(\nu+2)\gamma]$ corresponds to the centre of the explosion.

If $\nu = 3$ and $\gamma < 7$, then (11.14a) is also true; the range of variation of V for $\nu = 3$ and $\gamma > 7$ is determined by the inequalities:

$$\frac{4}{5(\gamma+1)} \leqslant V \leqslant \frac{2}{5}. \qquad (11.14b)$$

The value $V = 2/5$ in (11.14b) corresponds to the boundary of the expanding vacuum and $V = 4/[5(\gamma+1)]$ corresponds to the shock wave. Hence, a vacuum is obtained only in the spherical case for finite large γ.

When the calculations described are carried out, the complete solution is given by the formulas

$$\frac{r}{r_2} = \left[\frac{(\nu+2)(\gamma+1)}{4}V\right]^{-2/(2+\nu)}\left[\frac{\gamma+1}{\gamma-1}\left(\frac{(\nu+2)\gamma}{2}V-1\right)\right]^{-\alpha_2}$$

$$\times\left[\frac{(\nu+2)(\gamma+1)}{(\nu+2)(\gamma+1)-2[2+\nu(\gamma-1)]}\left(1-\frac{2+\nu(\gamma-1)}{2}V\right)\right]^{-\alpha_1},$$

$$\frac{r_0}{r_2} = \left[\frac{(\nu+2)(\gamma+1)}{4}V\right]^{-2/(2+\nu)}\left[\frac{\gamma+1}{\gamma-1}\left(\frac{(\nu+2)\gamma}{2}V-1\right)\right]^{\alpha_6}$$

$$\times\left[\frac{(\nu+2)(\gamma+1)}{(\nu+2)(\gamma+1)-2[2+\nu(\gamma-1)]}\left(1-\frac{2+\nu(\gamma-1)}{2}V\right)\right]^{\alpha_7}$$

$$\times\left[\frac{\gamma+1}{\gamma-1}\left(1-\frac{\nu+2}{2}V\right)\right]^{-[\alpha_6+\alpha_7]-[2/(2+\nu)]},$$

$$\frac{v}{v_2} = f = \frac{(\nu+2)(\gamma+1)}{4}V\frac{r}{r_2},$$

$$\frac{\rho}{\rho_2} = g = \left[\frac{\gamma+1}{\gamma-1}\left(\frac{(\nu+2)\gamma}{2}V-1\right)\right]^{\alpha_3}\left[\frac{\gamma+1}{\gamma-1}\left(1-\frac{\nu+2}{2}V\right)\right]^{\alpha_5}$$

$$\times\left[\frac{(\nu+2)(\gamma+1)}{(2+\nu)(\gamma+1)-2[2+\nu(\gamma-1)]}\left(1-\frac{2+\nu(\gamma-1)}{2}V\right)\right]^{\alpha_4},$$

$$\frac{p}{p_2} = h = \left[\frac{(\nu+2)(\gamma+1)}{4}V\right]^{2\nu/(2+\nu)}\left[\frac{\gamma+1}{\gamma-1}\left(1-\frac{\nu+2}{2}V\right)\right]^{\alpha_5+1}$$

$$\times\left[\frac{(\nu+2)(\gamma+1)}{(\nu+2)(\gamma+1)-2[2+\nu(\gamma-1)]}\left(1-\frac{2+\nu(\gamma-1)}{2}V\right)\right]^{\alpha_4-2\alpha_1},$$

$$\frac{T}{T_2} = \frac{p}{p_2}\frac{\rho_2}{\rho},$$

$$\left.\begin{matrix}\end{matrix}\right\} \quad (11.15)$$

where

$$\alpha_1 = \frac{(\nu+2)\gamma}{2+\nu(\gamma-1)}\left[\frac{2\nu(2-\gamma)}{\gamma(\nu+2)^2}-\alpha_2\right], \quad \alpha_2 = \frac{1-\gamma}{2(\gamma-1)+\nu},$$

$$\alpha_3 = \frac{\nu}{2(\gamma-1)+\nu}, \quad \alpha_4 = \frac{\alpha_1(\nu+2)}{2-\gamma},$$

$$\alpha_5 = \frac{2}{\gamma-2}, \quad \alpha_6 = \frac{\gamma}{2(\gamma-1)+\nu},$$

$$\alpha_7 = \frac{[2+\nu(\gamma-1)]\alpha_1}{\nu(2-\gamma)}.$$

$$\left.\begin{matrix}\end{matrix}\right\} \quad (11.16)$$

It is easy to derive asymptotic formulas for v, ρ, p and the temperature T near the centre of the explosion from (11.15) for $\gamma < 7$ as $V \to 2/[(2+\nu)\gamma]$ and $r \to 0$. We find

$$
\left.
\begin{aligned}
v &= \frac{2}{(2+\nu)\gamma}\frac{r}{t}, \\[2ex]
\rho &= k_1\rho_1\left(\frac{E}{\rho_1}\right)^{-\nu/(\nu+2)(\gamma-1)} t^{-2\nu/(\nu+2)(\gamma-1)}\, r^{\nu/(\gamma-1)}, \\[2ex]
p &= k_2\rho_1\left(\frac{E}{\rho_1}\right)^{2/(\nu+2)} t^{-2\nu/(\nu+2)}, \\[2ex]
T &= k_3\frac{1}{c_v}\left(\frac{E}{\rho_1}\right)^{[2(\gamma-1)+\nu]/(\gamma-1)(\nu+2)} t^{[2\nu(2-\gamma)]/(\nu+2)(\gamma-1)}\, r^{-\nu/(\gamma-1)},
\end{aligned}
\right\} \quad (11.17)
$$

where for spherical symmetry,

$$
\left.
\begin{aligned}
k_1 &= \frac{(\gamma+1)^{[3\gamma(\gamma+1)]/(\gamma-1)(3\gamma-1)}}{2^{6/[5(\gamma-1)]}\gamma^{(7\gamma-1)/(\gamma-1)(3\gamma-1)}}(\gamma-1)^{-1} \\[1ex]
&\qquad \times \left(\frac{2\gamma+1}{7-\gamma}\right)^{(13\gamma^2-7\gamma+12)/[5(\gamma-1)(2-\gamma)(3\gamma-1)]}, \\[2ex]
k_2 &= \frac{0\cdot 32}{2^{6/5}}\frac{(\gamma+1)^{(\gamma+1)/(3\gamma-1)}}{\gamma^{(4\gamma)/(3\gamma-1)}}\left(\frac{2\gamma+1}{7-\gamma}\right)^{(13\gamma^2-7\gamma+12)/[5(2-\gamma)(3\gamma-1)]}, \\[2ex]
k_3 &= \frac{k_2}{k_1(\gamma-1)}
\end{aligned}
\right\} \quad (11.18)
$$

for cylindrical symmetry,

$$
\left.
\begin{aligned}
k_1 &= \frac{(\gamma+1)^{(\gamma+1)/(\gamma-1)}}{\gamma-1}\frac{\gamma^{(3\gamma-4)/(\gamma-1)(2-\gamma)}}{2^{2/(\gamma-1)(2-\gamma)}}, \\[2ex]
k_2 &= \frac{\gamma^{[2(\gamma-1)]/(2-\gamma)}}{2^{(4-\gamma)/(2-\gamma)}}, \quad k_3 = \frac{k_2}{k_1(\gamma-1)}
\end{aligned}
\right\} \quad (11.19)
$$

for plane waves,

$$
\left.
\begin{aligned}
k_1 &= \frac{(2\gamma-1)^{(5\gamma-4)/[3(\gamma-1)(2-\gamma)]}(\gamma+1)^{(4+\gamma-3\gamma^2)/[3(\gamma-1)(2-\gamma)]}}{2^{2/[3(\gamma-1)]}\gamma^{1/(\gamma-1)}(\gamma-1)}, \\[2ex]
k_2 &= \frac{2^{7/3}}{9}\frac{(2\gamma-1)^{(5\gamma-4)/[3(2-\gamma)]}}{(\gamma+1)^{[2(\gamma+1)]/[3(2-\gamma)]}}, \quad k_3 = \frac{k_2}{k_1(\gamma-1)}
\end{aligned}
\right\} \quad (11.20)
$$

The variation of the constants k_1, k_2 and k_3 with γ is shown in Fig. 63.

The velocity is close to zero near the centre of symmetry; the pressure is not zero and is asymptotically constant in the r coordinate; the density approaches zero very rapidly and the temperature approaches infinity. It is easy to see that the entropy also approaches infinity. Large temperature gradients occur near the centre of the explosion; consequently, the heat conduction property becomes very important. If heat conduction is taken into account, then the temperature is finite at the centre of the explosion.

The mass of gas disperses from the centre of the explosion (we have $\rho \doteq 0$ for $r = 0$), the pressure is finite at the centre but approaches zero as time passes.

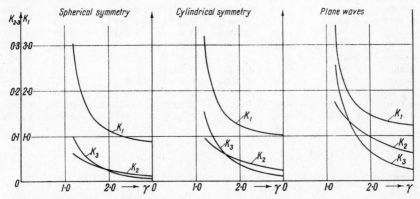

FIG. 63. The quantities $k_1(\gamma)$, $k_2(\gamma)$, $k_3(\gamma)$ in (11.17) which give asymptotic values of the density, pressure and temperature near the centre of the explosion.

Hence, it is clear that a reverse gas motion towards the centre of the explosion must occur in a violent explosion in a gas in which there is finite pressure prior to the explosion. We observe this effect well in such explosions in which repeated pulsations of the gas bubble arise.

An Archimedean lift, which causes a disturbance in the ambient atmosphere, results from the outward rush of gas from the centre and the reduction in pressure. During the atomic bomb explosion in New Mexico, the vertical lifting velocity of the luminous core was of the order of 35 m./sec. according to data on the photographs.

Graphs of the velocity, density, pressure and temperature distribution behind the wave front are given in Figs. 64–71: they define the motion and the changes of state in the gas in the three cases, $\nu = 1, 2, 3$.

Certain numerical values in practical applications are given in the tables.

$$\nu = 1, \quad \gamma = 1 \cdot 4$$

$\lambda = \dfrac{r}{r_2}$	$\dfrac{r_0}{r_2}$	$\dfrac{v}{v_2} = f$	$\dfrac{\rho}{\rho_2} = g$	$\dfrac{p}{p_2} = h$	$f'(\lambda)$	$g'(\lambda)$	$h'(\lambda)$
1	1	1	1	1	1·5	7·5	4·5
0·9797	0·8873	0·9699	0·8625	0·9162	1·4688	6·1575	3·8131
0·9420	0·7151	0·9156	0·6659	0·7915	1·4067	4·3615	2·8352
0·9013	0·5722	0·8599	0·5160	0·6923	1·3367	3·1109	2·0959
0·8565	0·4501	0·8017	0·3982	0·6120	1·2597	2·2183	1·5250
0·8050	0·3427	0·7390	0·3019	0·5457	1·1758	1·5638	1·0723
0·7419	0·2448	0·6678	0·2200	0·4904	1·0862	1·0738	0·7067
0·7029	0·1980	0·6263	0·1823	0·4661	1·0396	0·8726	0·5483
0·6553	0·1514	0·5780	0·1453	0·4437	0·9921	0·6919	0·4022
0·5925	0·1040	0·5172	0·1074	0·4229	0·9438	0·5232	0·2648
0·5396	0·0741	0·4682	0·0826	0·4116	0·9144	0·4213	0·1838
0·4912	0·0529	0·4244	0·0641	0·4038	0·8947	0·3473	0·1289
0·4589	0·0415	0·3957	0·0536	0·4001	0·8849	0·3020	0·1003
0·4161	0·0293	0·3580	0·0415	0·3964	0·8750	0·2551	0·0453
0·3480	0·0156	0·2988	0·0263	0·3929	0·8651	0·1905	0·0237
0·2810	0·0074	0·2410	0·0153	0·3911	0·8602	0·1370	0·0111
0·2320	0·0038	0·1989	0·0095	0·3905	0·8584	0·1025	0·0089
0·1680	0·0012	0·1441	0·0042	0·3901	0·8574	0·0630	0·0029
0·1040	0·0002	0·0891	0·0013	0·3900	0·8572	0·0307	0·0005
0·0000	0·0000	0·0000	0·0000	0·3900	0·8571	0·0000	0·0000

$$\nu = 3, \quad \gamma = 1 \cdot 4$$

$\lambda = \dfrac{r}{r_2}$	$\dfrac{r_0}{r_2}$	$\dfrac{v}{v_2} = f$	$\dfrac{\rho}{\rho_2} = g$	$\dfrac{p}{p_2} = h$	$f'(\lambda)$	$g'(\lambda)$	$h'(\lambda)$
1	1	1	1	1	2·1666	20·8333	11·1666
0·9913	0·9498	0·9814	0·8379	0·9109	2·0981	16·4159	9·2305
0·9773	0·8785	0·9529	0·6457	0·7993	1·9864	11·5610	6·9475
0·9622	0·8103	0·9237	0·4978	0·7078	1·8659	8·1607	5·2109
0·9342	0·7038	0·8744	0·3241	0·5923	1·6548	4·6349	3·2119
0·9080	0·6220	0·8335	0·2279	0·5241	1·4837	2·9536	2·1425
0·8747	0·5320	0·7872	0·1509	0·4674	1·3036	1·7841	1·3269
0·8359	0·4463	0·7397	0·0967	0·4272	1·1495	1·0716	0·7900
0·7950	0·3704	0·6952	0·0621	0·4021	1·0395	0·6690	0·4720
0·7493	0·2983	0·6496	0·0379	0·3856	0·9610	0·4090	0·2669
0·6788	0·2097	0·5844	0·0174	0·3732	0·8991	0·1981	0·1095
0·5794	0·1201	0·4971	0·0052	0·3672	0·8662	0·0679	0·0278
0·4560	0·0519	0·3909	0·0009	0·3656	0·8581	0·0142	0·0036
0·3600	0·0227	0·3086	0·0002	0·3655	0·8572	0·0030	0·0005
0·2960	0·0114	0·2538	0·0000	0·3655	0·8572	0·0008	0·0001
0·2000	0·0051	0·1714	0·0000	0·3655	0·8571	0·0001	0·0000
0·1040	0·0000	0·0892	0·0000	0·3655	0·8571	0·0000	0·0000
0·0000	0·0000	0·0000	0·0000	0·3655	0·8571	0·0000	0·0000

$$\nu = 2, \quad \gamma = 1\cdot 4$$

$\lambda = \dfrac{r}{r_2}$	$\dfrac{r_0}{r_2}$	$\dfrac{v}{v_2} = f$	$\dfrac{\rho}{\rho_2} = g$	$\dfrac{p}{p_2} = h$	$f'(\lambda)$	$g'(\lambda)$	$h'(\lambda)$
1	1	1	1	1	1·8333	14·1667	7·8333
0·9998	0·9954	0·9996	0·9973	0·9985	1·8325	14·1149	7·8102
0·9802	0·8911	0·9645	0·7653	0·8659	1·7516	9·8877	5·8478
0·9644	0·8164	0·9374	0·6285	0·7832	1·6848	7·5936	4·6993
0·9476	0·7460	0·9097	0·5164	0·7124	1·6132	5·8431	3·7665
0·9295	0·6789	0·8812	0·4234	0·6514	1·5371	4·4922	3·0009
0·9096	0·6140	0·8514	0·3451	0·5983	1·4569	3·4390	2·3668
0·8725	0·5115	0·7998	0·2427	0·5266	1·3206	2·1977	1·5608
0·8442	0·4461	0·7638	0·1892	0·4884	1·2313	1·6219	1·1574
0·8094	0·3695	0·7226	0·1414	0·4545	1·1394	1·1561	0·8136
0·7629	0·3009	0·6720	0·0975	0·4242	1·0455	0·7709	0·5170
0·7242	0·2480	0·6327	0·0718	0·4074	0·9885	0·5660	0·3567
0·6894	0·2071	0·5989	0·0545	0·3969	0·9503	0·4344	0·2551
0·6390	0·1577	0·5521	0·0362	0·3867	0·9120	0·2998	0·1556
0·5745	0·1081	0·4943	0·0208	0·3794	0·8832	0·1863	0·0798
0·5180	0·0748	0·4448	0·0123	0·3760	0·8698	0·1202	0·0423
0·4748	0·0552	0·4073	0·0079	0·3746	0·8640	0·0839	0·0249
0·4222	0·0370	0·3621	0·0044	0·3737	0·8602	0·0522	0·0123
0·3654	0·0220	0·3133	0·0021	0·3733	0·8583	0·0292	0·0052
0·3000	0·0111	0·2571	0·0008	0·3730	0·8574	0·0132	0·0016
0·2500	0·0058	0·2143	0·0003	0·3729	0·8572	0·0064	0·0005
0·2000	0·0027	0·1714	0·0001	0·3729	0·8572	0·0026	0·0001
0·1500	0·0010	0·1286	0·0000	0·3729	0·8571	0·0008	0·0000
0·1000	0·0002	0·0857	0·0000	0·3729	0·8571	0·0002	0·0000
0·0000	0·0000	0·0000	0·0000	0·3729	0·8571	0·0000	0·0000

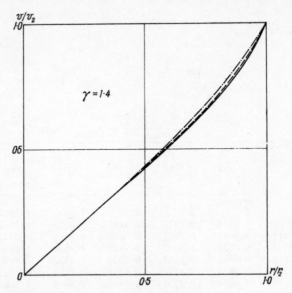

FIG. 64. Velocity distribution behind the shock wave ——— spherical case; ----- cylindrical case; —·—·—· plane case.

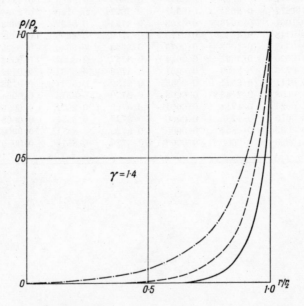

FIG. 65. Density distribution behind the shock wave.

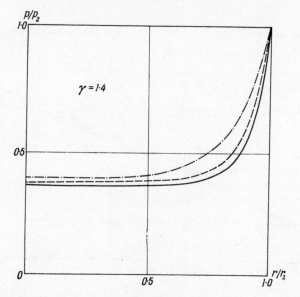

FIG. 66. Pressure distribution behind the shock wave ———— spherical case;
– – – – – cylindrical case; –·–·–·– plane case.

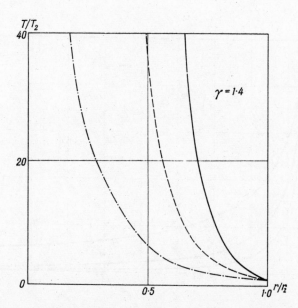

FIG. 67. Temperature distribution behind the shock wave.

8

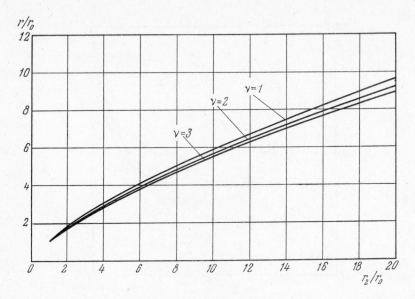

FIG. 68. Law of motion of a gas particle ($\gamma = 1\cdot4$).

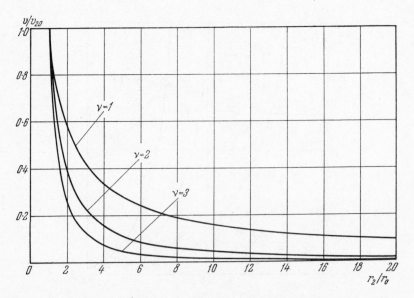

FIG. 69. Particle velocity as a function of shock wave position ($\gamma = 1\cdot4$).

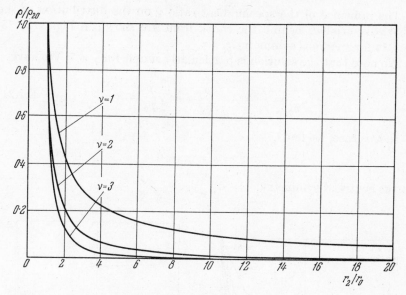

FIG. 70. Particle density as a function of shock wave position ($\gamma = 1 \cdot 4$).

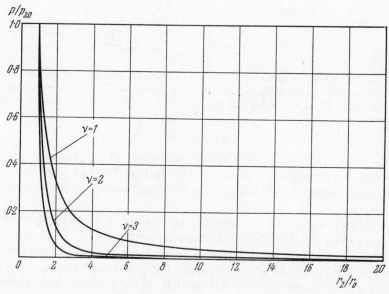

FIG. 71. Particle pressure as a function of shock wave position ($\gamma = 1 \cdot 4$).

The influence of the specific heat ratio γ on the distribution of the physical variables behind the shock front are shown in Figs. 72, 73, and 74 for spherical symmetry.

We note that the solution is particularly simple for $\gamma = 7$. We have

$$\frac{v}{v_2} = \frac{r}{r_2} = \lambda; \quad \frac{\rho}{\rho_2} = \lambda; \quad \frac{p}{p_2} = \lambda^3. \qquad (11.21)$$

In this case, we find

$$v = \rho = p = 0$$

at the centre of symmetry.

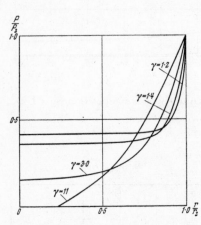

FIG. 72. Influence of the constant γ on the pressure distribution behind the shock wave front for spherical symmetry.

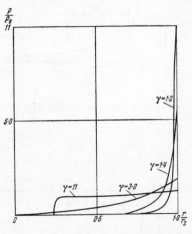

FIG. 73. Influence of the constant γ on the density distribution behind the shock wave front for spherical symmetry.

If $\gamma > 7$, then an empty sphere of radius r^* is produced at the centre; the pressure is zero within this sphere. The radius r^* increases with time so that the ratio r^*/r_2, which depends only on γ, remains constant.

The high values of the index γ arise when a violent explosion is initiated in a compressible medium such as water.

The problem of a point explosion formulated above is solved for any constant value of the abstract parameter γ. If $\gamma = \gamma^* = c_p/c_v$, then the process is adiabatic for each gas particle. If the constant γ is arbitrary and not equal to $\gamma^* = c_p/c_v$, then the process will be polytropic; in this case the jump front is similar to a detonation front or to a front of phase transition with heat absorption.

Actually, condition (2.6) can be rewritten as:

$$\frac{\gamma^* p_1}{(\gamma^*-1)\rho_1} + \tfrac{1}{2}(v_1-c)^2 + Q = \frac{\gamma^* p_2}{(\gamma^*-1)\rho_2} + \tfrac{1}{2}(v_2-c)^2, \qquad (11.22)$$

where

$$Q = \frac{\gamma-\gamma^*}{(\gamma-1)(\gamma^*-1)}\left(\frac{p_2}{\rho_2} - \frac{p_1}{\rho_1}\right);$$

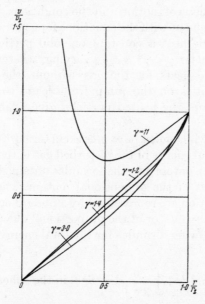

FIG. 74. Influence of the constant γ on the velocity distribution behind the shock wave front for spherical symmetry.

Since $p_2/\rho_2 = \mathbf{R}T_2 > \mathbf{R}T_1 = p_1/\rho_1$, then evidently, condition (11.22) is analogous to the condition on a detonation front for $\gamma > \gamma^* = c_p/c_v > 1$ when $Q > 0$. Heat absorption ($Q < 0$) will occur on the front if $\gamma > 1$ and $\gamma < \gamma^*$.

On the other hand, the following equation for a reversible process holds on each particle in the flow behind the jump front

$$dQ = T\,dS = Tc_v\,d\ln\frac{p}{\rho^{\gamma^*}} = Tc_v\,d\ln\frac{p}{\rho^\gamma}\rho^{\gamma-\gamma^*};$$

from which, since $p/\rho^\gamma = \text{const}$ and $p = \mathbf{R}\rho T$ on each particle, we obtain:

$$dQ = Tc_v(\gamma - \gamma^*)\frac{d\rho}{\rho} = \frac{\gamma - \gamma^*}{\gamma - 1}c_v\,dT. \qquad (11.23)$$

We would have $d\rho < 0$ and $dT < 0$ on fixed particles in the solutions considered; consequently, we have a process with heat release per particle $dQ > 0$ for $\gamma < \gamma^*$ and with heat absorption $dQ < 0$ for $\gamma > \gamma^*$.

The motion of a gas containing very fine solid or liquid particles (dusty atmosphere) can be considered. It can be shown (Sidorkina, 1957) that the adiabatic equations of motion and the conditions at jumps in such a mixture are in agreement with the equations for a gas without impurities in which the specific heat is constant on solid particles, but with the value of the parameter $\gamma < \gamma^* = c_p/c_v$ varying, where c_p and c_v are the appropriate specific heats for the gas without the impurities. The presence of impurities on the jump front leads to heat absorption $Q < 0$; in the stream, the heated particles give off heat to the gas so that $dQ > 0$.

The constant E, which must be expressed in terms of the charge energy E_0 (equal to the total energy of the disturbed gas in the present formulation of the problem), enters into the formulas obtained above giving the dependent variables of the dimensional motion. The nondimensional variables of the motion are represented by standard curves independent of the explosion energy E_0 or of the proportional quantity E.

We have the following formulas for the total energy:

$$E_0 = \int_0^{r_2} \frac{\rho v^2}{2} 4\pi r^2\,dr + \int_0^{r_2} \frac{p}{\gamma - 1} 4\pi r^2\,dr \quad \text{in the spherical case;}$$

$$E_0 = \int_0^{r_2} \frac{\rho v^2}{2} 2\pi r\,dr + \int_0^{r_2} \frac{p}{\gamma - 1} 2\pi r\,dr \quad \text{in the cylindrical case;}$$

$$\tfrac{1}{2}E_0 = \int_0^{r_2} \frac{\rho v^2}{2}\,dr + \int_0^{r_2} \frac{p}{\gamma - 1}\,dr \quad \text{for plane waves.}$$

The first term is the kinetic and the second is the thermal gas energy. Introducing nondimensional quantities, we find

$$E_0 = \alpha(\gamma)\,E, \qquad (11.24)$$

where

$$\alpha = 2\pi \int_0^1 R V^2 \lambda^4 \, d\lambda + \frac{4\pi}{(\gamma-1)} \int_0^1 P\lambda^4 \, d\lambda \quad \text{in the spherical case;}$$

$$\alpha = \pi \int_0^1 R V^2 \lambda^3 \, d\lambda + \frac{2\pi}{(\gamma-1)} \int_0^1 P\lambda^3 \, d\lambda \quad \text{in the cylindrical case;}$$

$$\alpha = 1 \int_0^1 R V^2 \lambda^2 \, d\lambda + \frac{2}{(\gamma-1)} \int_0^1 P\lambda^2 \, d\lambda \quad \text{for plane waves.}$$

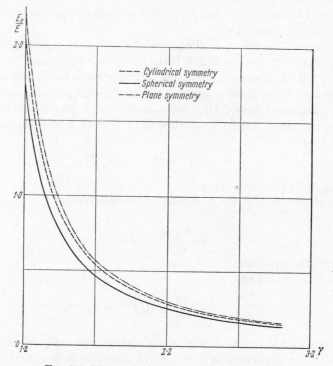

FIG. 75. The ratio $E_0/E = \alpha$ as a function of γ.

The function $\alpha(\gamma)$ is shown in Fig. 75, calculated for spherical, cylindrical and planar symmetry.

In particular, we have for $\gamma = 1\cdot4$, in the case of spherical symmetry.

$$E_0 = (0\cdot186 + 0\cdot665) E = 0\cdot851E \quad \text{or} \quad E = 1\cdot175E_0.$$

If we use the value of E given in (11.7) which was obtained from experimental results for the atomic bomb explosion in New Mexico, and if we take $\gamma = 1 \cdot 4$, then we obtain $E_0 = 7 \cdot 19 \times 10^{20}$ erg for the energy of the explosion,† which corresponds to the energy liberated by exploding 16,800 tons of TNT.

2. ON TAKING HEAT CONDUCTION INTO ACCOUNT

We noted above that the properties of viscosity and heat conduction can exert a definite influence on the gas motion near the centre of an explosion; if viscosity and heat conduction are not neglected, then the coefficient of viscosity μ and the coefficient of heat conduction κ enter into the equations of motion.

It should be noted that the coefficient of heat conduction enters as a factor of the temperature T.

When the temperature is eliminated by using the equation of state (for simplicity, we assume that it is the Clapeyron equation $T = p/[\mathbf{R}\rho]$), the ratio $\kappa/\mathbf{R}$ ($\mathbf{R}$ is the gas constant), rather than κ itself, will actually enter as a coefficient. As is known, the coefficients μ and κ depend on the temperature. Usually, they are assumed to be proportional to T to some power (most often, proportional to $T^{1/2}$ or a constant). Let us assume that

$$\mu = \mu_1 T^\alpha \quad \text{and} \quad \kappa = \kappa_1 T^\alpha$$

where μ_1 and κ_1 are constants.

After eliminating T, the new dimensional constants

$$\frac{\mu_1}{R^\alpha} \quad \text{and} \quad \frac{\kappa_1}{R^{\alpha+1}}$$

enter into the equations.

Their dimensions are

$$\left[\frac{\mu_1}{R^\alpha}\right] = \left[\frac{\kappa_1}{R^{\alpha+1}}\right] = M L^{-1-2\alpha} T^{-1+2\alpha}.$$

In order that the motion may be self-similar, it is sufficient that these dimensions be expressed in terms of the dimensions of ρ_1 and E_0. It is easy to see that it is necessary to put $\alpha = 1/6$ in the spherical case,‡ $\alpha = 0$ in the cylindrical case and $\alpha = -1/2$ in the plane case.

Hence, the problem of an intense explosion can be solved by integration

† In estimating the total energy liberated in an atomic bomb explosion, it should be kept in mind that a considerable part of the energy is expended in radiation.

‡ This result is obtained in Bam-Zelikovich (1949). We have $[\mu_1/R^\alpha] = [(E\rho_1^2)^{1/3}]$ for $\alpha = 1/6$.

of ordinary differential equations taking viscosity and heat conduction into account, if it is assumed that

$$\mu = \mu_1 T^{1/6}, \quad \kappa = \kappa_1 T^{1/6} \quad \text{for spherical waves;} \tag{11.25}$$

$$\mu = \text{const}, \quad \kappa = \text{const} \quad \text{for cylindrical waves;} \tag{11.26}$$

$$\mu = \frac{\mu_1}{\sqrt{T}}, \quad \kappa = \frac{\kappa_1}{\sqrt{T}} \quad \text{for plane waves.} \tag{11.27}$$

Approximate solutions of the problem of an intense point explosion taking heat conduction into account have been published recently (Korobeinikov, 1957).

It is not difficult to see that self-similarity of the problem of an intense explosion is retained for a perfect gas and for many other media in which the temperature is uniform in space, but varies with time in the disturbed region as a result of the intense heat exchange (due to very large heat conduction or to radiation and other processes).

These conditions apply when the adiabatic equation can be replaced by

$$\frac{\partial T}{\partial r} = 0.$$

The solution of the appropriate problems for spherical waves has been published by Korobeinikov (1956a).

3. SELF-SIMILARITY OF A POINT EXPLOSION IN IDEAL MEDIA

A further remark on solving the problem of an intense explosion by ideal fluid theory, with a more general kind of equation of state and an expression of the internal gas energy as a function of p and ρ, is appropriate.† The internal energy function $\epsilon(p, \rho)$ enters directly into the shock wave conditions and into the heat flow equation. In the general case, it can always be represented as

$$\epsilon = \frac{p}{\rho_1} \phi\left(\frac{p}{p^*}, \frac{\rho}{\rho_1}\right) + \text{const},$$

where ϕ is a nondimensional function of its arguments and p^* is any constant with the dimensions of pressure.

Since the dimensions of p^* cannot be expressed in terms of the dimensions of ρ_1 and E, then it is sufficient, for the motion to be self-similar, that ϵ should not contain p^*, i.e.

$$\epsilon = \frac{p}{\rho_1} \phi\left(\frac{\rho}{\rho_1}\right), \tag{11.28}$$

where ϕ is an arbitrary function of its argument.

† See Bam-Zelikovich (1949).

8*

No new physical dimensional constants can appear in the adiabatic equation

$$d\epsilon + pd\frac{1}{\rho} = 0.$$

Condition (11.28) imposes a certain limitation on the equation of state. Actually, since

$$\frac{d\epsilon + pd\dfrac{1}{\rho}}{T} = dS$$

is a total differential, T and ϵ must satisfy the following partial differential equation

$$T + \frac{\partial T}{\partial p}\left(\rho^2\frac{\partial\epsilon}{\partial\rho} - p\right) - \rho^2\frac{\partial T}{\partial\rho}\frac{\partial\epsilon}{\partial p} = 0.$$

Substituting $\epsilon = p\phi(\rho)$ (for simplicity, we can put $\rho_1 = 1$), we obtain:

$$T + \frac{\partial T}{\partial p}p(\phi'(\rho)\rho^2 - 1) - \rho^2\phi(\rho)\frac{\partial T}{\partial\rho} = 0.$$

An equivalent system of ordinary differential equations can be written:

$$\frac{dT}{T} = -\frac{dp}{p[\phi'(\rho)\rho^2 - 1]} = \frac{d\rho}{\phi(\rho)\rho^2}.$$

It has two integrals

$$T\exp\left(-\int\frac{d\rho}{\rho^2\phi(\rho)}\right) = C_1 \quad\text{and}\quad p\phi(\rho)\exp\left(-\int\frac{d\rho}{\rho^2\phi(\rho)}\right) = C_2,$$

therefore,

$$p\phi(\rho)\exp\left(-\int\frac{d\rho}{\rho^2\phi(\rho)}\right) = \Phi\left[T\exp\left(-\int\frac{d\rho}{\rho^2\phi(\rho)}\right)\right], \qquad (11.29)$$

where ϕ and Φ are arbitrary functions of their arguments. Equation (11.29) is satisfied by all equations of state in which the temperature is proportional to the pressure and depends on the density arbitrarily. However, in spite of the presence of two arbitrary functions, many interesting equations of state (for example, the van der Waal) cannot be written in the form (11.29).

If the function Φ reduces to a constant, the pressure and, therefore, the internal energy depend only on the density; a corresponding dimensional constant, equal to Φ, enters into the internal energy expression. In this case, the system depends on one parameter; consequently, the preceding formulation of the problem becomes impossible.

The function Φ in (11.29) in the general case depend on a number of dimensional constants which cannot violate the self-similarity of the problem formulated in terms of v, p, ρ in §1. However, the presence of these constants can lead to a non-self-similar relation between the temperature T and r and t in the disturbed flow of the medium.

We note that (11.28) gives

$$p\phi(\rho)\exp\left(-\int\frac{d\rho}{\rho^2\phi(\rho)}\right) = \psi(S), \quad T\exp\left(-\int\frac{d\rho}{\rho^2\phi(\rho)}\right) = \psi'(S),$$

where $\psi(S)$ is a function of the entropy which depends on the type of function Φ in (11.29) and which satisfies the equation

$$\psi(S) = \Phi(\psi'(S)).$$

4. POINT EXPLOSION IN AN IDEAL INCOMPRESSIBLE FLUID

The problem of the point explosion can be analysed under the assumption that the medium is incompressible. The adiabatic equation can be replaced by

$$\rho = \rho_1 = \text{const.} \tag{11.30}$$

The disturbance is propagated with an infinitely high velocity in this case; consequently, a solution without a shock wave is possible. The case of incompressible fluid motion during a point explosion can be obtained as the limit of adiabatic gas motion as $\gamma \to +\infty$. As in the general case, if $p_1 = 0$, the fluid motion is self-similar and, as is easily verified, the following formula is true for the velocity field:

$$v = \frac{2}{5}\left(\frac{E}{2\pi\rho_1}\right)^{3/5}\frac{t^{1/5}}{r^2}, \tag{11.31}$$

since the motion must correspond to a source of variable intensity dependent only on E, ρ_1, t. The constant factor is determined by assigning the fluid kinetic energy which equals $4E/25$. It follows from the Lagrange integral that

$$\frac{1}{\rho_1}p = \frac{2}{25}\left(\frac{E}{2\pi\rho_1}\right)^{2/5}t^{-6/5}\frac{r^*}{r}\left[1-\left(\frac{r^*}{r}\right)^3\right]. \tag{11.32}$$

A spherical vacuum with increasing radius r^* is formed at the centre, for which we find

$$r^* = \left(\frac{E}{2\pi\rho_1}\right)^{1/5}t^{2/5}. \tag{11.33}$$

The pressure is zero inside a sphere of radius r^*.

Curve 1 on Fig. 76 gives the pressure distribution in the fluid which is a universal curve in the variables

$$\frac{pt^{6/5}}{\rho_1\left(\dfrac{E}{2\pi\rho_1}\right)^{2/5}}; \quad \frac{r}{r^*}.$$

The peak pressure drops in inverse proportion to $t^{6/5}$, while the peak pressure is reached independently of the time at $r^*/(r_{p_{\max}}) = 4^{-1/3}$, from which

$$r_{p_{\max}} = \sqrt[3]{(4)}\, r^* = 4^{1/3}\left(\frac{E}{2\pi\rho_1}\right)^{1/5} t^{2/5}. \qquad (11.34)$$

Only the ratio E/ρ_1 is essential in the motion found.

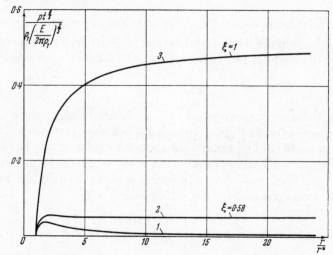

FIG. 76. Pressure distribution in an incompressible fluid for a point explosion. 1, self-similar solution; 2, solution taking counter-pressure into account for small $\xi = r^*/r^*_{\max}$; 3, pressure when the internal cavity has the maximum radius.

If $p_1 \neq 0$, the motion of an incompressible fluid is not self-similar, just as for a gas. However, the complete solution is easily obtained in simple analytic form in this case.

The following formulas can be written for the velocity potential and for the magnitude of the velocity:

$$\phi = -\frac{r^{*2}\dfrac{dr^*}{dt}}{r} \quad \text{and} \quad v = \frac{r^{*2}}{r^2}\frac{dr^*}{dt}, \qquad (11.35)$$

which are true for any law $r^*(t)$ according to which the internal cavity expands.

We obtain from the Lagrange integral, if the internal pressure is $p^* = 0$ and the external pressure at infinity is $p_1 = \text{const}$:

$$-\frac{p_1}{\rho} = r^* \frac{d^2 r^*}{dt^2} + \frac{3}{2}\left(\frac{dr^*}{dt}\right)^2. \tag{11.36}$$

Hence

$$\left(\frac{dr^*}{dt}\right)^2 = -\frac{2}{3}\frac{p_1}{\rho_1} + Cr^{*-3}. \tag{11.37}$$

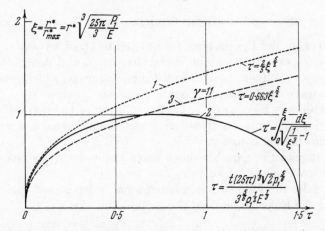

FIG. 77. Expansion law of the internal cavity; 1, self-similar solution; 2, incompressible fluid motion taking counter-pressure into account; 3, compressible medium with $\gamma = 11$.

We find from the solution of the self-similar problem using (11.33):

$$C = \frac{4}{25}\left(\frac{E}{2\pi\rho_1}\right).$$

If the following notation is introduced

$$\tau = t\frac{(25\pi)^{1/3}\sqrt{(2)}\,p_1^{5/6}}{3^{5/6}\rho_1^{1/2}E^{1/3}}, \tag{11.38}$$

$$\xi = \frac{r^*}{r^*_{\max}}, \quad r^*_{\max} = \left(\frac{3}{25\pi}\frac{E}{p_1}\right)^{1/3},$$

then the rate of expansion of the internal cavity is given by the universal formula

$$\tau = \int_0^\xi \frac{d\xi}{\sqrt{\left(\frac{1}{\xi^3} - 1\right)}}. \tag{11.39}$$

It is evident that r_{max} will equal the greatest radius of the internal cavity.

The expansion law given by (11.39) is shown on the graph (Fig. 77). The expansion law for self-similar motion in the same variables is given by

$$\tau = \tfrac{2}{5} \zeta^{5/2}; \tag{11.40}$$

corresponding to the dashed curve on Fig. 77. Also superposed on Fig. 77 is the curve for the self-similar solution for a gas with $\gamma = 11$.

§12. Point Explosion taking Counter-Pressure into Account

Now let us consider the problem formulated in §1, of a point explosion in a perfect gas with constant initial density ρ_1 and constant initial pressure $p_1 \neq 0$. Self-similarity is lost if the pressure p_1, which enters into the shock wave conditions, is taken into account.

As was already mentioned in §1, the solution of the non-self-similar problem depends only on two independent nondimensional variables and on one abstract constant γ.

Let us denote the variable shock wave radius by r_2, as before, and the shock wave velocity by $c = dr_2/dt$.

Let us take the following nondimensional independent variables in describing the Eulerian motion:

$$\lambda = \frac{r}{r_2} \quad \text{and} \quad q = \frac{a_1^2}{c^2} \tag{12.1}$$

where $a_1^2 = \gamma p_1/\rho_1$; a_1 is the speed of sound in the undisturbed gas. The ranges of variation of q and λ are determined by the inequalities:

$$0 \leqslant \lambda \leqslant 1; \quad 0 \leqslant q \leqslant 1.$$

One of the following nondimensional variables can be used instead of the variable q:

$$\tau = \frac{t}{t^0} \quad \text{where} \quad t^0 = E_0^{1/\nu} \sqrt{(\rho_1)}\, p_1^{-(\nu+2)/2\nu} \tag{12.2}$$

and τ is nondimensional time or

$$l = \frac{r_2}{r^0} \quad \text{where} \quad r^0 = \left(\frac{E_0}{p_1}\right)^{1/\nu} \tag{12.3}$$

l is the nondimensional shock wave coordinate; t^0 is a characteristic time; r^0 is a characteristic length.

Graphs are given in Fig. 78 to determine r^0 in metres and t^0 in seconds in the spherical case, depending on the explosive energy and on the

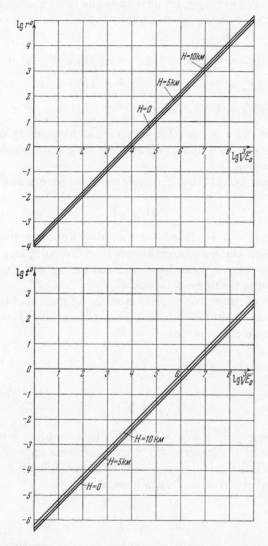

FIG. 78. Scale constants: $r^0 = (E_0/p_1)^{1/3}$ is a characteristic linear dimension in metres; $t^0 = E_0^{1/3}\rho_1^{1/2}p_1^{-5/6}$ is a characteristic time in seconds; H is the height of a standard atmosphere; E_0 is the explosive energy in ergs.

height of a standard atmosphere which determine the pressure p_1 and the density ρ_1. For example, let us take as the basic energy the E_0 calculated on p. 230. To clarify ideas we take the following numerical values at a height $H = 5$ km.

$$1/10{,}000 \; E_0 = 7 \cdot 19 \times 10^{16} \text{ erg}; \; r^0 \simeq 51 \text{ m.}, \; t^0 \simeq 0 \cdot 002 \text{ sec.};$$

$$E_0 = 7 \cdot 19 \times 10^{20} \text{ erg}; \; r^0 \simeq 1100 \text{ m.}, \; t^0 \simeq 4 \cdot 1 \text{ sec.};$$

$$1000 \; E_0 = 7 \cdot 19 \times 20^{23} \text{ erg}; \; r^0 \simeq 11{,}000 \text{ m.}, \; t^0 \simeq 41 \text{ sec.}$$

We have $q = 0$, $l = 0$, $\tau = 0$ for $p_1 = 0$ for an intense explosion.

We have $\tau \to 0$, $l \to 0$ and $q \to 0$ as $t \to 0$ for a point explosion taking $p_1 \neq 0$ into account.

Later, we can assume that if nondimensional functions of the motion of the type

$$f(\lambda, q, \gamma)$$

remain finite as $q \to 0$, $\tau \to 0$ or $l \to 0$, then they approach the appropriate functions for self-similar motion known from the solution given in §11.

Finite values of q, τ, l can be considered for the self-similar solution independent of p_1 for finite p_1, ρ_1 and E_0.

Since the shock wave velocity is known as a function of the time t and of the coordinate r_2 in self-similar motions (formulas (11.4), (11.5) and (11.6)) we obtain the relations

$$\left.\begin{aligned} q &= \gamma \left(\frac{2+\nu}{2}\right)^2 \alpha^{2/(2+\nu)} \tau^{2\nu/(2+\nu)} \\ q &= \gamma \left(\frac{2+\nu}{2}\right)^2 \alpha l^\nu. \end{aligned}\right\} \qquad (12.4)$$

The functions $q(\tau)$ or $q(l)$ in non-self-similar motion are determined by the motion of the shock wave; these functions approach the corresponding function in (12.4) as $q \to 0$. It is not difficult to see that for constant sound speed $a_1 = \sqrt{(\gamma p_1/\rho_1)}$, we have

$$\frac{dl}{dq} = \sqrt{\left(\frac{\gamma}{q}\right)} \frac{d\tau}{dq}. \qquad (12.5)$$

As before, we denote the values of the velocity, density and pressure behind the explosion wave front by v_2, ρ_2 and p_2. Then from the shock wave conditions (11.1), we obtain

$$v_2 = \frac{2a_1}{\gamma+1} \frac{1-q}{\sqrt{q}}; \quad \rho_2 = \frac{\rho_1(\gamma+1)}{\gamma-1+2q}; \quad p_2 = \frac{p_1}{\gamma+1} \frac{2\gamma-(\gamma-1)q}{q}. \qquad (12.6)$$

These define all the physical variables at the shock wave front as explicit functions of q. The relations between v_2, ρ_2, p_2 and τ or l are not known in advance; finding these relations is equivalent to finding the function $q(\tau)$ or the function $q(l)$.

If we take as unknown functions

$$f(\lambda, q) = \frac{v}{v_2}; \quad q(\lambda, q) = \frac{\rho}{\rho_2}; \quad h(\lambda, q) = \frac{p}{p_2} \qquad (12.7)$$

then the exact conditions at the shock wave are represented in the simple form

$$\lambda = 1 \quad f(1, q) = g(1, q) = h(1, q) = 1. \qquad (12.8a)$$

The conditions that the velocity becomes zero at the centre of symmetry give

$$f(0, q) = 0. \qquad (12.8b)$$

The initial conditions in the problem of a point explosion can be written as

$$f(\lambda, 0) = f_0(\lambda); \quad g(\lambda, 0) = g_0(\lambda); \quad h(\lambda, 0) = h_0(\lambda), \qquad (12.9)$$

where $f_0(\lambda)$, $g_0(\lambda)$ and $h_0(\lambda)$ are the corresponding functions for self-similar motion which we determined and discussed in §11.

The system of equations for one-dimensional unsteady adiabatic gas motion (1.3) in these variables is transformed into

$$\left[\frac{2(1-q)}{\gamma+1}f - \lambda\right](1-q)\frac{\partial f}{\partial \lambda}$$
$$+ \frac{[\gamma+1+(\gamma-1)(1-q)][\gamma+1-2(1-q)]}{2\gamma(\gamma+1)} \cdot \frac{1}{g}\frac{\partial h}{\partial \lambda}$$
$$+ \left[(1-q)\frac{\partial f}{\partial q} - \frac{2-(1-q)}{2q}f\right]r_2\frac{dq}{dr_2} = 0,$$

$$\frac{2(1-q)}{\gamma+1}\frac{\partial f}{\partial \lambda} + \left[\frac{2(1-q)}{\gamma+1}f - \lambda\right]\frac{1}{g}\frac{\partial g}{\partial \lambda} + \frac{2(1-q)}{\gamma+1}\cdot\frac{(\nu-1)}{\lambda}f$$
$$+ \left[\frac{1}{g}\frac{\partial g}{\partial q} - \frac{2}{\gamma+1-2(1-q)}\right]r_2\frac{dq}{dr_2} = 0,$$

$$\left[\frac{2(1-q)}{\gamma+1}f - \lambda\right]\frac{1}{h}\frac{\partial h}{\partial \lambda} - \left[\frac{2(1-q)}{\gamma+1}f - \lambda\right]\frac{\gamma}{g}\frac{\partial g}{\partial \lambda}$$
$$+ \left[\frac{1}{h}\frac{\partial h}{\partial q} - \frac{\gamma}{g}\frac{\partial g}{\partial q} - \frac{2\gamma(\gamma-1)(1-q)^2}{[\gamma+1-2(1-q)][\gamma+1+(\gamma-1)(1-q)]\cdot q}\right]$$
$$\times r_2\frac{dq}{dr_2} = 0.$$

$$\left.\rule{0pt}{40pt}\right\} \quad (12.10)$$

Equation (12.10) and conditions (12.8) and (12.9) can be satisfied by putting:

$$\left.\begin{aligned}
f(\lambda, q) &= f_0(\lambda) + q f_1(\lambda) + \dots, \\
g(\lambda, q) &= g_0(\lambda) + q g_1(\lambda) + \dots, \\
h(\lambda, q) &= h_0(\lambda) + q h_1(\lambda) + \dots, \\
l^\nu &= \left(\frac{2}{2+\nu}\right)^2 \frac{q}{\gamma^\alpha(\gamma)} e^{Aq +} \dots
\end{aligned}\right\} \quad (12.11)$$

Using (12.10) and boundary conditions (12.8a) and (12.8b), the functions $f_1(\lambda)$, $g_1(\lambda)$, $h_1(\lambda)$ and the value of A can be determined.†

The boundary conditions yield:

$$f_1(0) = 0; \quad f_1(1) = 0; \quad g_1(1) = 0; \quad h_1(1) = 0. \quad (12.12)$$

Equations (12.4) reduce to the following system of linear ordinary differential equations containing the unknown constant A:

$$\left.\begin{aligned}
&\left[\frac{2}{\gamma+1}f_0 - \lambda\right] g_0 \frac{df_1}{d\lambda} + \frac{\gamma-1}{\gamma+1}\frac{dh_1}{d\lambda} + \left[\frac{2}{\gamma+1}f_0' + \frac{\nu}{2}\right]g_0 f_1 \\
&+ \left[\frac{2}{\gamma+1}f_0 f_0' - \lambda f_0' - \frac{\nu}{2}f_0\right]g_1 \\
&+ \left[\frac{\gamma^2 + 4\gamma - 1}{2\gamma(\gamma+1)}h_0' - \nu f_0 g_0 - \frac{2}{\gamma+1}f_0 f_0' g_0 + \frac{\nu}{2}f_0 g_0 A\right] = 0, \\[2mm]
&\frac{2g_0}{\gamma+1}\frac{df_1}{d\lambda} + \left[\frac{2}{\gamma+1}f_0 - \lambda\right]\frac{dg_1}{d\lambda} + \frac{2}{\gamma+1}\left[\frac{\nu-1}{\lambda}g_0 + g_0'\right]f_1 \\
&+ \left[\frac{2}{\gamma+1}\cdot\frac{\nu-1}{\lambda}\cdot f_0 + \frac{2}{\gamma+1}f_0' + \nu\right]g_1 - \lambda g_0' - \frac{2}{\gamma-1}\nu g_0 = 0, \\[2mm]
&\left[\frac{2f_0}{\gamma+1} - \lambda\right]g_0 \frac{dh_1}{d\lambda} - \left[\frac{2f_0}{\gamma+1} - \lambda\right]\gamma h_0 \frac{dg_1}{d\lambda} + \frac{2}{\gamma+1}[g_0 h_0' - \gamma h_0 g_0']f_1 \\
&+ \left[\frac{2f_0}{\gamma+1} - \lambda\right]h_0' g_1 - \nu(\gamma+1)h_0 g_1 - \gamma\left[\frac{2f_0}{\gamma+1} - \lambda\right]g_0' h_1 \\
&- \frac{2}{\gamma+1}(g_0 h_0' - \gamma h_0 g_0')f_0 + \nu h_0 g_0\left[A + \frac{4\gamma^2 - (\gamma-1)^2}{2\gamma(\gamma-1)}\right] = 0.
\end{aligned}\right\} \quad (12.13)$$

† The solution of this problem is given in Mel'nikova (1954). A similar linearized problem of a point explosion taking counter pressure into account was studied somewhat later, independently, and by another method by Sakurai (1953, 1954).

The detailed investigation and numerical results obtained by N. S. Mel'nikova were published in the third edition of this book in 1954.

The coefficients of these equations are unknown functions of λ, expressed in terms of the functions f_0, g_0, h_0 of the appropriate self-similar solution. The constant A only enters into the free term linearly, consequently, the desired solution of the system (12.10) can be written as

$$\left.\begin{aligned}
f_1 &= f_{11} + A f_{12}, \\
g_1 &= g_{11} + A g_{12}, \\
h_1 &= h_{11} + A h_{12}.
\end{aligned}\right\} \quad (12.14)$$

We determined the functions $f_{11}(\lambda)$, $g_{11}(\lambda)$, $h_{11}(\lambda)$ as the solution of the Cauchy problem for the system (12.13) at $A = 0$ and with the initial data

$$f_{11}(1) = 0; \quad g_{11}(1) = 0; \quad h_{11}(1) = 0.$$

We define the functions $f_{12}(\lambda)$, $g_{12}(\lambda)$ and $h_{12}(\lambda)$ as the solution of the Cauchy problem for a similar system of equations with appropriately altered free terms and with the initial data

$$f_{12}(1) = 0; \quad g_{12}(1) = 0; \quad h_{12}(1) = 0.$$

Condition (12.12) at the centre will be satisfied if the constant A is determined from the condition

$$\lim_{\lambda \to 0} (f_{11} + A f_{12}) = 0. \qquad (12.15)$$

The solution of these Cauchy problems can be carried out numerically.

After replacing the variable λ by V with use of formulas (11.10) and (11.15), the system of ordinary linear differential equations (12.13) is transformed into a system of linear differential equations with simple rational functions of V as coefficients.

Certain coefficients have poles at points corresponding to the centre of symmetry for $V = 2/[(2+\nu)\gamma]$. The required solutions of the transformed system can be constructed as series which converge for all values of V in the interval $2/[(2+\nu)\gamma] < V < 4/(\gamma+1)(2+\nu)$ (for $\nu = 3$ with $\gamma < 7$) in which the required solution must be constructed.

The numerical calculations yield a value of A close to 2 in all three cases $\nu = 1, 2, 3$.†

Condition (12.8) and condition (12.15) in the linearized case can be replaced by the requirement that the energy of the disturbed motion is finite. This is equivalent to the condition (12.15) for the linearized equations. After the constant A has been determined from formula (12.11) by discarding higher order terms not shown, we obtain a more

† The numerical calculations in $\nu = 1$ and $\nu = 2$ cases are given by Brushlinskii and Solomakhova (1956).

precise relation between l and q for small q. From (12.5), the relation between q and τ for small q is

$$\tau = \left(\frac{2}{2+\nu}\right)^{2/\nu} \frac{q^{2+\nu/2\nu}}{\nu(\gamma\alpha)^{1/\nu}\gamma^{1/2}} \left[\frac{2\nu}{2+\nu} + A\frac{2(\nu+1)}{2+3\nu}q\right]. \qquad (12.16)$$

The corresponding curves are given in Figs. 79 and 80. The dashed lines correspond to the self-similar law (12.4), the chain dotted lines correspond to the improved law taking linear terms into account, the solid lines are obtained from numerical results using electronic computers (see the last paragraph on this page).

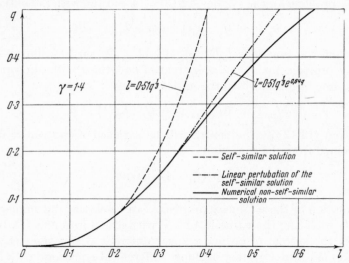

FIG. 79. Comparison of analytical and numerical methods of calculating the function $q(l)$ $(q = a_1^2/c^2, l = r_2/r^0)$.

The results of numerical computations of the functions $f_1(\lambda)$, $g_1(\lambda)$ and $h_1(\lambda)$ are given in Fig. 81 for $\nu = 3$. The variation of characteristics of the gas motion with time can be estimated from (12.11) for small q by using these graphs.

Papers (Neumann and Goldstine, 1955†; Brode, 1955‡; Okhotsimskii, Kondrasheva, Vlasova and Kasakova, 1957§) published, from 1955 onwards, show the results of numerical computations of the solution of the non-self-similar problem of a point explosion (spherical case: $\nu = 3$) taking counter-pressure into account.

† The computation is made in the range $100 \geqslant p_2/p_1 \geqslant 1 \cdot 017$.
‡ The computation in the range $1600 \geqslant p_2/p_1 \geqslant 1 \cdot 06$.
§ Computational range is $1740 \geqslant p_2/p_1 \geqslant 1 \cdot 008$.

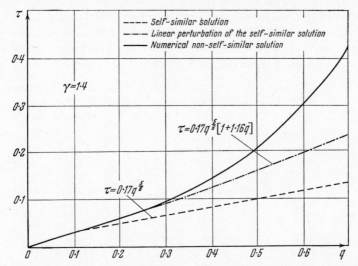

FIG. 80. Comparison of analytical and numerical methods of calculating the function $\tau(q)$ $(\tau = t/t^0)$.

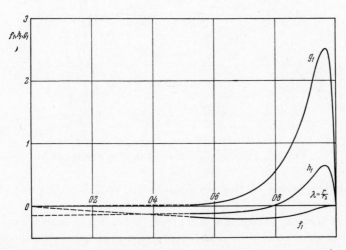

FIG. 81. Graphs of the derivatives $g_1(\lambda) = [\partial(\rho/\rho_2)/\partial q]_{q=0}$, $h_1(\lambda) = [\partial(p/p_2)/\partial q]_{q=0}$ and $f_1(\lambda) = [\partial(v/v_2)/\partial q]_{q=0}$ which determine the influence of the counter-pressure in the initial stages of a point explosion.

The results of all the work referred to were compared and expressed in terms of the nondimensional variables introduced above.†

The velocity, density and pressure distributions behind the shock wave are given in Figs. 82–84; the most sensitive variable $l = r_2/r^0$ which defines the distance from the blast wave to the centre of the explosion,

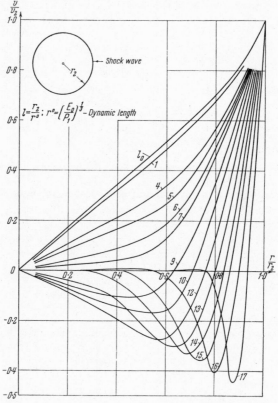

FIG. 82. Velocity distributions in a point explosion.

measured in terms of the dynamic length $r^0 = (E_0/p_1)^{1/3}$, is indicated on the graphs.

The numerical values of l_k are given in the table shown with corresponding values of $\tau = t/t^0$ and the pressure drop across the shock wave front p_2/p_1.

All the graphs shown apply to a point explosion for any p_1, ρ_1 and E_0 when $\gamma = 1\cdot4$.

† The numerical computations given below were carried out by N. S. Mel'nikova. V. P. Korobeinikov, E. V. Riazanov.

$l = \dfrac{r_2}{r^0}$	$\tau = \dfrac{t}{t^0}$	$\dfrac{p_2}{p_1}$	Notation for l
Self-similar curves		∞	l_0
0·1867	0·01403	21·1	l_1
0·2669	0·03230	10·31	l_2
0·3342	0·05431	4·967	l_3
0·4890	0·1231	3·056	l_4
0·6003	0·1842	2·321	l_5
0·6812	0·2333	2·010	l_6
0·7566	0·2807	1·835	l_7
0·9566	0·3667	1·615	l_8
0·9801	0·4323	1·521	l_9
1·2524	0·6296	1·338	l_{10}
1·3210	0·6811	1·315	l_{11}
1·5171	0·8299	1·255	l_{12}
1·8751	1·1080	1·185	l_{13}
1·3222	1·4636	1·136	l_{14}
2·8641	1·8973	1·102	l_{15}
3·6983	2·7763	1·066	l_{16}
7·0791	5·3764	1·032	l_{17}
9·6424	7·5180	1·022	l_{18}

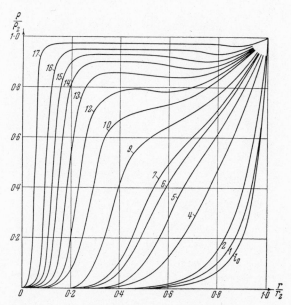

FIG. 83. Density distribution in a point explosion.

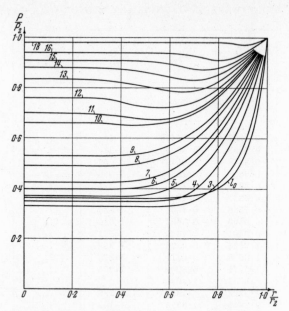

FIG. 84. Pressure distribution in a point explosion.

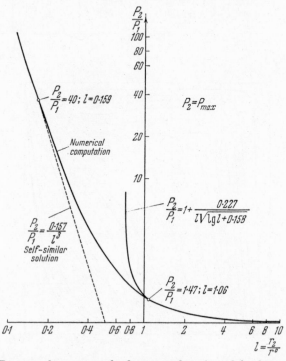

FIG. 85. Pressure drop across the front as a function of shock wave radius.

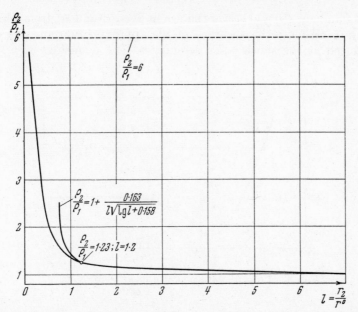

FIG. 86. Density behind the wave front at a function of shock wave radius.

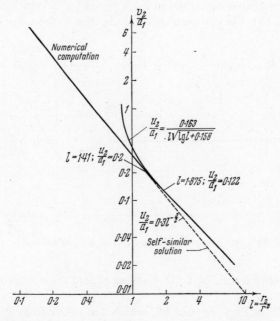

FIG. 87. Velocity behind the wave front as a function of shock wave radius.

The functions $q(l)$ and $q(\tau)$ are shown in Figs. 79 and 80, from which it is easy to estimate the accuracy of the approximation and the error introduced as the shock wave attenuates. The quantities p_2/p_1, ρ_2/ρ_1

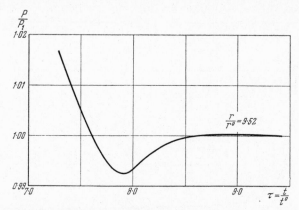

FIG. 88. Pressure variation at a fixed point of space.

and v_2/a_1 behind the shock, can easily be constructed as functions of l and τ from the graphs in Figs. 79 and 80 and from formula (12.6). The corresponding graphs are given in Figs. 85–87 (on a logarithmic scale in Figs. 85 and 87). Curves corresponding to the self-similar motion and

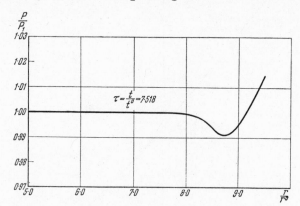

FIG. 89. Pressure distribution at a fixed instant.

the asymptotic formula of L. D. Landau (1945) are also given on these graphs.

The graphs in Figs. 88 and 89, which are taken from the paper by von Neumann and Goldstine give, respectively, the pressure p/p_1 as a function of time at a fixed point of space and as a function of the co-ordinate r/r^0 at a fixed instant.

The graphs were derived in the spherical case with $\gamma = 1 \cdot 4$. Approximate formulas based on theoretical arguments give the motion of the shock wave and the motion behind the wave front for other values of γ, and also cover the cylindrical and plane wave cases (Korobeinikov, 1956).

§13. On Simulation and on Formulas for the Peak Pressure and Impulse in Explosions

The nature of the formulas determining the peak pressure, impulse, etc., as a function of the distance from the explosion, the energy of the explosion and other variables could have been given at the very beginning of the book as one of the examples in the series discussed in section 6 of Chapter II.

However, the explanation of this question, as well as the rule for modelling the behaviour and effect of an explosion is given more conveniently after the development of the theory of a point explosion in a gas; this simplifies the problem and allows us to make use of the formulation of the appropriate problems described above.

In order to estimate the effect of an explosion on a body which happens to interact with the field of disturbed gas motion, the peak pressure $p_{\max}$ or the magnitude of the impulse I per unit area acting on the body surface must be considered.

When only pressure stresses act on the body surface, the impulse I can be determined from the formula

$$ I = \int\limits_{t_0}^{t_1} (p - p_1)\, dt, $$

where p_1 is the initial undisturbed pressure; t_0 and t_1 are characteristic times (for example, t_0 is the time the interaction starts, $t_1 = \infty$ or t_1 corresponds to the moment the difference $p - p_1$ changes sign, etc.). Sometimes, it is necessary to take the total impulse vector J acting on the body during a characteristic time interval.

Interaction between disturbances in the medium surrounding the body and the body itself occurs in an explosion. The magnitude of $p_{\max}$ at various points of the body surface, the impulse I, the total impulse vector J and the other interaction variables depend partly on the properties of the charge, its location relative to the body, and the properties of the ambient atmosphere. But they are also influenced considerably by the properties of the body itself, its shape, structure, mass distribution and other dynamic characteristics, on its deformability and internal cohesion, the conditions of its location relative to other bodies, etc.

When the system of characteristic parameters and the similarity conditions are being singled out, it is necessary to pick out the dimensional and nondimensional parameters, the characteristics of the body itself which determine its size and shape, the mass and its distribution over the body volume, the parameters which characterize the internal cohesion, the body inertia, and so on.

The main difficulty in the general case is to isolate only the essential factors: these differ from case to case and depend on the character of the problems to be solved.

Additional physical constants characterizing the body do not arise in the limiting cases when the body is absolutely solid and is rigidly

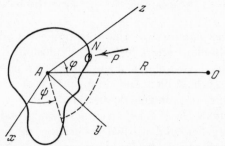

FIG. 90. The point O is the centre of the explosion.

supported (this can be a body with an infinite mass, in the limit) or when the body is very small and its influence on the explosive field can be neglected. Actually, an absolutely solid, rigid body of arbitrary, fixed, geometric shape can be assigned only one dimensional parameter, the characteristic linear dimension D; the other parameters can be considered abstract and they remain constant for a series of geometrically similar cases. If we assume that the effect of the explosion is determined only by the charge energy E_0 (it is common practice to characterize the charge effect only by its weight which is proportional to the charge energy and to neglect the shape, chemical or physical nature of the charge, etc.), then the idea of a point, instantaneous explosion, which is determined completely by the magnitude of the energy E_0 liberated, can be used. As is known, such an assumption is not correct in many cases,† in particular, when the question of the effect of cumulative charges arises.

If we consider an intense point explosion in a perfect gas occupying the whole space outside an absolutely solid body, then according to the

† The conclusions which follow require extensive review in such cases.

formulation of the problem given in §1, we see that the system of parameters defining the phenomenon as a whole is

$$\gamma,\ E_0,\ \rho_1,\ \xi = \frac{D}{R},\ R,\ \phi,\ \psi, \tag{13.1}$$

where R, ϕ, ψ are the polar coordinates of the centre of the explosion in a certain coordinate system fixed in the body (Fig. 90).

Let us consider a certain point N on the body and let us denote the peak pressure and impulse at this point by p_{max} and I, respectively.

It is easy to verify from dimensional analysis that

$$p_{max} = \frac{E_0}{R^3} f_1(\xi,\ \phi,\ \psi,\ N,\ \gamma), \tag{13.2}$$

$$I = \sqrt{\left(\frac{\rho_1 E_0}{R}\right)} f_2(\xi,\ \phi,\ \psi,\ N,\ \gamma). \tag{13.3}$$

The coordinates of the point N can be considered as abstract numbers. Without loss of generality, we can place the origin A at the fixed point N.

For the total impulse summed over all the points of the body surface we have

$$J = \sqrt{\left(\frac{\rho_1 E_0}{R}\right)} D^2 f_3(\xi,\ \phi,\ \psi,\ \gamma). \tag{13.4}$$

A complex three-dimensional problem must be solved to calculate these functions theoretically and this is not always possible in practice.

The relations between p_{max}, I and J and E_0, ρ_1 and R in geometrically similar cases are given explicitly for an intense explosion ($p_1 = 0$) from (13.2), (13.3) and (13.4), without solving this complex problem.

The dynamic similarity conditions reduce to the parameters $\xi(= D/R)$, ϕ, ψ being constant in this case and do not impose any other limitations on ρ_1, R and E_0.

The quantities p_{max}, I remain finite as $\xi = D/R \rightarrow 0$, when the body shrinks to a point, consequently, the functions f_1, f_2 and f_3 retain finite values in the limit as $\xi \rightarrow 0$.

The parameter ξ can be put equal to zero for bodies of small dimensions. Formulas (13.2), (13.3) and (13.4) define p_{max}, I and J as functions of the distance R from the body to the centre of the explosion in this case.

If the initial pressure ($p_1 \neq 0$) is taken into account, then the non-dimensional argument

$$\chi = \frac{R}{\sqrt[3]{\left(\dfrac{E_0}{p_1}\right)}} = \frac{R}{r^0}$$

is added to the right sides of (13.2), (13.3) and (13.4).

This parameter is almost zero for large E_0 and small R; it is large for any E_0 and large R; consequently, the similarity conditions given above and the relation between $p_{\max}$, I and $\boldsymbol{J}$ and the energy E_0 for large R must be modified. The similarity conditions when counter-pressure is taken into account are:

$$\frac{R}{\sqrt[3]{\left(\dfrac{E_0}{p_1}\right)}} = \text{const}; \quad \frac{D}{R} = \text{const}; \quad \phi = \text{const} \quad \text{and} \quad \psi = \text{const}. \quad (13.5)$$

Such conditions are satisfied in setting up experiments if there is geometric similarity and if the relation

$$E_0 = \text{const}\, p_1\, R^3$$

holds. These requirements can easily be fulfilled in practice, by a reduction in scale. If a relation between any of the quantities $p_{\max}$, I or $\boldsymbol{J}$ and the explosion energy E_0 is derived from additional data (experimental, say), then we can determine f_1, f_2, f_3 as functions of the parameter χ, and therefore, as functions of the distance R for $\xi = 0$.

The similarity conditions (13.5) can be extended to more general cases when the body used can be considered as a continuum with the constant density ρ^*, either in the form of a weightless incompressible fluid or of a weightless elastic body which is characterized by Young's modulus ϵ and Poisson's ratio σ. The following conditions must be added to conditions (13.5) to accomplish this:

$$\frac{\rho^*}{\rho_1} = \text{const} \quad \text{and} \quad \frac{\mathscr{E}}{p_1} = \text{const}, \quad \sigma = \text{const}. \quad (13.6)$$

The similarity conditions (13.5) and (13.6) still retain their form for a system of elastic bodies in liquids with perfectly rigid boundaries. In the more general case, conditions of type (13.6), together with those of geometric similarity (13.5) will be satisfied for any body with appropriate constant ratios.

It is evident that the density ρ^* may vary with position but the density distributions in such cases must be similar.

All the preceding conclusions are obtained without taking gravitational forces into account.

If the intrinsic weight of different bodies and also of gases is taken into account, then the constant acceleration due to gravity g must be added to the system of characteristic parameters. In this case the similarity conditions are supplemented by the additional requirement that:

$$\frac{p_1}{\rho_1 g D} = \text{const}. \quad (13.7)$$

Condition (13.7) can be satisfied by decreasing p_1, increasing ρ_1 or increasing g for a large decrease in the linear dimension D. If the elastic properties are fixed and it is impossible to alter the Young's modulus $\mathscr{E}$ substantially, then the condition $p_1/\mathscr{E} = $ const leads to the requirement that the magnitude of p_1 be conserved. The air density can vary within broad limits but the density of elastic bodies only varies within narrow limits, consequently, the condition

$$\frac{\rho^*}{\rho_1} = \text{const},$$

fixes the density ρ_1, generally speaking.

Therefore, it is necessary to alter g, which can be accomplished with a centrifuge or by using an accelerated moving platform (§6, Chapter II), in order to satisfy (13.7). However, both these methods complicate the experimental set-up radically, due to the use of small models with very large linear scale ratios, the additional influence of the moving air flowing around the model, and the influence of processes of short duration on the accelerated platform, etc. The influence of a number of other factors may appear with large-scale reductions (the method of preparing the model, the impossibility of guaranteeing total similarity of the structure of the materials, viscous and capillary forces and many others) but these clearly have no physical significance.

The condition $p_1/\mathscr{E}$ drops out for an incompressible fluid ($\mathscr{E} = \infty$); consequently, condition (13.7) can be satisfied by a decrease in the characteristic pressure p_1 (experiment in a vacuum). We can give examples in which the intrinsic weight plays an essential part: the effect of an explosion which leads to the jarring of large structures, the stability of a structure when tilted, the phenomenon of crater formation for explosions in the ground and of base surge and other surface wave phenomena in underwater explosions and so on.

Inertial and strength properties can be very important in the study of destruction of a local character on the surface and within different kinds of metal structures. However, the effect of weight is generally insignificant.

Let us consider the question of magnitudes of craters due to explosions in the ground in more detail. In many cases, especially for explosions in sandy soil, the internal stresses specified by the physical properties of the soil (by friction or shear stress) and the atmospheric pressure, are small. The internal pressures, determined by the effect of gravity and which have the character of hydrostatic pressure, can be large and we take their influence into account.

Let us consider the explosions which are shown in Fig. 91.

We assume that the phenomenon of soil ejection, for a definite type of explosive, depends essentially on the density and weight of the soil, just as in the phenomenon of water ejection during explosions. Following usual practice, we assume that the dimensions of the craters depend only on the charge energy E (charge weight) and on the depth h.† Moreover, let us neglect the pressure p_1 at the soil surface. Then, we obtain a system of characteristic parameters ρ_1, g, h, R. For similarity, we must have

$$\frac{E}{\rho_1 g h^4} = \text{const} = c. \tag{13.8}$$

A crater is not obtained for large depths h, i.e. small c. It is clear from (13.8) that the craters will be similar if

$$E = c\rho_1 g h^4. \tag{13.9}$$

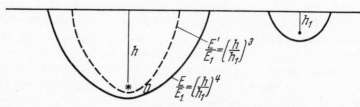

FIG. 91. Point explosion within the ground; O is the centre of the explosion; h is the depth. In similar cases we shall have $(E/E_1) = (h/h_1)^4$.

The charge energy and, therefore, the charge weight for similar craters must be proportional to the fourth power of the scale. Since the volume ejected is proportional to the scale cubed, it is evident that the relation between the charge energy and the volume of ejected soil Q will be, for similarity,

$$E = c_1(c)Q^{4/3}\rho_1 g. \tag{13.10}$$

If the explosion is on the surface, $h = 0$, and the similarity condition (13.8) drops out. In this case, (13.10) remains valid under the assumptions made above, where the constant $c_1 = c_1(\infty)$ is a definite numerical constant independent of the soil density and the charge energy.

Formula (13.10) can be rewritten in a more practical form if we replace the charge energy E by the charge weight and the volume Q by the weight of soil.

† More precisely, the quantity E equals the mechanical energy transferred from the charge to the soil. Part of the charge energy can be lost by radiation, fusion and evaporation of the soil, etc.

For explosions following the law $E'/E = (h/h')^3$ similarity is violated and the craters are not similar; the character of the crater in this case is shown by dashes on Fig. 91.

We neglected the effect of the external atmospheric pressure p_1 on the soil surface in the preceding discussions. This effect, as well as the effect of internal elastic forces in identical materials, exert an influence through the parameters $\mu = p_1/\rho_1 gh$ and $p_1/\mathscr{E}$.

The quantity μ decreases for identical p_1, ρ_1 and g as h increases; a decrease in μ corresponds to a decrease in the pressure p_1 for ρ, g, h constant.

It can be assumed that the crater dimensions increase for constant ρ_1, g, h, $\mathscr{E}$ and a decrease in the external pressure p_1.

Therefore, a decrease in the parameter $\mu = p_1/\rho_1 gh$ because of an increase in h, with the remaining nondimensional characteristic parameters kept constant, violates similarity. The amount of soil ejected is increased as compared with the amount calculated using similarity.

Hence, it is clear that a smaller increase in the charge energy than that given by (13.10) is required to obtain similarity in the crater volume when making the transition to full-scale conditions.

Hence, the role of the external pressure and of the elastic cohesive forces is reduced as the linear dimensions increase if the explosion energy increases in proportion to the fourth power of the linear dimensions.

The relative displacement and deformation in full-scale conditions will be larger than in the model if the energy increases in proportion to the fourth power of the scale.

On the other hand, if weight is not important, then similarity is guaranteed by conditions (13.5) and (13.6) from which it follows that the energy is proportional to the scale cubed. The additional influence of gravity violates similarity by decreasing the relative deformation and displacement when the transition is made from the model to full-scale conditions.

Therefore, the actual, relative deformation at full scale is larger than on small models for constant ρ_1, p_1 and g if $E \sim h^4$; the actual, relative deformation at full scale is less than on small models if $E \sim h^3$.

It is useful to keep these points in mind when setting up experiments to investigate effects of explosions.

It also follows from the above that the use of very large, concentrated charges is unsuitable from the energy viewpoint, apparently, when using explosions to eject soil. The questions of dispersion of the charge and of suitable methods for carrying out the explosion have serious, and important, aspects.

9

We have already discussed the general formulas for p_{max}, I and J when unsteady gas motion interacts with a solid body. If the body is nonexistent, a certain imaginary surface can be fixed in the gas, which does not interact with the gas motion, and similar quantities p^0_{max}, I^0 and J^0, which define the explosion field, can be considered; evidently, they will not equal the corresponding above quantities p_{max}, I, J which depend essentially on the interaction of the gas motion and the solid body.

In this case, we obtain for each element of the surface Σ the relations

$$p^0_{max} = \frac{E_0}{r^3} f_1\left(\frac{r}{r^0}, \gamma\right) \quad \text{and} \quad I^0 = \sqrt{\left(\frac{\rho_1 E_0}{r}\right)} f_2\left(\frac{r}{r^0}, \gamma\right), \quad (13.11)$$

where r is the distance of the element from the centre of the explosion. The functions f_1 and f_2 depend only on the parameters r/r^0 ($r^0 = \sqrt[3]{(E_0/p_1)}$ and γ. Since the pressure in an ideal gas is independent of the direction of the surface area on which it acts, then p^0_{max} will equal the value of the pressure behind the shock front at the moment it passes through the point considered. For a violent explosion $r/r^0 \doteq 0$ and, therefore, the functions f_1 and f_2 reduce to constants which can easily be calculated by using the solutions found in §11.

The values of these constants for $\gamma = 1 \cdot 4$,[†] are

$$f_1 = 0 \cdot 157, \quad f_2 = 0 \cdot 486. \quad (13.12)$$

A graph of the function $f_1(r/r^0, \gamma)(r^0/r)^3 = p_{max}/p_1$ for $\gamma = 1 \cdot 4$ is given in Fig. 85.

The pressure can be calculated as a function of time at each value of r by using the solution described in §12 for a point explosion with counter-pressure; this pressure is always larger at first than the initial unperturbed pressure p_1, but later, after a certain time $\bar{t}$, the pressure becomes less than p_1, reaches a minimum and then increases monotonically to the value p_1 (see Fig. 88).

We determine the positive impulse for each point from

$$I^+ = \int_{t_0}^{\bar{t}} (p - p_1)\, dt = \sqrt{\left(\frac{\rho_1 E_0}{r}\right)} f_2^+\left(\frac{r}{r^0}\right) \quad (13.13)$$

and the negative impulse from

$$I^- = -\int_{\bar{t}}^{\infty} (p - p_1)\, dt = \sqrt{\left(\frac{\rho_1 E_0}{r}\right)} f_2^-\left(\frac{r}{r^0}\right).$$

† The impulse I^0 is defined by the formula $I^0 = \int_{t_0}^{\infty} (p - p_1)\, dt$; t_0 the time at which the shock wave passes through the point considered.

The functions $f_2^+(r/r^0)$ and $f_2^-(r/r^0)$ were determined from a numerical solution. The corresponding graphs are given in Fig. 92. It follows from these graphs that the negative impulse is larger in magnitude than the positive impulse for $r/r^0 > 0 \cdot 3$.

The dimensions of E_0 are changed for a line explosion, the cylindrical wave case; consequently, formulas (13.11) are replaced by

$$p_{\max}^0 = \frac{E_0}{r^2} f_1\left(\frac{r}{r^0}\right) \quad \text{and} \quad I^0 = \sqrt{(\rho_1 E_0)} f_2\left(\frac{r}{r^0}\right), \quad r^0 = \sqrt{\left(\frac{E_0}{p_1}\right)}, \quad (13.14)$$

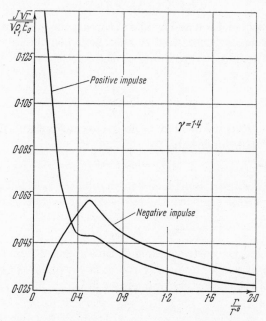

FIG. 92. Distribution of total positive and negative impulses for a point explosion as a function of distance from the centre of the explosion.

and for an explosion along a plane, by:

$$p_{\max}^0 = \frac{E_0}{r} f_1\left(\frac{r}{r^0}\right) \quad \text{and} \quad I^0 = \sqrt{(\rho_1 E_0 r)} f_2\left(\frac{r}{r^0}\right), \quad r^0 = \frac{E_0}{p_1} \quad (13.15)$$

where r is the distance to the line or plane of the explosion.

If the impulse I^0 for an intense explosion is defined as

$$I^0 = \int_{t_0}^{\infty} p \, dt = \int_{r_2 = r_0}^{\infty} p \frac{dr_2}{c},$$

then we obtain $I^0 = \infty$ in the plane and cylindrical cases. Finite limit values are obtained for I^0 in these cases if the following definition is used

$$I^0 = \int_r^{kr} p \frac{dr_2}{c}$$

where k is a certain number larger than one.

§14. Problem of an Intense Explosion in a Medium with Variable Density

The formulation, method and results of solving the problem of an intense explosion are easily generalized to the case when the initial density ρ_1 depends on the initial coordinate of the particle r_0 according to the relation,

$$\rho_1 = \frac{A}{r_0^\omega} \tag{14.1}$$

where ω is an arbitrary positive or negative constant and A is a positive constant with the dimensions

$$[A] = ML^{\omega-3}.$$

If the initial pressure p_1 is neglected in an intense explosion and adiabatic motion of a perfect gas is considered, then the system of characteristic parameters is

$$\omega, \gamma, A, E_0, r, t. \tag{14.2}$$

It follows that the corresponding disturbed gas motion is self-similar. Furthermore, if we consider only the case of spherical symmetry, the dimensions of the characteristic energy will be:

$$[E_0] = ML^2 T^{-2}.$$

All the nondimensional quantities can be considered as functions of the parameters

$$\omega, \gamma, \lambda = \left(\frac{A\alpha}{E_0}\right)^{1/(5-\omega)} \frac{r}{2/t^{(5-\omega)}} \tag{14.3}$$

where α is a constant to be chosen, $(\delta = 2/(5-\omega))$.

It follows from the formulation of the problem that in states with the same phase ($\lambda = \text{const}$) and, in particular, on the shock wave:

$$\left.\begin{array}{l} \lambda = \lambda_2 = \text{const}, \quad r_2 = \lambda_2\left(\dfrac{E_0}{A\alpha}\right)^{1/(5-\omega)} t^{2/(5-\omega)}, \\[3mm] c = \dfrac{dr_2}{dt} = \dfrac{2}{5-\omega}\dfrac{r_2}{t} = \dfrac{2\lambda_2^{(5-\omega)/2}}{5-\omega}\sqrt{\left(\dfrac{E_0}{A\alpha}\right)} r_2^{(\omega-3)/2}. \end{array}\right\} \tag{14.4}$$

We determine the constant α from the condition that $\lambda_2 = 1$ on the shock wave. With this definition of α we shall have

$$\lambda = \frac{r}{r_2}.$$

The shock wave will decelerate if $\omega > 3$ and will accelerate if $\omega < 3$. The $\omega < 3$ case corresponds to a finite spherical mass containing the centre of symmetry; this mass will be infinite for $\omega \geqslant 3$.

It follows from (14.4) that the shock wave which forms near the centre of symmetry for $t = 0$ is propagated with finite velocity to the periphery, for $t > 0$, only when $\omega < 5$. In what follows, we shall assume $\omega < 5$.

It has been shown in §5 that the problem in this case reduces to the integration of the ordinary differential equation (5.10).

It follows from an analysis of the field of integral curves drawn for different ω in Figs. 34 and 35 that, as in the $\rho_1 = \text{const}$ case, the region of disturbed motion is separated from the region by a spherical shock wave which is propagated through the undisturbed region with the variable velocity c determined by (14.4).

From the condition on the strong shock wave (11.2), and Equations (14.4) and (1.1), we find that the quantities V, R and $z = \gamma P/R$ have the following constant values behind the shock:

$$V_2 = \frac{4}{(5-\omega)(\gamma+1)} = \frac{2\delta}{\gamma+1}, \quad R_2 = \frac{\gamma+1}{\gamma-1}, \quad z_2 = \frac{8\gamma(\gamma-1)}{(5-\omega)^2(\gamma+1)^2}. \quad (14.5)$$

As we have shown in §3, the energy integral (3.11) holds in this case.

After substituting the values on the shock front (14.5) into the left side of the energy integral (3.11), we find that the constant on the right is zero. Hence, it follows that the equation of the integral curve passing through the point V_2, z_2 in the z, V plane is:

$$z = \frac{(\gamma-1)\, V^2\left(V - \dfrac{2}{5-\omega}\right)}{2\left[\dfrac{2}{\gamma(5-\omega)} - V\right]} = \frac{(\gamma-1)\, V^2(V-\delta)}{2\left(\dfrac{\delta}{\gamma} - V\right)}. \quad (14.6)$$

It is clear from physical considerations that $z \geqslant 0$, consequently, admissible values of V must lie on the segment

$$\frac{\delta}{\gamma} = \frac{2}{\gamma(5-\omega)} \leqslant V \leqslant \frac{2}{5-\omega} = \delta. \quad (14.7)$$

Since $\gamma > 1$, we have

$$\frac{1}{\gamma} < \frac{2}{\gamma+1} < 1;$$

consequently, the value of V_2 determined from (14.5) lies in the interval (14.7). The end points of the integral curve (14.6)

$$\left(z = 0, \quad V = \frac{2}{5-\omega} \quad \text{and} \quad z = \infty, \quad V = \frac{1}{\gamma}\frac{2}{5-\omega} \right)$$

are singular points of the differential equation (5.10).

Further, it is easy to verify that the singular point† with the coordinates

$$\frac{2}{3\gamma-1} = V^*, \quad \frac{2(\gamma-1)\left(\dfrac{2}{3\gamma-1}-\delta\right)}{(3\gamma-1)^2\left(\dfrac{\delta}{\gamma}-\dfrac{2}{3\gamma-1}\right)} = z^*$$

which belongs to the integral curve (14.6) for the values $\gamma = 1\cdot 4$ or $5/3$ and ω close to zero, lies outside the interval (14.7). This singular point coincides with the upper limit of the interval (14.7) for $\omega = 6-3\gamma$ and moves upwards along the integral curve (14.6) as ω increases further. The singular point corresponds to values (V_2, z_2) on the shock front for values of ω determined from the equation

$$\frac{4}{(5-\omega)(\gamma+1)} = \frac{2}{3\gamma-1}$$

or

$$\omega = \frac{7-\gamma}{\gamma+1} \tag{14.8}$$

(for $\gamma = 1\cdot 4$, $\omega = 7/3$ and $\omega = 2$ for $\gamma = 5/3$).

It is easy to see that the following simple exact solution of the system of ordinary differential equations (2.1)–(2.3) corresponds to the singular point V^*, z^*:

$$V = V^* = \text{const}, \quad z = z^* = \text{const}, \quad R = B\lambda^{[(\omega-3)V^*]/(V^*-\delta)},$$

$$P = B\frac{z^*}{\gamma}\lambda^{[(\omega-3)V^*]/(V^*-\delta)}, \tag{14.9}$$

This corresponds to the exact solution of the partial differential equation in the dimensional variables

$$v = V^*\frac{r}{t}, \quad \rho = \frac{AB}{r^\omega}\lambda^{[(\omega-3)V^*]/(V^*-\delta)},$$

$$p = \frac{ABz^*}{\gamma r^{\omega-2}t^2}\lambda^{[(\omega-3)V^*]/(V^*-\delta)}. \tag{14.10}$$

† See §5 and Fig. 35.

Here B is an arbitrary constant. Formulas (14.10) give the solution of the problem[†] of an intense explosion for

$$\omega = \frac{7-\gamma}{\gamma+1}.$$

Determining the constant B from the conditions on the shock wave in this case, we obtain:

$$v = \frac{2}{3\gamma-1}\frac{r}{t}, \quad \rho = \frac{A(\gamma+1)}{r^\omega(\gamma-1)}\lambda^{8/(\gamma+1)}, \quad p = \frac{A}{r^{\omega-2}t^2}\frac{2(\gamma+1)}{(3\gamma-1)^2}\lambda^{8/(\gamma+1)}$$

or, expressing the velocity, density and pressure in terms of appropriate shock wave quantities, we find

$$\frac{v}{v_2} = \lambda, \quad \frac{\rho}{\rho_2} = \lambda, \quad \frac{p}{p_2} = \lambda^3. \tag{14.11}$$

It is clear from these formulas that in this case the motion can be continued to the centre of symmetry, where the density and pressure are zero.

We note that the derivative $d\ln\lambda/dV$ at the point V_2, z_2 changes sign as ω varies when the point V^*, z^* passes through the point V_2, z_2 along the integral curve (14.6). Hence, it follows that it is necessary to move upward along the integral curve (14.6) when moving from the shock wave to the centre of symmetry, as λ decreases, if $V_2 < V^*$ and downward if $V_2 > V^*$ (see Fig. 35). In the first case, λ approaches zero as the singular point $z = \infty$, $V = (1/\gamma)2/(5-\omega)$ is approached, and the motion is determined up to the centre of symmetry; in the second case, it is necessary to move downward along the integral curve to the singular point $V = 2/(5-\omega)$, $z = 0$ at which the parameter λ takes a finite value.[‡]

It is easy to determine the combination p/ρ^γ, related to the entropy, as a function of the origin r_0. We obtain from the shock wave conditions (11.2) and formula (14.4)

$$\frac{p}{\rho^\gamma} = \frac{2(\gamma-1)^\gamma}{(\gamma+1)^{\gamma+1}}\rho_1^{1-\gamma} \quad c^2 = \frac{8(\gamma-1)^\gamma A^{-\gamma}E_0}{\alpha(\gamma+1)^{\gamma+1}(5-\omega)^2}r_0^{\gamma\omega-3}. \tag{14.12}$$

We use (14.12) to determine r_0/r_2 as a function of the parametric variable V.

It is curious to note that the particle entropy behind the shock front decreases as the shock wave is propagated over the particles if $\omega < 3/\gamma$, remains constant for $\omega = 3/\gamma$ and increases if $\omega > 3/\gamma$.

---

† The appropriate relation between ω and γ for cylindrical waves at $\nu = 2$ is $\omega = 4/(\gamma+1)$ and for plane waves with $\nu = 1$ is $\omega = 1$ if $\gamma \neq -1$.

‡ This is seen directly from formulas (14.14), (14.15) which give the solution.

The shock wave is retarded in the range

$$\frac{3}{\gamma} < \omega < 3$$

while the entropy behind the wave front increases. This is explained by the fact that, in spite of the shock wave attenuation and the pressure drop behind the wave front, the density drops so sharply as to cause the entropy to increase. For such an effect, if

$$\frac{3}{\gamma} < \omega \leqslant \frac{7-\gamma}{\gamma+1}$$

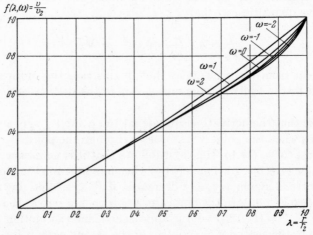

FIG. 93. Velocity field for a point explosion in a gas with the variable initial density distribution $\rho_1 = A/r_0^\omega$; $\gamma = 1\cdot4$. The motion extends to the centre of symmetry when $\omega \leqslant 7/3$.

the disturbance occupies the whole interior of the shock wave sphere, right up to the centre of symmetry; if

$$\frac{7-\gamma}{\gamma+1} < \omega < 3,$$

then a vacuum is formed near the centre.

Let us consider the behaviour of the temperature T_2 behind the shock wave front. Taking the conditions on the shock (11.2) into account, we obtain:

$$\mathbf{R}T_2 = \frac{p_2}{\rho_2} = \frac{2(\gamma-1)}{(\gamma+1)^2}c^2. \tag{14.13}$$

Here $\mathbf{R}$ is the gas constant. The temperature T_2 is proportional to c^2, therefore, the temperature behind the shock front can increase only as c^2 increases for the gas motions under consideration, i.e. for $\omega > 3$.

Using the integral (14.6) of Equation (5.10), $\lambda(V)$ and $R(V)$ can be determined from (5.11) and (5.12) by simple quadratures, which permits the complete solution of the problem to be obtained in a simple closed form.

This solution is given by the formulas

$$
\left.\begin{aligned}
\frac{r}{r_2} &= \left[\frac{(5-\omega)(\gamma+1)}{4}\,V\right]^{-2/(5-\omega)}\left[\frac{\gamma+1}{\gamma-1}\left(\frac{\gamma(5-\omega)}{2}\,V-1\right)\right]^{-\alpha_2} \\
&\quad \times\left[\frac{(5-\omega)(\gamma+1)}{7-\gamma-(\gamma+1)\omega}\left(1-\frac{3\gamma-1}{2}\,V\right)\right]^{-\alpha_1}, \\[6pt]
\frac{r_0}{r_2} &= \left[\frac{(5-\omega)(\gamma+1)}{4}\,V\right]^{-2/(5-\omega)}\left[\frac{\gamma+1}{\gamma-1}\left(\frac{\gamma(5-\omega)}{2}\,V-1\right)\right]^{\alpha_6} \\
&\quad \times\left[\frac{(5-\omega)(\gamma+1)}{7-\gamma-(\gamma+1)\omega}\left(1-\frac{3\gamma-1}{2}\,V\right)\right]^{\alpha_7} \\
&\quad \times\left[\frac{\gamma+1}{\gamma-1}\left(1-\frac{5-\omega}{2}\,V\right)\right]^{-\{\alpha_6+\alpha_7-[2/(5-\omega)]\}}, \\[6pt]
\frac{v}{v_2} &= f = \frac{(5-\omega)(\gamma+1)}{4}\,V\frac{r}{r_2}, \\[6pt]
\frac{\rho}{\rho_2} &= g = \left[\frac{(5-\omega)(\gamma+1)}{4}\,V\right]^{2\omega/(5-\omega)}\left[\frac{\gamma+1}{\gamma-1}\left(1-\frac{5-\omega}{2}\,V\right)\right]^{\alpha_5} \\
&\quad \times\left[\frac{\gamma+1}{\gamma-1}\left(\frac{\gamma(5-\omega)}{2}\,V-1\right)\right]^{\alpha_3+\omega\alpha_2} \\
&\quad \times\left[\frac{(5-\omega)(\gamma+1)}{7-\gamma-(\gamma+1)\omega}\left(1-\frac{3\gamma-1}{2}\,V\right)\right]^{\alpha_4+\omega\alpha_1}, \\[6pt]
\frac{p}{p_2} &= h = \left[\frac{(5-\omega)(\gamma+1)}{4}\,V\right]^{6/(5-\omega)}\left[\frac{\gamma+1}{\gamma-1}\left(1-\frac{5-\omega}{2}\,V\right)\right]^{\alpha_5+1} \\
&\quad \times\left[\frac{(5-\omega)(\gamma+1)}{7-\gamma-(\gamma+1)\omega}\left(1-\frac{3\gamma-1}{2}\,V\right)\right]^{\alpha_4+\alpha_1(\omega-2)}, \\[6pt]
\frac{T}{T_2} &= \frac{p}{p_2}\frac{\rho_2}{\rho},
\end{aligned}\right\} \quad (14.14)
$$

9*

where

$$\alpha_1 = \frac{(5-\omega)\gamma}{3\gamma-1}\left[\frac{2(6-3\gamma-\omega)}{\gamma(5-\omega)^2} - \alpha_2\right],$$

$$\alpha_2 = \frac{1-\gamma}{(2\gamma+1-\gamma\omega)}, \qquad \alpha_3 = \frac{-\omega+3}{(2\gamma+1-\gamma\omega)},$$

$$\alpha_4 = \frac{(5-\omega)(3-\omega)\alpha_1}{(6-3\gamma-\omega)}, \qquad \alpha_5 = \frac{\omega(1+\gamma)-6}{(6-3\gamma-\omega)},$$

$$\alpha_6 = \frac{\gamma}{(2\gamma+1-\gamma\omega)}, \qquad \alpha_7 = \frac{(3\gamma-1)\alpha_1}{(6-3\gamma-\omega)}.$$

$$\left.\right\} \quad (14.15)$$

Formulas (14.14) give the complete solution from the Eulerian and Lagrange viewpoints. These formulas show that the distribution of the nondimensional variables in the motion will be identical for different values of the explosion energy E_0.

The curves $v/v_2 = f(\lambda, \omega)$; $\rho/\rho_2 = g(\lambda, \omega)$; $p/p_2 = h(\lambda, \omega)$ are shown in Figs. 93, 94 and 95 for $\gamma = 1\cdot4$ and for $\omega = 2, 1, 0, -1, -2, 7/3$. The gas motion extends right up to the centre of symmetry in all these cases.

The curves for $\omega = 7/3, 2\cdot5, 3, 3\cdot5$ with $\gamma = 7/5$ and $\omega = 2, 2\cdot5, 3, 3\cdot5$ for $\gamma = 5/3$ are shown in Figs. 96, 97 and 98. A vacuum (a region without mass whose motion is taken into account) is formed near the centre of the explosion for each γ in the last three cases.

The following asymptotic formulas hold near the centre of the explosion for $\omega < (7-\gamma)/(\gamma+1)$

$$\frac{v}{v_2} = \frac{\gamma+1}{2\gamma}\frac{r}{r_2},$$

$$\frac{p}{p_2} = \left(\frac{\gamma+1}{2\gamma}\right)^{6/(5-\omega)}\left(\frac{\gamma+1}{\gamma}\right)^{\alpha_5+1}\left(\frac{\omega-\dfrac{2\gamma+1}{\gamma}}{\omega-\dfrac{7-\gamma}{\gamma+1}}\right)^{\alpha_1+\alpha_1(\omega-2)},$$

$$\frac{\rho}{\rho_2} = \left(\frac{r}{r_2}\right)^{-(\alpha_3/\alpha_2)-\omega}\left(\frac{\gamma+1}{\gamma}\right)^{\alpha_5}\left(\frac{\gamma+1}{2\gamma}\right)^{-[2/(5-\omega)](\alpha_3/\alpha_2)}$$

$$\times\left[\frac{\omega-\dfrac{2\gamma+1}{2}}{\omega-\dfrac{7-\gamma}{\gamma+1}}\right]^{\alpha_4-(\alpha_1/\alpha_2)\alpha_3}.$$

$$\left.\right\} \quad (14.16)$$

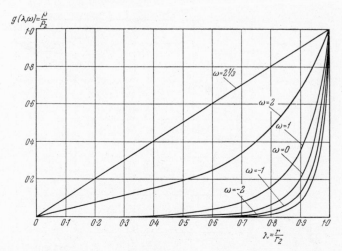

FIG. 94. Density field for a point explosion in a gas with variable initial density distribution $\rho_1 = A/r_0^\omega$, $\gamma = 1 \cdot 4$. The motion extends to the centre of symmetry for $\omega \leqslant 7/3$.

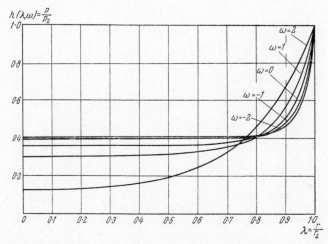

FIG. 95. Pressure field for a point explosion in a gas with the variable initial density distribution $\rho_1 = A/r_0^\omega$, $\gamma = 1 \cdot 4$. The motion extends to the centre of symmetry for $\omega \leqslant 7/3$.

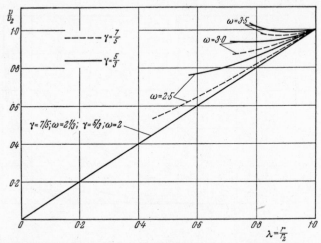

FIG. 96. Velocity field for a point explosion in a gas with the variable initial density distribution $\rho_1 = A/r_0^\omega$, $\gamma = 7/5$ and $\omega \geqslant 7/3$; $\omega \geqslant 2$ for $\gamma = 5/3$.

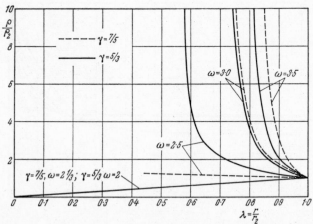

FIG. 97. Density field for a point explosion in a gas with the variable density distribution $\rho_1 = A/r_0^\omega$; $\omega \geqslant 7/3$ for $\gamma = 1\cdot4$ and $\omega \geqslant 2$ for $\gamma = 5/3$.

It follows from (14.16) that if $\omega > 0$, the density approaches zero rapidly as $r \to 0$ while the pressure is not zero; however, the pressure approaches zero at the centre as $\omega \to (7 - \gamma)/(\gamma + 1)$.

The asymptotic and exact formulas are in agreement and are given by (14.11) for $\omega = (7 - \gamma)/(\gamma + 1)$.

If $\omega > (7 - \gamma)/(\gamma + 1)$, then a vacuum extending from the centre of the explosion is formed with boundary $V = 2/(5 - \omega)$, where the following asymptotic formulas hold:

$$\left.\begin{aligned}
\frac{v}{v_2} &= \frac{\gamma + 1}{2}\frac{r^*}{r_2}, \\[2mm]
\frac{\rho}{\rho_2} &= A\left(\frac{r}{r_2} - \frac{r^*}{r_2}\right)^{\alpha_5}, \\[2mm]
\frac{p}{p_2} &= B\left(\frac{r}{r_2} - \frac{r^*}{r_2}\right)^{\alpha_5 + 1}
\end{aligned}\right\} \quad (14.17)$$

where

$$\frac{r^*}{r_2} = \left(\frac{\gamma + 1}{2}\right)^{-2/(5-\omega)}(\gamma + 1)^{-\alpha_2}\left(\frac{3\gamma + \omega - 6}{\omega - \dfrac{7 - \gamma}{\gamma + 1}}\right)^{-\alpha_1},$$

$$A = \left(\frac{\gamma + 1}{2}\right)^{2\omega/(5-\omega)}(\gamma + 1)^{\alpha_3 + \omega\alpha_2}\left(\frac{3\gamma + \omega - 6}{\omega - \dfrac{7 - \gamma}{\gamma + 1}}\right)^{\alpha_4 + \omega\alpha_1}$$

$$\times \left[\frac{\gamma(1 - \gamma)}{(\gamma + 1)(3\gamma + \omega - 6)}\right]^{-\alpha_5}\left(\frac{r^*}{r_2}\right)^{-\alpha_5},$$

$$B = \left(\frac{\gamma + 1}{2}\right)^{6/(5-\omega)}\left(\frac{3\gamma + \omega - 6}{\omega - \dfrac{7 - \gamma}{\gamma + 1}}\right)^{\alpha_4 + \alpha_1(\omega - 2)}$$

$$\times \left[\frac{\gamma(1 - \gamma)}{(\gamma + 1)(3\gamma + \omega - 6)}\right]^{-(\alpha_5 + 1)}\left(\frac{r^*}{r_2}\right)^{-(\alpha_5 + 1)}$$

in which α_1, α_2, α_3, α_4, α_5 are determined from (14.15). Here r^* is the radius at $V = 2/(5 - \omega)$; a vacuum is obtained within the corresponding sphere of radius r^*. The ratio r^*/r_2 is shown in Fig. 99 as a function of ω.

The pressure equals zero for $\omega < 3$ at $r = r^*$, the density equals zero for $\omega > 6/(1+\gamma)$ and is infinite for $\omega > 6/(\gamma+1)$. The pressure at the internal boundary is infinite for $\omega > 3$.

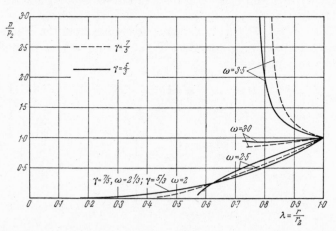

FIG. 98. Pressure field for a point explosion in a gas with variable density distributed as $\rho_1 = A/r_0^\omega$; $\omega \geqslant 7/3$ for $\gamma = 1 \cdot 4$ and $\omega \geqslant 2$ for $\gamma = 5/3$.

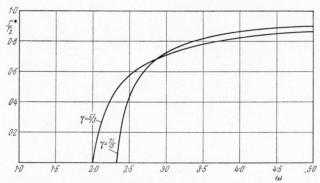

FIG. 99. r^* is the radius of the interior spherical boundary: r_2 is the radius of the shock wave sphere. The ratio r^*/r_2 is shown as a function of the exponent ω when $\rho_1 = A/r_0^\omega$.

It is clear from the asymptotic formulas established that the energy of the disturbed motion is finite for $\omega < 3$. It can also be seen that this energy is infinite for $\omega > 3$ because the pressure is always infinite at $r = r^*$ and, consequently, the work being transmitted by the gas to this boundary becomes infinite.

§15. Unsteady Motion of a Gas when the Velocity is Proportional to Distance from the Centre of Symmetry[†]

We consider the adiabatic unsteady motion of a perfect gas in the spherical, cylindrical and plane symmetry cases ($\nu = 1, 2, 3$) when the velocity distribution can be represented as[‡]:

$$v = r\Phi(t) \tag{15.1}$$

where r is the distance from the centre of symmetry.

The formula (15.1) is correct for self-similar motions if the independent dimensions of the characteristic constants are:

$$[a] = ML^k T^s \quad \text{and} \quad [b] = T \quad (m = 0)$$

and, therefore, the variable $\lambda = t/b$. The general class of self-similar solutions for arbitrary values of a and b include the solutions (15.1) as special cases. In particular, the solutions, corresponding to $V = V^* = \text{const}$, have the form of (15.1),

$$v = (r/t) V^*. \tag{15.2}$$

Hence, the particular solutions corresponding to the singular points in the z, V plane with constant values of z^* and V^* belong to the type being considered.

We recall that solutions of the problem of propagation of a detonation wave in a medium with variable density $\rho_1 = A/r^\omega$ for $\omega = [3(\gamma + 1)]/(3\gamma - 1)$ [formula (8.4)], the problem of an intense point explosion for $\rho_1 = \text{const}$ and $\gamma = 7$ [formula (11.21)] and the problem of an intense point explosion in a medium with variable initial density for $\omega = (7 - \gamma)/(\gamma + 1)$ [formula (14.11)] all have velocity distributions of the type (15.1).

We note that in all these problems Equation (15.1) is a limit solution separating the solution which can be continued to the centre of symmetry from that in which an extensive vacuum is formed near the centre of symmetry.

If we ignore Newtonian gravitational forces then we find from Equations (1.3) and (15.1), that the required solution for adiabatic

---

[†] The formulas for this solution were first published in Sedov (1953). The classification and detailed investigation of these motions were published in the 3rd ed. of this book (1954).

[‡] It can be deduced from the equations of motion that (15.1) follows from the assumption $v = f_1(r)f_2(t)$ if $f_2 \neq \text{const.}$

motion of a perfect gas is

$$
\left.
\begin{aligned}
dt &= \pm \frac{d\mu}{\mu^2 [A + B\mu^{\nu(\gamma-1)}]^{1/2}}, \\[2mm]
v &= \frac{dr}{dt} = -\frac{1}{\mu}\frac{d\mu}{dt} r = \mp \mu [A + B\mu^{\nu(\gamma-1)}]^{1/2} r, \\[2mm]
\rho &= \frac{\mu^{\nu-1}}{r} \phi'(\mu r), \\[2mm]
p &= \mu^{\gamma\nu}\left\{ C + \frac{(\gamma-1)\nu}{2} B\phi(\mu r) \right\}
\end{aligned}
\right\} \quad (15.3)
$$

where A, B and C are arbitrary constants, and $\phi(\mu r)$ is an arbitrary function. Without loss of generality, the function $\mu(t)$ can always be considered positive.

Hence, formulas (15.3) determine the exact solution of (1.3) depending on the arbitrary function $\phi(\mu r)$ and on the three arbitrary constants A, B, C.

Since the density ρ is positive, the function $\phi'(\mu r)$ must satisfy the condition

$$
\phi'(\mu r) > 0.
$$

It follows from (15.3) that

$$
\frac{p}{\rho^\gamma} = \Phi(r\mu) = \frac{\left[C + \dfrac{\gamma-1}{2}\nu B\phi(\mu r) \right](\mu r)^\gamma}{\phi'^\gamma(\mu r)}. \tag{15.4}
$$

Since the entropy is constant on each particle, the solution is defined by

$$
r = \frac{F(\xi)}{\mu(t)}
$$

where ξ is the Lagrangian coordinate. The function $\mu(t)$ is determined by the first of Equations (15.3); this function completely determines the motion of all the gas particles, depending on the values of the constants A and B.

Let us denote the coordinate r of a gas particle at a certain instant t_0, corresponding to the value $\mu(t_0) = \mu_0$ by ξ. Then

$$
F(\xi) = \mu_0 \xi
$$

and, the motion of the different gas particles can therefore be written

$$
r = \frac{\mu_0}{\mu(t)}\xi; \quad v = \frac{dr}{dt} = -\frac{\mu_0 \mu'(t)}{\mu^2(t)}\xi. \tag{15.5}
$$

Formula (15.5) is also obtained directly from the second of Equations (15.3).

It follows from (15.4) that the arbitrary function $\phi(\mu r)$ is directly related to the entropy distribution through the gas.

From the last two of Equations (15.3) we obtain for the pressure and density at any instant

$$\frac{\partial p}{\partial r} = B\frac{\gamma - 1}{2}\nu\mu^{\nu(\gamma-1)+2}r\rho. \tag{15.6}$$

The pressure gradient per particle at a given point of space depends on the time through μ. The pressure increases with r for $B > 0$, for $B < 0$ it decreases with distance from the bounding sphere.

The $r(t)$ relation gives rise to three main cases

I. $A > 0, B > 0$ ⎫ Pressure increases with distance from the
II. $A < 0, B > 0$ ⎭ centre of symmetry.
III. $A > 0, B < 0$ Pressure decreases with distance from centre
        of symmetry.

In the first case the first of the formulas in (15.3) and (15.5) can be written

$$\tau = \sqrt{(A)}\mu_0 t = \mp \int_0^y \sqrt{\left(\frac{y^{\nu(\gamma-1)}}{1+y^{\nu(\gamma-1)}}\right)} dy,$$

$$y = \frac{\mu_0}{\mu} = \frac{r}{\xi}, \quad \mu_0 = \left(\frac{A}{B}\right)^{1/[\nu(\gamma-1)]}. \tag{15.7}$$

From these formulas we see that the universal curves $y(\tau)$ shown in Fig. 100 for different ν enable us to obtain the motion of each particle easily for any $A > 0$ and $B > 0$.

All the gas particles are concentrated at the centre of symmetry for $\tau = t = 0$, an explosion occurs as t increases, which starts with infinitely large velocity. On each particle the coordinate r equals ξ for $y = 1$. The corresponding time is found from

$$\tau = \int_0^1 \sqrt{\left(\frac{y^{\nu(\gamma-1)}}{1+y^{\nu(\gamma-1)}}\right)} dy.$$

As $\tau \to \infty$; $y \to \infty$ and $\mu \to 0$; here, the velocity of each particle approaches a limiting value.

Since $dy/d\tau \underset{\tau \to \infty}{= 1}$ then $dr/dt = \xi\sqrt{(A)}\mu_0.$

For each particle $r\mu = \text{const}$; consequently, from (15.3), the density and pressure at $t = 0$ and $\mu = \infty$ become infinite and approach zero when $\mu \to 0$ and $t \to \infty$.

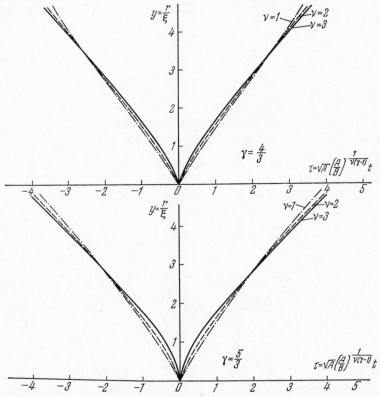

FIG. 100. Motion of gas particles in case I: $A > 0$, $B > 0$. For $t < 0$, implosion; for $t > 0$, explosion at infinity.

In case II the first of formulas (15.3) and (15.5) can be written in the analogous form

$$\left. \begin{aligned} \tau &= \sqrt{(|A|)}\,\mu_0 t = \mp \int_0^y \sqrt{\left(\frac{y^{\nu(\gamma-1)}}{1 - y^{\nu(\gamma-1)}} \right)}\, dy \\ y &= \frac{\mu_0}{\mu} = \frac{r}{\xi}, \quad \mu_0 = \left(\frac{|A|}{B} \right)^{1/[\nu(\gamma-1)]}. \end{aligned} \right\} \quad (15.8)$$

The corresponding universal curves for different ν are shown in Fig. 101. Cases I and II are completely analogous near time $t = 0$.

However, in contrast to case I, as the disturbance progresses coordinate r has a maximum value equal to $\xi(y = 1)$. This position is attained by all the particles simultaneously at τ^*, determined by

$$\tau^* = \sqrt{(|A|)}\,\mu_0 t^* = \int_0^1 \sqrt{\left(\frac{y^{\nu(\gamma-1)}}{1 - y^{\nu(\gamma-1)}}\right)}\,dy.$$

All the particles come to rest at the instant τ^* and a reverse motion towards the centre follows immediately under the influence of a pressure gradient with the result that all the particles arrive at the centre of symmetry simultaneously at the instant $2\tau^*$.

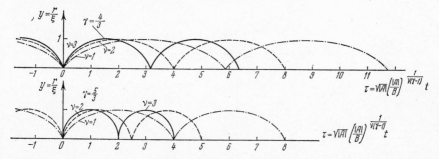

FIG. 101. Gas particle motion in case II: $A < 0$, $B > 0$. Explosion from the centre starting at time $t = 0$, then a state of rest at an infinite distance followed by implosion.

This process is repeated as the time increases further, and we get a pulsating periodic gas motion.

Finally, in case III, the first of formulas (15.3) and (15.5) become

$$\left. \begin{aligned} \tau &= \sqrt{(A)}\left(\frac{A}{|B|}\right)^{1/[\nu(\gamma-1)]} t = \mp \int_1^y \sqrt{\left(\frac{y^{\nu(\gamma-1)}}{y^{\nu(\gamma-1)} - 1}\right)}\,dy, \\ y &= \frac{\mu_0}{\mu} = \frac{r}{\xi}, \quad \mu_0 = \left(\frac{A}{|B|}\right)^{1/[\nu(\gamma-1)]}. \end{aligned} \right\} \quad (15.9)$$

The corresponding universal curves $y(\tau)$ are shown in Fig. 102 for different ν.

We have $y = 1$ and $d\mu/dt = 0$ for $\tau = t = 0$. The Lagrangian coordinate ξ is the value of the coordinate r at the instant $t = 0$, when all the gas particles are at rest. The particles move towards the centre of symmetry for $t < 0$, all are at rest at $t = 0$ and then an explosion starts. The explosion of gas concentrated at the centre of symmetry at $t = 0$ occurs in cases I and II.

Case III corresponds to a distributed mass of gas at rest which is exploded at $t = 0$ since the pressure decreases as r increases, according to (15.6).

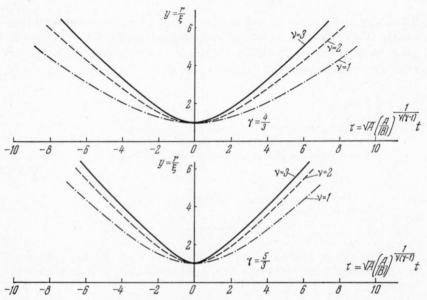

FIG. 102. Gas particle motion for case III: $A > 0$, $B < 0$. Motion from infinity toward the centre with all the particles stationary at $t = 0$, and at an infinite distance from the centre for $t < 0$. Explosion to infinite for $t > 0$.

If the expression

$$\left\{ C + \frac{(\gamma - 1)\nu}{2} B\phi(\mu r) \right\}$$

becomes zero on a certain sphere for $\nu = 3$ (on a cylinder for $\nu = 2$ and on a plane for $\nu = 1$), then the pressure will always vanish on this sphere, which, as it moves, contains the same gas particles. This sphere is the external boundary of the disturbance due to an exploding mass of gas in a vacuum.

We can thus find the exact solution of the problem of a gas explosion in a vacuum in a simple analytic form when the initial density and pressure distributions are related by means of (15.6).

As an example, we take as our function $\phi(r\mu)$ the form

$$\phi(r\mu) = \frac{\rho_0}{\mu_0^\nu} \frac{(r\mu)^2}{2}.$$

Then for the density and pressure distributions, we have

$$\rho = \rho_0 \left(\frac{\mu}{\mu_0}\right)^\nu = \frac{\rho_0}{y^\nu},$$

$$p = \left[C - A\rho_0 \mu_0^{-\nu\gamma} \frac{(\gamma-1)^\nu}{4} (r\mu)^2\right] \mu^{\nu\gamma}.$$

The density is independent of the geometric coordinates and it decreases with time. The pressure is distributed according to a parabolic law. The radius of the expanding sphere bounding the vacuum is determined by the formula

$$r^* = \sqrt{\left(\frac{4C}{A\rho_0\nu(\gamma-1)\mu_0^{-\nu\gamma}}\right)} \frac{1}{\mu}.$$

Let us consider the case $A = 0$ or $B = 0$. If $A = 0$ and $B > 0$, then

$$\mu = \left[\pm \sqrt{(B)}t\left(1 + \frac{\nu(\gamma-1)}{2}\right)\right]^{-1/[1+\nu(\gamma-1)/2]} = \mu_0 \left(\frac{t}{t_0}\right)^{-2/[2+\nu(\gamma-1)]}. \quad (15.10)$$

Therefore, $r\mu = \xi\mu_0$, $r = \xi(t/t_0)^{2/[2+\nu(\gamma-1)]}$ where μ_0 and t_0 are constants. The gas motion is self-similar and corresponds to case I. We have an explosion for $t_0 > 0$ and motion toward the centre of symmetry for $t_0 < 0$. If $B = 0$ and $A > 0$, then

$$\mu = \mp \frac{1}{\sqrt{A}} \frac{1}{t} = \mu_0 \frac{t_0}{t}, \quad r\mu = \xi\mu_0, \quad r = \xi\frac{t}{t_0}. \quad (15.11)$$

The corresponding motion is also self-similar and belongs to case III, however, all the gas particles are concentrated at the centre of symmetry in this case. Each gas particle moves at a constant velocity equal to ξ/t_0.

According to (15.3), we have in this case:

$$p = c_{-1}\left(\frac{\mu_0 t_0}{t}\right)^{\nu\gamma}, \quad \rho = \left(\frac{\mu_0 t_0}{t}\right)^\nu \Phi(\mu_0\xi). \quad (15.12)$$

The pressure is constant through space and falls with time, the density falls in inverse proportion to t^ν.

Exact solutions analogous to (15.3) can be constructed (Lidov, 1954) in the problem of the motion of a perfect gas taking into account Newtonian gravitational forces between the gas particles. If we put $\gamma = 4/3$, we have the following solution for the system (3.1), (3.2), (3.3) and (3.4) with $\nu = 3$:

$$
\left.
\begin{aligned}
dt &= \pm \frac{d\mu}{\mu^2 (A + B\mu)^{1/2}}, \\
v &= -\frac{1}{\mu}\frac{d\mu}{dt} r, \qquad r = \frac{\xi}{\mu(t)}, \\
\rho &= \mu^3(t)\,\xi \Psi(\xi), \quad p = \mu^4(t)\,\Phi(\xi).
\end{aligned}
\right\} \quad (15.13)
$$

Formulas (15.13) and (15.3) are completely similar when $\gamma = 4/3$ and $\nu = 3$; a difference occurs only because of the relation between the functions $\Psi(\xi)$ and $\Phi(\xi)$, which in the presence of gravitation is:

$$
\frac{d\Phi}{d\xi} \xi + 4\pi f \Psi(\xi) \int_0^\xi \eta^3 \Psi(\eta)\, d\eta = \frac{B}{2} \xi^3 \Psi(\xi), \qquad (15.14)
$$

where A and B are arbitrary constants and $\Psi(\xi)$ is an arbitrary function.

It is evident that the solution (15.13) is a direct generalization of the solution (15.3) considered above. The classification of the solution given above is retained in this case. The presence of gravitational forces does not alter the general qualitative type of gas motion.

If γ is a number, then self-similar motion in which the velocity is a linear function of the radius (Staniukovich, 1949; Rosseland, 1952) can be obtained from the equations of motion (3.1), (3.2), (3.3) and (3.4) taking gravitation into account. We can, for example, find a self-similar motion which depends on only two independent dimensional constants. These have the dimensions of the gravitational constant $[f] = M^{-1} L^3 T^{-2}$, which enters into the equations of motion and the constant t_0, a certain characteristic time. From general considerations it follows that this solution must be

$$
v = \frac{r}{t} V\left(\frac{t}{t_0}\right), \quad \rho = \frac{1}{ft^2} R\left(\frac{t}{t_0}\right), \quad p = \frac{r^2}{ft^4} P\left(\frac{t}{t_0}\right) \quad \text{and} \quad \mathcal{M} = \frac{r^3}{ft^2} M\left(\frac{t}{t_0}\right).
$$
$$(15.15)$$

After substitution of (15.15) into the equations of motion, we find

$$
\left.
\begin{aligned}
v &= -\frac{1}{\mu}\frac{d\mu}{d\tau}\frac{r}{t_0}, \qquad && \tau = \frac{t}{t_0}, \\
\rho &= \frac{1}{ft_0^2}\mu^3(\tau), \qquad && \mathcal{M} = \frac{4\pi r^3}{3ft_0^2}\mu^3(\tau), \\
p &= \kappa \frac{r^2}{ft_0^4}\mu^{3\gamma+2}(\tau),
\end{aligned}
\right\} \quad (15.16)
$$

where κ is an arbitrary constant and the function $\mu(\tau)$ is determined from the integral

$$\tau = \int \frac{d\mu}{\mu^2 \sqrt{\left(\chi + \frac{8\pi}{3} \mu + \frac{4\kappa}{3(\gamma - 1)} \mu^{3(\gamma - 1)} \right)}} . \qquad (15.17)$$

Here χ is an arbitrary constant. The family of solutions (15.16) depends essentially on two nondimensional constants χ and κ.

It is easy to see that the family of solutions (15.16) can be generalized if the formula for the pressure in (15.16) is replaced by the new formula

$$p = \kappa \frac{r^2}{ft_0^4} \mu^{3\gamma + 2} + \beta \frac{r_0^2}{ft_0^4} \mu^{3\gamma}, \qquad (15.18)$$

in which βr_0^2 is a new arbitrary constant with the dimensions of length squared. Actually, the pressure enters into the momentum and adiabatic equations. The additional term in (15.18) depends only on time and, consequently, is eliminated in the momentum equation; the combination p/ρ^γ in the abiabatic equation is altered only by an additive constant when the term is introduced, so that the adiabatic equation is also satisfied.

The motion of each particle in the family of solutions obtained is represented by

$$r = \frac{\mu(\tau_0)}{\mu(\tau)} \xi, \qquad (15.19)$$

where ξ is the Lagrangian coordinate; $\xi = r$ at $\tau = \tau_0$. The variation of the coordinate r with time for each particle is defined by means of the identical function $\mu(\tau)$. The function $\mu(\tau)$ can change character for different values of χ and κ. In particular, $\mu(\tau)$ can be a positive periodic function oscillating between two positive values μ_1 and μ_2 which make the radical

$$f_1(\mu, \kappa, \chi, \gamma) = \chi + \frac{8\pi}{3} \mu + \frac{4\kappa}{3(\gamma - 1)} \mu^{3(\gamma - 1)}$$

zero in the integral of (15.17).

We now examine the behaviour of the function $f_1(\mu, \kappa, \chi, \gamma)$. If $\kappa > 0$ when $\gamma > 4/3$, $f_1(\mu, \kappa, 0, \gamma)$ increases monotonically as μ increases from zero to infinity, consequently, there is only one positive root for $\chi < 0$. If $\kappa < 0$ and $\chi = 0$, then there are two positive roots: $\mu_1^* = 0$ and $\mu_2^* = \{ -[2\pi(\gamma - 1)/\kappa] \}^{1/(3\gamma - 4)}$.

The function f_1 has a maximum in the μ_1^*, μ_2^* interval for $\kappa < 0$ which equals $\chi + 8\pi/9[(3\gamma - 4)/(\gamma - 1)](-2\pi/3\kappa)^{1/3\gamma - 4}$. Therefore, if $\gamma > 4/3$,

$\kappa < 0$ and $0 > \chi > -8\pi/9[(3\gamma-4)/(\gamma-1)](-2\pi/3\kappa)^{1/3\gamma-4}$, then the function $f(\chi, \kappa, \mu)$ has two positive roots μ_1 and μ_2 within the interval μ_1^*, μ_2^* and the corresponding motion is periodic.

In the case $\chi = \kappa = 0$ for $\gamma \neq 4/3$ or $\chi = 0$ and $\kappa \neq -2\pi(\gamma-1)$ for $\gamma = 4/3$ formula (15.17) gives

$$\tau = \frac{1}{\sqrt{3}\sqrt{\left(2\pi + \dfrac{\kappa}{\gamma-1}\right)}}\frac{1}{\mu^{3/2}}. \tag{15.20}$$

The additive constant and the sign in front of the root are not essential since they determine the origin and the direction of the time variation. If $\gamma = 4/3$, $\kappa = -2\pi(\gamma-1)$ and $\chi > 0$, then we shall have, instead of (15.20):

$$\tau = \frac{1}{\sqrt{\chi}}\frac{1}{\mu}. \tag{15.21}$$

All the appropriate formulas for $\gamma = 4/3$ are particular cases of the more general solution defined by (15.13) and (15.14).

Formulas (15.16), when (15.20) and (15.21) exist, determine all the variables of the gas motion in terms of power laws, giving monotonic rates of change.

This solution is characterized by a density which is constant along the radius.

According to (15.19), formula (15.18) written in terms of Lagrange variables for $\kappa = -\beta/\mu_0^2$ $(\beta > 0)$ is

$$p = \frac{\beta}{ft_0^4}(r_0^2 - \xi^2)\mu^{3\gamma}(\tau). \tag{15.22}$$

If we consider a gas sphere with radius r_0 at $\tau = \tau_0$, then it follows from (15.22) that the pressure is always zero on the surface of the sphere of radius r_0 when $\tau = \tau_0$. As the motion proceeds the pressure increases toward the centre of the sphere according to a parabolic law.

Hence, nonlinear periodic motions of a homogeneous gas sphere under the action of gravitational forces can be constructed by using the solution found. This motion is continuous, without shock waves. The maximum and minimum values of $\mu(\tau)$ are determined by the values of the square root in the integral (15.17). The corresponding values of the period of oscillation, τ^*, are determined by the formula

$$\tau^* = 2\int_{\mu_1}^{\mu_2}\frac{d\mu}{\mu^2\sqrt{\left(\chi + \dfrac{4\kappa}{3(\gamma-1)}\mu^{3(\gamma-1)} + \dfrac{8\pi}{3}\mu\right)}} = \Phi(\chi, \beta, \gamma).$$

Let us denote the whole mass of the gas sphere by $\mathfrak{M}$ and the gas density at time τ_0 and $\mu_0 = \mu(\tau_0)$ by ρ_0. Since $\rho_0 = \mu_0^3/ft_0^2$ and $\mathfrak{M} = [(4\pi r_0^3)/(3ft_0^2)]\mu_0^3$, the formula for the dimensional period of oscillation, t^*, can be written

$$t^* = \frac{\mu_0^{3/2}}{\sqrt{(f\rho_0)}}\,\Phi = \frac{\mu_0^{3/2}}{\sqrt{\left(f\,\dfrac{3\mathfrak{M}}{4\pi r_0^3}\right)}}\,\Phi(\chi, \beta, \gamma). \qquad (15.23)$$

The parameters χ, β, γ determine the amplitude and the form of the oscillations. The dependence of the period of the oscillations on the gravitational constant f, on the mass $\mathfrak{M}$ and on the radius r_0 is obtained in explicit form. If $r_0 = r_{\min}$, then $\mu_0 = \mu_1$ is a known function of χ, β, γ.

This solution of gas sphere oscillations was used by different authors to analyse star pulsations or cepheids (see Chapter V).

When (15.20) is applicable, we shall have

$$\left. \begin{aligned} v &= \frac{2}{3}\frac{r}{t}, \quad \rho = \frac{R_0}{ft^2}, \quad \mathcal{M} = \frac{4\pi r^3}{3}\frac{R_0}{ft^2} \\[2mm] p &= \frac{1}{ft_0^4}\left[\kappa r^2\left(\frac{R_0}{\tau^2}\right)^{2/3} + \beta r_0^2\right]\left(\frac{R_0}{\tau^2}\right)^{\gamma} \end{aligned} \right\} \qquad (15.24)$$

Here, for brevity, we introduced the notation

$$R_0 = \frac{1}{3\left(2\pi + \dfrac{\kappa}{\gamma-1}\right)}.$$

We use the solution (15.24) in Chapter V to construct the unsteady motion of gravitating gas masses of explosive type with shock waves present. The pressure is only time dependent for $\kappa = 0$. A solution of this type with the law of motion $r = \xi/[\mu(t)]$ can be found for gases, compressible media with more general physical properties, and for the magnetogasdynamical equations in certain cases.

§16. On the General Theory of One-Dimensional Motion of a Gas

The solution of boundary problems for the nonlinear system of Equations (1.3) is very difficult when self-similarity does not exist.

In order to obtain the results required, it is usually necessary to use numerical methods which rely on modern computers. In order to work out such methods and to construct analytic approximate methods of solution, various general relations, which occur for certain general classes of gas motion, appear to be useful (Sedov, 1952).

1. BEHAVIOUR OF THE SOLUTIONS OF THE EQUATIONS OF ONE-DIMENSIONAL UNSTEADY GAS MOTION NEAR THE CENTRE OF SYMMETRY

Let us consider the one-dimensional unsteady motion of a perfect gas by using the equations obtained by a simple transformation of Equations (1.3)

$$\frac{\partial \rho r^{\nu-1}}{\partial t} + \frac{\partial \rho v r^{\nu-1}}{\partial r} = 0, \tag{16.1}$$

$$\frac{\partial \rho v}{\partial t} + \frac{\partial \rho v^2}{\partial r} + (\nu-1)\frac{\rho v^2}{r} + \frac{\partial p}{\partial r} = 0, \tag{16.2}$$

$$\frac{\partial p}{\partial t} + v\frac{\partial p}{\partial r} + \gamma p\left[\frac{\partial v}{\partial r} + (\nu-1)\frac{v}{r}\right] = 0. \tag{16.3}$$

We can express $v(r, t)$ and $p(r, t)$ in terms of the density $\rho(r, t)$ from (16.1) and (16.2).

In fact, integrating Equation (16.1) under the condition $v = 0$ when $r = 0$, we have

$$v = -\frac{1}{r^{\nu-1}\rho}\int_0^r r^{\nu-1}\frac{\partial \rho}{\partial t}\,dr. \tag{16.4}$$

It then follows from (16.2) that

$$p(r, t) = p(0, t) - \rho v^2 - (\nu-1)\int_0^r \frac{\rho v^2}{r}\,dr + \int_0^r \frac{1}{\xi^{\nu-1}}\int_0^\xi \eta^{\nu-1}\frac{\partial^2 \rho(\eta, t)}{\partial t^2}\,d\eta\,d\xi. \tag{16.5}$$

We consider the class of gas motions in which the density can be expanded in the following power series near the centre of symmetry

$$\rho(r, t) = r^s[\omega_0(t) + \omega_1(t)r^m + \ldots + \omega_n(t)r^{nm} + \ldots], \tag{16.6}$$

where $\omega_0 \neq 0$, s and m are constants satisfying the inequalities

$$s+\nu+nm \neq 0, \quad s+2+nm \neq 0, \quad (n = 0, 1, 2, 3, \ldots). \tag{16.7}$$

Substituting the series (16.6) into (16.4), and integrating, we obtain

$$v = -r\frac{\dfrac{\omega_0'}{s+\nu} + \dfrac{\omega_1'}{s+\nu+m}r^m + \ldots + \dfrac{\omega_n'}{s+\nu+nm}r^{nm} + \ldots}{\omega_0 + \omega_1 r^m + \ldots + \omega_n r^{nm} + \ldots}. \tag{16.8}$$

The logarithmic term does not appear in the integration because of condition (16.7).

It is evident that the velocity $v(r, t)$ can be represented near $r = 0$ by the power series

$$v = r[\phi_0(t) + \phi_1(t)r^m + \ldots + \phi_n(t)r^{nm} + \ldots]. \tag{16.9}$$

Comparison of (16.8) and (16.9) leads to the following system of equations for the coefficients ϕ_i

$$\sum_{i=0}^{n} \phi_i \omega_{n-i} + \frac{\omega_n'}{s+\nu+nm} = 0. \qquad (16.10)$$

Evidently, from the system of Equations (16.10), the function $\phi_n(t)$ can be expressed in terms of the functions $\omega_0(t)$, $\omega_1(t)$, ..., $\omega_n(t)$ and their first derivatives by the determinant

$$\phi_n = \frac{1}{\omega_0^{n+1}} \begin{vmatrix} -\dfrac{\omega_n'}{s+\nu+nm}, & \omega_1 & \omega_2 ... \omega_{n-1} & \omega_n \\[2ex] -\dfrac{\omega_{n-1}'}{s+\nu+(n-1)m}, & \omega_0 & \omega_1 ... \omega_{n-2} & \omega_{n-1} \\[2ex] 0 & \omega_0 ... \omega_{n-3} & \omega_{n-2} \\[1ex] . & 0 & . & . \\[1ex] . & . & . & . \\[1ex] . & . & \omega_0 & \omega_1 \\[2ex] -\dfrac{\omega_0'}{s+\nu}, & 0 & 0 \quad 0 & \omega_0 \end{vmatrix} \qquad (16.11)$$

Substituting (16.6) and (16.9) into (16.5), we find, after integration, [logarithmic terms do not appear because of conditions (16.7)] that the pressure $p(r, t)$ is given by the power series

$$p(r, t) = \psi_{-1}(t) + r^{s+2}[\psi_0(t) + \psi_1(t)\, r^m + ... + \psi_n(t)\, r^{nm} + ...], \qquad (16.12)$$

where the functions $\psi_i(t)$ are given by the relations

$$\left.\begin{aligned}
\psi_{-1} &= p(0, t), \\[2ex]
\psi_0 &= \frac{\omega_0''}{(s+\nu)(s+2)} - \frac{(\nu+s+1)}{s+2}\phi_0^2\,\omega_0, \\[2ex]
\psi_1 &= \frac{\omega_1''}{(s+\nu+m)(s+2+m)} - \frac{\nu+s+1+m}{s+2+m}(2\phi_0\phi_1\omega_0 + \phi_0^2\,\omega_1), \\[2ex]
&\cdots\cdots\cdots\cdots\cdots\cdots\cdots\cdots\cdots \\[2ex]
\psi_n &= \frac{\omega_n''}{(s+\nu+nm)(s+2+nm)} - \frac{\nu+s+1+nm}{s+2+nm}\sum_{j=0}^{n}\omega_j\sum_{i=0}^{n-j}\phi_i\phi_{n-j-i}, \\[2ex]
&\cdots\cdots\cdots\cdots\cdots\cdots\cdots\cdots\cdots
\end{aligned}\right\} \qquad (16.13)$$

The order of summation in the general formula for ψ_n can be changed as follows

$$\sum_{j=0}^{n} \omega_j \sum_{i=0}^{n-j} \phi_i \phi_{n-j-i} = \sum_{k=0}^{n} \phi_{n-k} \sum_{i=0}^{i=k} \omega_i \phi_{k-i}.$$

Taking this and (16.10) into account, we obtain a general formula for ψ_n in the following simplified form

$$\psi_n = \frac{\omega_n''}{(s+\nu+nm)(s+2+nm)} + \frac{\nu+s+1+nm}{s+2+nm} \sum_{k=0}^{n} \frac{\phi_{n-k} \omega_k'}{s+\nu+km}. \quad (16.14)$$

Hence, if ϕ_n is determined from (16.11) and ψ_n from (16.14) the series (16.6), (16.9) and (16.12) satisfy Equations (16.1) and (16.2).

If we now substitute these series into (16.3), we obtain

$$\left\{ \psi_{-1}' + \gamma \psi_{-1} \sum_{n=0}^{\infty} \phi_n (\nu+nm) r^{nm} \right\} + r^{s+2}$$

$$\times \sum_{n=0}^{\infty} \left[\psi_n' + \sum_{i=0}^{n} \phi_i \psi_{n-i} (\gamma\nu + s + 2 + im(\gamma-1) + nm) \right] r^{nm} = 0. \quad (16.15)$$

This relation, together with (16.11) and (16.14), can be used to derive equations for the functions $p(0, t)$, $\omega_0(t)$, $\omega_1(t)$, ... Here we need to use the relations between s and m.

Equation (16.15) is satisfied for arbitrary s if both expressions in brackets vanish. The vanishing of the first bracketed expression is equivalent to the system of equations

$$\left. \begin{aligned} \psi_{-1}' + \nu\gamma\psi_{-1}\phi_0 &= 0, \\ (m+\nu)\psi_{-1}\phi_1 &= 0, \\ (2m+\nu)\psi_{-1}\phi_2 &= 0, \\ \cdot \quad \cdot \quad \cdot \quad \cdot \quad \cdot \quad \cdot \quad \cdot \quad & \\ (nm+\nu)\psi_{-1}\phi_n &= 0. \end{aligned} \right\} \quad (16.16)$$

Hence if $\psi_{-1} = p(0, t) \neq 0$, the system (16.16), with use of (16.11) for $km+\nu \neq 0$ $(k=1, 2, ...)$, yields

$$\phi_0 = -\frac{\omega_0'}{(s+\nu)\omega_0}, \quad \phi_1 = \phi_2 = ... = \phi_n = ... = 0. \quad (16.17)$$

Thus, in this case, the velocities vary linearly with r

$$v = \phi_0 r.$$

From (16.17) and (16.10), it follows that

$$\frac{\omega_n'}{s+\nu+nm} = \omega_n \frac{\omega_0'}{(s+\nu)\,\omega_0},$$

consequently,

$$\omega_0 = c_0\,\omega(t), \quad \omega_n = c_n\,\omega^{[(nm)/(s+\nu)]+1} \quad (n = 1, 2, 3, \ldots) \quad (16.18)$$

and, therefore, we have from (16.6)

$$\rho = r^s\,\omega[c_0 + c_1(r\omega^{1/(s+\nu)})^m + c_2(r\omega^{1/(s+\nu)})^{2m} + \ldots] = r^s\,\omega P'[(r\omega^{1/(s+\nu)})^{s+2}], \tag{16.19}$$

where

$$P(x) = c_0 x + (s+2)\left[c_1 \frac{x^{(m+s+2)/(s+2)}}{m+s+2} + c_2 \frac{x^{(2m+s+2)/(s+2)}}{2m+s+2} + \ldots\right]. \tag{16.20}$$

Since the coefficients $c_0, c_1, \ldots, c_n, \ldots$ and the index m are constant in value, $P(x)$ is to a certain extent an arbitrary function of its argument. Furthermore, it follows from the first of Equations (16.6), using the relation $\omega_0 = c_0\omega$, that

$$\psi_{-1} = p(0, t) = c_{-1}\omega^{\gamma\nu/(s+\nu)}. \tag{16.21}$$

From (16.14) we find

$$\psi_n = \frac{\omega_n''}{(s+\nu+nm)(s+2+nm)} - \frac{\nu+s+1+nm}{(s+2+nm)}\frac{\omega'}{(s+\nu)\,\omega}\frac{\omega_n'}{s+\nu+nm}.$$

Hence, from (16.18)

$$\psi_n = \frac{c_n\,\omega^{(nm)/(s+\nu)}}{(s+\nu)(s+2+nm)}\left[\omega'' - \frac{\nu+s+1}{s+\nu}\frac{\omega'^2}{\omega}\right] \quad (n = 0, 1, 2, \ldots). \tag{16.22}$$

By equating the first bracketed expression to zero in (16.15) we obtain expressions for all the functions $\omega_n(t)$, $\psi_n(t)$ and $\phi_n(t)$ in terms of the single function $\omega(t)$, which can be arbitrary, given by (16.17), (16.18), (16.21) and (16.22).

We now show that the solution obtained leads to the vanishing of the second expression when the function $\omega(t)$ is suitably chosen.

In fact, the appropriate infinite system of equations is, from (16.17),

$$\psi_n' - \frac{\omega'}{(s+\nu)\,\omega}\psi_n(\gamma\nu+s+2+nm) = 0 \quad (n = 0, 1, 2, \ldots).$$

It follows from (16.22) that all these equations are satisfied if the function $\omega(t)$ is a solution of the ordinary differential equation

$$\omega'' - \left(\frac{\nu+s+1}{\nu+s}\right)\frac{\omega'^2}{\omega} = \frac{(\gamma-1)\nu(s+\nu)}{2} B\omega^{(s+2+\nu\gamma)/(s+\nu)}, \quad (16.23)$$

where B is an arbitrary constant.

The general solution of (16.23) can be given as

$$t = \pm\frac{1}{s+\nu}\int\frac{d\omega}{\omega^{1+[1/(\nu+s)]}\sqrt{(A+B\omega^{[(\gamma-1)\nu/(s+\nu)]})}} = \pm\int\frac{d\mu}{\mu^2\sqrt{(A+B\mu^{\nu(\gamma-1)})}}, \quad (16.24)$$

where A is an arbitrary constant and $\mu = \omega^{1/(\nu+s)}$ is a quantity which can be considered positive for $c_0 > 0$ since $\omega_0 > 0$ on physical grounds. The choice of the lower limit in the integral in (16.24) is unimportant since this only affects the origin from which the time is measured.

Using (16.21), (16.22) and (16.23), the expansion (16.12) can be written

$$
\left.
\begin{aligned}
p = \omega^{(\gamma\nu)/(s+2)}\Bigg\{ & c_{-1} + \frac{(\gamma-1)\nu}{2}B \\
& \times\bigg[\frac{c_0(r\omega^{1/(s+\nu)})^{s+2}}{s+2} + \frac{c_1(r^{s+2}\omega^{(s+2)/(s+\nu)})^{(s+2+m)/(s+2)}}{s+2+m} \\
& \quad + \frac{c_2(r^{s+2}\omega^{-(s+2)/(s+\nu)})^{(s+2+2m)/(s+2)}}{s+2+2m} + \dots\bigg]\Bigg\}
\end{aligned}
\right\} \quad (16.25)
$$

which, with use of (16.20) becomes

$$p = \omega^{(\gamma\nu)/(s+2)}\left\{c_{-1} + \frac{(\gamma-1)\nu}{2(s+2)} BP[(r\omega^{1/(s+\nu)})^{s+2}]\right\}. \quad (16.25')$$

Replacing ω by ν, the solution obtained can be written

$$
\left.
\begin{aligned}
& dt = \pm\frac{d\mu}{\mu^2(A+B\mu^{\nu(\gamma-1)})^{1/2}}, \\
& \frac{dr}{dt} = v = -\frac{1}{\mu}\frac{d\mu}{dt}\cdot r = \mp\mu\sqrt{(A+B\mu^{\nu(\gamma-1)})}\,r, \\
& \rho = \mu^\nu(r\mu)^s\,P'[(r\mu)^{s+2}], \\
& p = \mu^{\gamma\nu}\left\{c_{-1} + \frac{(\gamma-1)\nu}{2(s+2)} BP[(r\mu)^{s+2}]\right\}.
\end{aligned}
\right\} \quad (16.26)
$$

The solution (16.26) agrees with the solution determined by formulas (15.3) in §15†, if we put

$$\phi(r\mu) = \frac{1}{s+2} P[(r\mu)^{s+2}].$$

We now return to (16.15). The solution considered above holds for any values of m and s.

If conditions (16.7) are satisfied, then it is evident that (16.26) is a general solution of (16.15) for $s+2 < 0$. This conclusion is valid even in the case when $s+2 > 0$ provided that $s+2 \neq km$ where k is a certain positive integer.

Terms containing $\ln r$ can appear in formulas (16.9) and (16.12) in cases when conditions (16.7) are not satisfied and, in particular, when $s = -2$.

If $s+2 = km$, where $k \geqslant 1$ is an integer, then corresponding first order terms appear in the first and second bracketed expressions starting with the term of order r^{km}. Equations (16.16) are valid for $n = -1, 0, ..., (k-1)$ and, consequently, formulas (16.17), (16.18), (16.21) and (16.22) are correct in this case for $n < k$.

The following formulas are obtained for $n \geqslant k$

$$\left.\begin{array}{l}\psi_0' + (km + \gamma\nu)\,\phi_0\psi_0 + (km+\nu)\,\gamma\psi_{-1}\phi_k = 0, \\[4pt] \cdots\cdots\cdots\cdots\cdots\cdots\cdots\cdots\cdots\cdots\cdots\cdots\cdots \\[4pt] \psi_{n-k}' + \sum_{i=0}^{n-k} \phi_i\psi_{n-k-i}(nm-im+i\gamma m+\gamma\nu) + (nm+\nu)\,\gamma\psi_{-1}\phi_n = 0.\end{array}\right\} \quad (16.27)$$

When combined with relations (16.11) and (16.14), these formulas can be considered as differential equations for the successive determination of $\omega_k, \omega_{k+1}, ..., \omega_n, ...$ when the function $\omega_0(t)$ is given.

Hence, all the coefficients $\omega_n(t)$, $\phi_n(t)$ and $\psi_n(t)$ can be expressed in terms of the single function $\omega_0(t)$, the constants γ, k, m and the constants of integration. The constants mentioned and the function $\omega_0(t)$ can be given arbitrarily.

If $\psi_{-1} = p(0, t) = 0$, then the first bracketed expression in (16.15) becomes zero. The vanishing of the second expression is equivalent to the system of equations

$$\psi_n' + \sum_{i=0}^{n} \phi_i\psi_{n-i}(s+2+\gamma\nu+nm+im(\gamma-1)) = 0, \qquad (16.28)$$

which can be integrated successively to determine the functions $\omega_0(t), \omega_1(t), ..., \omega_n(t)$ with the aid of (16.11) and (16.24). The solution

† This solution was first obtained by this approach.

is only arbitrary in respect of the constants of integration and the value of the exponent m. The series (16.16) for the density ρ_0, (16.9) for the velocity v and (16.12) for the pressure p can be written

$$
\left.
\begin{aligned}
g &= \frac{\rho}{\rho_2} = \lambda^s(\alpha_0 + \alpha_1 \lambda^m + \alpha_2 \lambda^{2m} + \ldots), \\[2ex]
f &= \frac{v}{v_2} = \lambda(\beta_0 + \beta_1 \lambda^m + \beta_2 \lambda^{2m} + \ldots), \\[2ex]
h &= \frac{p}{p_2} = \gamma_{-1} + \lambda^{s+2}(\gamma_0 + \gamma_1 \lambda^m + \gamma_2 \lambda^{2m} + \ldots),
\end{aligned}
\right\} \quad (16.29)
$$

where $\lambda = r/r_2$ and r_2, ρ_2, v_2, p_2 are certain characteristic quantities generally dependent on time, with the dimensions of length, density, velocity and pressure respectively. Nondimensional quantities α_i, β_i and γ_i replacing ω_i, ϕ_i and ψ_i can be considered as nondimensional functions of a certain nondimensional parameter τ, which can be introduced in place of the time t.

According to the above, all the coefficients of the series (16.29) can be expressed in terms of the functions $r_2(\tau)$, $\rho_2(\tau)$, $p_2(\tau)$, $v_2(\tau)$, $\alpha_0(\tau)$ for $s + 2 = km$ ($k > 0$). In particular, if there is a shock wave in the unsteady gas motion, the values of the dependent variables on the shock wave can be taken as r_2, ρ_2, v_2 and p_2.

Consider a self-similar motion of a gas in which the dependent variables, including r_2, ρ_2, v_2, p_2 are defined as functions of the time t and of constants, only two of which have independent dimensions

$$
a = M L^\chi T^\zeta \quad \text{and} \quad b = L T^\eta
$$

(where χ, ζ and η are certain real constants). Then relations of the following type are satisfied

$$
\left.
\begin{aligned}
r_2 &= \delta_1 \frac{b}{t^\eta}, \qquad \rho_2 = \delta_2 a b^{-\chi-3} t^{\eta(\chi+3)-\zeta}, \\[2ex]
v_2 &= \delta_3 \frac{b}{t^{\eta+1}}, \qquad p_2 = \delta_4 a b^{-\chi-1} t^{\eta(\chi+1)-\zeta-2}.
\end{aligned}
\right\} \quad (16.30)
$$

(Here δ_1, δ_2, δ_3 and δ_4 are abstract constants.) In this case, all the coefficients α_i, β_i and γ_i are independent of time and therefore are arbitrary constants.

For given δ_1, δ_2, δ_3 and δ_4 Equation (16.15) reduces to algebraic equations determining s, m and giving the coefficients α_i, β_i, γ_i in terms of the constants α_0 and γ_{-1} (when $\gamma_{-1} \neq 0$). Actually, the abstract constants δ_1, δ_2, δ_3 and δ_4 can be put equal to unity without loss of generality if the definitions of the constant coefficients α_i, β_i and γ_i are modified correspondingly.†

Using this, we find that for the self-similar motion (16.10) and (16.14) become

$$\beta_n = -\frac{f_n}{s+\nu+nm}\frac{\alpha_n}{a_0} - \frac{1}{a_0}\sum_{i=0}^{n-1}\alpha_{n-i}\beta_i \qquad (16.31)$$

and

$$\gamma_n = \frac{f_n(f_n-1)\alpha_n}{(s+\nu+nm)(s+2+nm)} + \frac{(\nu+s+1+nm)}{s+2+nm}\sum_{k=0}^{n}\frac{f_k\beta_{n-k}\alpha_k}{s+\nu+km} \qquad (16.32)$$

where

$$f_n = \eta(3+s+nm+\chi)-\zeta \quad (n = 0, 1, 2, \ldots). \qquad (16.33)$$

In this case the fundamental equation (16.15) becomes

$$\gamma_{-1}\Big\{\eta(\chi+1)-\zeta-2+\gamma\sum_{n=0}^{\infty}\beta_n(\nu+nm)\lambda^{nm}\Big\} + \lambda^{s+2}$$

$$\times\Big\{\sum_{n=0}^{\infty}\Big[(f_n-2)\gamma_n + \sum_{i=0}^{n}\beta_i\gamma_{n-i}(\gamma\nu+s+2+im(\gamma-1)+nm)\Big]\lambda^{nm}\Big\} = 0. \qquad (16.34)$$

Let $\gamma_{-1} \neq 0$ and $s+2 = km$ ($k \geqslant 1$). Equation (16.34) is equivalent to the system

$$\eta(\chi+1)-\zeta-2+\gamma\nu\beta_0 = 0,$$

$$\beta_1 = 0,$$

$$\beta_2 = 0,$$

. . . .

$$\beta_{k-1} = 0.$$

$$\gamma\gamma_{-1}\beta_k(km+\nu)+(f_0-2)\gamma_0+\beta_0\gamma_0(km+\gamma\nu) = 0.$$

. . . .

$$\gamma\gamma_{-1}(nm+\nu)+(f_{n-k}-2)\gamma_{n-k}$$
$$+ \sum_{i=0}^{n-k}\beta_i\gamma_{n-k-i}(nm-im+i\gamma m+\gamma\nu) = 0.$$

. . . .

$\left.\begin{array}{c}\\ \\ \\ \\ \\ \\ \\ \\ \\ \\ \\ \\ \\ \end{array}\right\}$ (16.35)

† This is equivalent to modifying the meaning of the dimensional parameters r_2, ρ_2, v_2 and p_2 while retaining the choice of the constants a and b as arbitrary.

From (16.31) and the first of Equations (16.35), we obtain

$$\beta_0 = -\frac{f_0}{s+\nu} = -\frac{[\eta(3+s+\chi)-\zeta]}{s+\nu}, \tag{16.36}$$

$$\beta_0 = \frac{2+\zeta-\eta(\chi+1)}{\gamma\nu}. \tag{16.37}$$

Since $\beta_1 = \beta_2 = \ldots = \beta_{k-1} = 0$ from (16.35), (16.31) will give, for $n = 0, 1, 2, \ldots, k-1$

$$\beta_0 = -\frac{[\eta(3+s+nm+\chi)-\zeta]}{s+\nu+nm}. \tag{16.38}$$

Hence it follows that for $k > 1$,

$$\zeta = (3-\nu+\chi)\eta \tag{16.39}$$

which is an additional condition imposed on the exponents in the formulas for the dimensions of the constants a and b. The exponent s remains arbitrary in this case.

If condition (16.39) is not satisfied, we must have $k = 1$ and, therefore

$$m = s+2 = \frac{\eta[\nu(\gamma-1)+2](\chi+1)+(\nu-2)(\zeta+2)-\gamma\nu\zeta}{\eta(\chi+1-\gamma\nu)-\zeta-2}. \tag{16.40}$$

Evidently, the constants α_0 and γ_{-1} remain arbitrary for $k = 1$, the rest of the constant coefficients in the expansion (16.29) are easily determined in succession by using (16.31), (16.32) and (16.35).

If condition (16.39) is satisfied, then $m = (s+2)/k$ where k, s, α_0, α_1, ..., α_{k-1} and γ_{-1} remain arbitrary, $\beta_1 = \beta_2 = \ldots = \beta_{k-1} = 0$; the rest of the coefficients are determined from Equations (16.31), (16.32) and (16.35).

If $\gamma_{-1} \neq 0$ and $s+2 \neq km$, where k is an integer, then we find from Equations (16.34)

$$\beta_0 = -\frac{f_0}{s+\nu} = \frac{\zeta-\eta(3+s+\chi)}{s+\nu}, \quad \beta_1 = \beta_2 = \ldots = \beta_n = \ldots = 0$$

and, moreover, we obtain

$$\gamma_n\left\{\frac{\zeta-n(3-\nu+\chi)}{s+\nu}nm-\frac{[-\zeta+\eta(3+s+\chi)][2+\nu(\gamma-1)]+2(s+\nu)}{s+\nu}\right\} = 0$$
$$(n = 0, 1, 2, \ldots). \tag{16.41}$$

Formulas (16.32) become in this case

$$\gamma_n = \frac{\alpha_n f_n}{(s+2+nm)(s+\nu+nm)}\left[f_n-1-\frac{(\nu+s+1+nm)f_0}{s+\nu}\cdot\right].$$

Equations (16.41) can be satisfied in the general case for arbitrary s, m and α_n $(n = 0, 1, 2, ...)$ if the exponents ζ, η satisfy the relations†

$$\eta = -\frac{1}{1 + \dfrac{\nu(\gamma - 1)}{2}} \quad \text{and} \quad \zeta = \eta(3 - \nu + \chi) \qquad (16.42)$$

the constant γ_{-1} can also be arbitrary. The corresponding solution is a particular case of self-similar motion in the exact solution (16.26).

If $\gamma_{-1} = 0$ and the pressure $p \not\equiv 0$, then the exponent s is determined from the formula

$$s = \frac{-2\nu - [\eta(3 + \chi) - \zeta][2 + \nu(\gamma - 1)]}{\eta[2 + (\gamma - 1)\nu] + 2}.$$

The coefficients α_0, α_1 and the exponent m remain undetermined, all the remaining coefficients are defined by formulas (16.31), (16.32) and (16.34).

2. ON THE DERIVATIVES OF v, ρ, p, T, S WITH RESPECT TO THE
    CHARACTERISTIC COORDINATES OF ONE-DIMENSIONAL UNSTEADY GAS
    MOTION.

We consider a certain point M, moving with velocity c in fixed space. Using Equations (1.3), the first derivatives of v, ρ, p and the temperature T and the entropy S with respect to the coordinate r can easily be expressed in terms of their values and the values of their derivatives with respect to time t at the moving point M.

We have, for an arbitrary function $F(r, t)$

$$\frac{dF}{dt} = \frac{d}{dt} F(r_M(t), t) = \frac{\partial F}{\partial t} + c\frac{\partial F}{\partial r}, \quad \text{where } c = \frac{dr_M}{dt}.$$

Using this relation, Equations (1.3) can be written

$$\left.\begin{aligned}
\frac{\partial v}{\partial r}\rho + \frac{\partial \rho}{\partial r}(v - c) &= -\frac{(\nu - 1)\rho v}{r} - \frac{d\rho}{dt}, \\[2mm]
\frac{\partial v}{\partial r}(v - c) + \frac{1}{\rho}\frac{\partial p}{\partial r} &= -\frac{dv}{dt}, \\[2mm]
\frac{\partial S}{\partial r}(v - c) &= -\frac{dS}{dt}.
\end{aligned}\right\} \qquad (16.43)$$

---

† It is evident that the system (16.41) can be satisfied in certain special cases when relations (16.42) are not satisfied.

In addition to these relations, we have, from the equation of state in the forms $p = p(\rho, S)$ and $T = T(\rho, S)$

$$\left.\begin{aligned}
\frac{\partial p}{\partial r} &= \left(\frac{\partial p}{\partial \rho}\right)_S \frac{\partial \rho}{\partial r} + \left(\frac{\partial p}{\partial S}\right)_\rho \frac{\partial S}{\partial r}, \\
\frac{\partial T}{\partial r} &= \left(\frac{\partial T}{\partial \rho}\right)_S \frac{\partial \rho}{\partial r} + \left(\frac{\partial T}{\partial S}\right)_\rho \frac{\partial S}{\partial r}.
\end{aligned}\right\} \quad (16.44)$$

When the linear systems of Equations (16.43) and (16.44) are solved for the partial derivatives with respect to r, we arrive at formulas which are correct at a moving point M for the adiabatic motion of any gas

$$\left.\begin{aligned}
\frac{\partial v}{\partial r} &= -\frac{(\nu-1)\rho v \left(\frac{\partial p}{\partial \rho}\right)_S}{r\Delta} - \frac{1}{\Delta}\left[(c-v)\rho\frac{dv}{dt} + \frac{dp}{dt}\right], \\[2ex]
\frac{\partial \rho}{\partial r} &= -\frac{(\nu-1)\rho^2 v(c-v)}{r\Delta} \\
&\quad - \frac{1}{\Delta}\left[(c-v)\rho\frac{d\rho}{dt} + \rho^2\frac{dv}{dt} + \left(\frac{\partial p}{\partial S}\right)_\rho \frac{\rho}{c-v}\frac{dS}{dt}\right], \\[2ex]
\frac{\partial p}{\partial r} &= -\frac{(\nu-1)\rho^2 v(c-v)\left(\frac{\partial p}{\partial \rho}\right)_S}{r\Delta} \\
&\quad - \frac{1}{\Delta}\left[(c-v)\rho\frac{d\rho}{dt} + \rho^2\frac{dv}{dt} - (c-v)\rho\left(\frac{\partial \rho}{\partial S}\right)_p \frac{dS}{dt}\right]\left(\frac{\partial p}{\partial \rho}\right)_S, \\[2ex]
\frac{\partial T}{\partial r} &= -\frac{(\nu-1)\rho^2 v(c-v)\left(\frac{\partial T}{\partial \rho}\right)_S}{r\Delta} - \frac{1}{\Delta}\left[\rho^2\left(\frac{\partial T}{\partial \rho}\right)_S \frac{dv}{dt} + (c-v)\rho\frac{dT}{dt}\right. \\
&\quad \left. - \frac{\rho}{c-v}\left(\frac{\partial p}{\partial \rho}\right)_T \left(\frac{\partial T}{\partial S}\right)_\rho \frac{dS}{dt}\right], \\[2ex]
\frac{\partial S}{\partial r} &= \frac{1}{(c-v)}\frac{dS}{dt}, \quad \text{when } \Delta = \rho\left[\left(\frac{\partial p}{\partial \rho}\right)_S - (c-v)^2\right].
\end{aligned}\right\} \quad (16.45)$$

The following well-known identities for the Jacobian are used in deriving (16.45)

$$\left(\frac{\partial p}{\partial S}\right)_\rho = \frac{D(\rho, p)}{D(\rho, S)} = \frac{D(\rho, p)}{D(p, S)}\frac{D(p, S)}{D(\rho, S)} = -\left(\frac{\partial \rho}{\partial S}\right)_p \left(\frac{\partial p}{\partial \rho}\right)_S$$

and

$$\frac{D(T,\,\rho)}{D(\rho,\,S)} = \frac{D(T,\,p)}{D(T,\,\rho)}\frac{D(T,\,\rho)}{D(\rho,\,S)} = \left(\frac{\partial p}{\partial \rho}\right)_T\left(\frac{\partial T}{\partial S}\right)_\rho.$$

If the relative velocity of the particle motion of the point M is $u = c - v$, which equals the speed of sound $\pm (\partial p/\partial \rho)_S^{1/2}$, then $\varDelta = 0$ and, therefore, the determination of the desired derivatives becomes possible only if the corresponding numerators become zero. The condition $\varDelta = 0$ is satisfied on the characteristics of the system of Equations (16.43) and (16.44).

Formulas (16.45) can be used in the case when the point M coincides with the front of an intense explosion. From the Hugoniot conditions, all the quantities behind the shock can be expressed in terms of the parameters defining the state of the region into which the shock propagates and the velocity $c(r_2) = dr_2/dt$ defined by the motions of the shock motion (r_2 is the shock coordinate).

In particular, let us consider the case when the shock wave is propagated from the centre $(c > 0)$ through a perfect gas at rest with the density ρ_1 and pressure p_1. As before, let us denote

$$\frac{a_1^2}{c^2} = \frac{\gamma p_1}{\rho_1 c^2} = q, \quad \frac{v}{v_2} = f, \quad \frac{\rho}{\rho_2} = g, \quad \frac{p}{p_2} = h.$$

Using the conditions on the shock wave (12.6), formulas (16.45) can be transformed into

$$r_2\frac{\partial f}{\partial r} = -\frac{(\nu - 1)\left[1 + \dfrac{\gamma - 1}{\gamma + 1}(1 - q)\right]}{1 - q} + \frac{3 + q}{2q(1 - q)^2}r_2\frac{dq}{dr_2},$$

$$r_2\frac{\partial g}{\partial r} = -\frac{2(\nu - 1)}{\gamma + 1} - \frac{1}{\gamma + 1 - 2(1 - q)}$$

$$\times \left[\frac{2\left[1 + \dfrac{\gamma - 1}{\gamma + 1}(1 - q)\right]}{1 - \dfrac{2}{\gamma + 1}(1 - q)} - \frac{3(1 + q)}{q}\right]\frac{r_2}{1 - q}\frac{dq}{dr_2},$$

$$r_2\frac{\partial h}{\partial r} = -\frac{2\gamma(\nu - 1)}{\gamma + 1}$$

$$+ \frac{\gamma}{q}\left[\frac{q + 1}{\gamma + 1 - 2(1 - q)} + \frac{2}{\gamma + 1 + (\gamma - 1)(1 - q)}\right]\frac{r_2}{1 - q}\frac{dq}{dr_2},$$

$$r_2\frac{\partial S}{\partial r} = -\frac{2c_p(1 - q)^2(\gamma^2 - 1)}{[\gamma + 1 - 2(1 - q)]^2[\gamma + 1 + (\gamma - 1)(1 - q)].q}r_2\frac{dq}{dr_2}.$$

$$\left.\right\}\ (16.46)$$

Formulas for the second derivatives, similar to formulas (16.45) and (16.46), can be obtained by the same means.

It follows from (16.46) that if the shock moves with the constant velocity c into an undisturbed medium with the constant temperature T_1, then $q = \text{const}$ and, therefore,

$$\left.\begin{array}{l}
\dfrac{\partial f}{\partial \lambda} = -(\nu-1)\dfrac{\left[1+\dfrac{\gamma-1}{\gamma+1}(1-q)\right]}{1-q} \leqslant 0, \\[4mm]
\dfrac{\partial g}{\partial \lambda} = -\dfrac{2(\nu-1)}{\gamma+1} \leqslant 0, \\[4mm]
\dfrac{\partial h}{\partial \lambda} = -\dfrac{2\gamma(\nu-1)}{\gamma+1} \leqslant 0, \quad \dfrac{\partial S}{\partial \lambda} = 0, \quad \lambda = \dfrac{r}{r_2}.
\end{array}\right\} \quad (16.47)$$

Equality in this condition is attained only for plane waves $\nu = 1$. The derivatives of the variables defining the disturbed gas motion behind the front equal zero for a plane shock wave at constant speed.

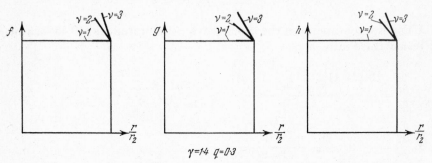

$\gamma=1\!\cdot\!4 \quad q=0\!\cdot\!3$

Fig. 103. Elements of curves of the gas velocity distribution behind the shock wave front for constant front speed in the plane, cylindrical and spherical cases.

In the spherical and cylindrical wave cases, only $\partial S/\partial r = 0$ when $q = \text{const}$, the derivatives $\partial v/\partial r$, $\partial \rho/\partial r$, $\partial p/\partial r$ are not zero but approach zero like A/r_2 (A is a suitable constant; see (16.47)).

If the shock wave velocity c decreases, then the quantity $q = a_1^2/2$ increases and, consequently, $dq/dr_2 > 0$. In this case the direction of the tangents to the corresponding curves on Fig. 103, shown for constant velocity c, rotate counter-clockwise.

It is evident that the intensity of shock wave damping is amplified for given q as the values of the derivatives $\partial f/\partial \lambda$, $\partial g/\partial \lambda$, $\partial h/\partial \lambda$ increase. As the shock wave moves away from the centre of symmetry near which a finite energy distribution was formed in a finite time interval, shock damping occurs in which the shock propagation velocity c approaches the speed of sound and, therefore $q \to 1$.

The asymptotic law of shock wave attenuation can be determined by using (16.46) if the limiting behaviour of the expression on the left of these equations is known as $q \to 1$.

Approximate solutions (Sedov, 1952) of the problems of unsteady gas motion within a shock wave $(0 \leqslant r \leqslant r_2)$ can be constructed in certain cases by using interpolation formulas for f, g, h or for any other functions which can be expressed in terms of f, g, h.

Parametric representations of the unknown functions $\omega_0(t)$ and $q(t)$ can be introduced in the interpolation formulas to determine the behaviour of the solution near the centre of symmetry and near the shock wave.

A different kind of integral relation, similar to those in the Ritz and Galerkin methods and in approximate boundary layer theory, can be used to determine these functions.

The conditions on the shock wave (formula (1.46)) and their generalizations for derivatives with respect to the radius r higher than the first order permit the gas motion behind the shock front to be calculated if the law of shock motion is known. (The shock motion—the shock wave—can be determined experimentally or on the basis of additional assumptions.)

§17. Asymptotic Laws of Shock Wave Decay

In the general case, when a disturbance bounded by a shock wave is propagated into a gas at rest, the asymptotic laws of the behaviour of the shock wave velocity as a function of the shock wave coordinate r_2 and, therefore, of the variation of the shock wave intensity, can be very complicated and depend essentially on the conditions determining the gas motion within the shock wave.

The laws of plane shock wave attenuation were analysed and established by Crussard (1913)† under the assumption that the disturbed gas motion behind the shock front is a Riemann travelling wave containing a point at which the gas velocity is zero.

Riemann (1953) has shown (for one-dimensional gas motions with plane waves when the gas fills all space) that if the initial disturbance is continuous and distributed along a finite segment of the x axis, then the initial disturbance would develop into two travelling waves which are

† A bibliography of works devoted to the question of plane shock wave damping is given in the book by Y. B. Zel'dovich (1948).

propagated in opposite directions after the motion has existed for a certain finite time. If the gas motion is continuous in the travelling wave moving in the positive x direction, and if there are regions in which the pressure decreases with increasing x at a given time, then shock waves or compression shocks, form in the travelling wave. These result from the overtaking of one part of the wave by another.

L. D. Landau (1945) was the first to describe the damping of spherical and cylindrical shock waves under the assumption that the disturbed gas motion behind the shock wave front is attenuated, and that this motion approaches a travelling wave in which the disturbance at a fixed instant is confined to an interval of finite length. The motion differs from the acoustic motion only in the use of a more exact value of the speed of sound.

Subsequently, much work on this question appeared in which the damping of spherical and cylindrical waves was studied under the same or analogous assumptions as in the Landau paper. It was also shown in the Landau paper that corresponding methods, reasoning and results carry over directly into the case of damping of curved shock waves formed in a plane or axially symmetrical supersonic flow past a body.

A theory of shock wave damping is given below which is based on Equation (16.46), determining the derivatives behind the shock wave front.

The differential relation

$$dr - (a+v)\,dt = 0 \tag{17.1}$$

is satisfied along characteristics of Equations (1.3) which carry disturbances in the direction of increasing r. The equation of these characteristics in the r, t plane can be written as

$$\xi(r, t) = \text{const}$$

where the function $\xi(r, t)$ is determined from the relation

$$\mu\,d\xi = dr - (a+v)\,dt.$$

Hence

$$dt = \frac{dr - \mu\,d\xi}{a+v} \tag{17.2}$$

where μ is a certain integrating factor.

Let us replace the independent variables r, t by the new independent variables r, ξ. The formulas defining the transformation of the partial derivatives are

$$\frac{\partial f(r, t)}{\partial t} = -\frac{\partial f(\xi, r)}{d\xi}\frac{(a+v)}{\mu}, \quad \frac{\partial f(r, t)}{\partial r} = \frac{\partial f(\xi, r)}{\partial r} + \frac{\partial f(\xi, r)}{\partial \xi}\frac{1}{\mu}.$$

Using these, the equations of motion (1.3) can be rewritten in the form

$$a\frac{\partial v}{\partial \xi} - \frac{1}{\rho}\frac{\partial p}{\partial \xi} = \mu\left(v\frac{\partial v}{\partial r} + \frac{1}{\rho}\frac{\partial p}{\partial r}\right),$$

$$a\frac{\partial \rho}{\partial \xi} - \rho\frac{\partial v}{\partial \xi} = \frac{\mu}{r^{\nu-1}}\frac{\partial r^{\nu-1}\rho v}{\partial r},$$

$$a\frac{\partial}{\partial \xi}\left(\frac{p}{\rho^{\gamma}}\right) = \mu v\frac{\partial}{\partial r}\frac{p}{\rho^{\gamma}}.$$

$$(17.3)$$

The integrability condition (17.2) yields one equation

$$\frac{\partial(a+v)}{\partial \xi} = (a+v)\frac{\partial \mu}{\partial r} - \mu\frac{\partial(a+v)}{\partial r}. \tag{17.4}$$

It is easy to see that the system (17.3) and Equation (17.4) admit a solution for $\nu = 1$ (in the plane wave case) in which v, ρ, p depend only on ξ. This solution is a Riemann travelling wave: it can be represented in the form

$$p = \frac{p_1}{\rho_1^{\gamma}}\rho^{\gamma}, \quad a^2 = \frac{\gamma p_1}{\rho_1^{\gamma}}\rho^{\gamma-1}, \quad v = \frac{2}{\gamma-1}(a-a_1)$$

$$\mu = \frac{d\ln(a+v)}{d\xi}r + \left[\frac{F(v)}{a+v}\right]'_{\xi}(a+v) = \frac{d\ln(a+v)}{d\xi}r + \left[\frac{\Omega(\xi)}{a+v}\right]'_{\xi}(a+v)$$

and

$$r - (a+v)t = F(v) = \Omega(\xi) \quad \text{where } v = \mathcal{G}[r - (a+v)t] = \Phi(\xi).$$

$$(17.5)$$

where $F(v) = \Omega(\xi)$ and, therefore, $\mathcal{G}[r-(a+v)t] = \Phi(\xi)$ are arbitrary functions of their arguments (p_1 and ρ_1 are constants).

In particular, without loss of generality, we can take

$$\Omega(\xi) = \xi = r - (a+v)t, \tag{17.6}$$

keeping $F(v)$ and the inverse function $\Phi(\xi)$, respectively, arbitrary. From (17.5) and (17.6), we shall have, for the derivative $(\partial v/\partial r)_{t=\text{const}}$.

$$\frac{\partial v}{\partial t} = \Phi'(\xi)\left[1 - \frac{\partial(a+v)}{\partial r}\Big|_{t=\text{const}}t\right] = \Phi'(\xi)\left[1 - \frac{\gamma+1}{2}\frac{\partial v}{\partial r}t\right].$$

Hence

$$r\frac{\partial v}{\partial r} = \frac{\Phi'(\xi)r}{1 + \dfrac{\gamma+1}{2}t\Phi'(\xi)} \tag{17.7}$$

10*

as $r_2 \to \infty$ the shock wave degenerates into a sound wave and the gas motion behind the shock front degenerates into a Riemann flow, with $q \to 1$ and $(a_1 t / r_2) \to 1$.

Let us denote the limiting value of ξ on the shock wave as $r_2 \to \infty$ by ξ_0. Since the shock wave degenerates into a sound wave by hypothesis it is evident that we must have $v \to 0$ as $\xi \to \xi_0$ and, therefore

$$\Phi(\xi_0) = 0.$$

Furthermore, we find the asymptotic law of shock wave damping under the assumption that

$$\lim_{r_2 \to \infty} r_2 \Phi'(\xi) \to \infty \tag{17.8}$$

If $\Phi'(\xi) \neq 0$, the limiting relation (17.8) is not admissible.

If

$$v = \Phi(\xi) = c(\xi - \xi_0)^n + \dots,$$

where c and $n > 1$ are constants, then $\Phi'(\xi) = c(v/c)^{(n-1)/n} + \dots$ consequently, the assumption (17.8) is equivalent to the condition

$$r_2 v^{(n-1)/n} \to \infty$$

which will be satisfied if the velocity behind the shock front v decreases more slowly than

$$\frac{k}{r_2^{1+[1/(n-1)]}}.$$

We show further that the assumption (17.8) leads to a law in which the velocity decreases more slowly than $k/\{r^{1+[1/(n-1)]}\}$; consequently, the assumption (17.8) is also not essential in this case.

We obtain the following asymptotic equation from (16.46) and (17.7) for $\nu = 1$

$$\frac{2a_1}{\gamma+1} = \frac{4a_1}{\gamma+1} \frac{r_2}{1-q} \frac{dq}{dr_2}.$$

Hence, we find after integration

$$1 - q = \sqrt{\left(\frac{r_0}{r_2}\right)} \quad \text{or} \quad a_1^2 \left(\frac{dt}{dr_2}\right)^2 = 1 - \sqrt{\left(\frac{r_0}{r_2}\right)} \tag{17.9}$$

where r_0 is a certain constant.

From (17.9) and the shock conditions (12.6) we obtain the following asymptotic formulas for the law of shock motion† and for the dependent variables of the gas motion

$$
\left.\begin{array}{l}
a_1(t-t_0) = r_2\left[1 - \sqrt{\left(\dfrac{r_0}{r_2}\right)} - \dfrac{1}{8}\dfrac{r_0\ln r_2/r_0}{r_2} + \ldots\right], \\[2em]
v_2 = \dfrac{2a_1}{\gamma+1}\sqrt{\left(\dfrac{r_0}{r_2}\right)} + \ldots, \\[2em]
\rho_2 = \rho_1 + \dfrac{2\rho_1}{\gamma+1}\sqrt{\left(\dfrac{r_0}{r_2}\right)} + \ldots, \\[2em]
p_2 = p_1 + \dfrac{2p_1}{\gamma+1}\sqrt{\left(\dfrac{r_0}{r_2}\right)} + \ldots.
\end{array}\right\} \quad (17.10)
$$

Formulas (17.10) are in agreement with the solution (17.5) and give the desired asymptotic laws for plane waves. It is evident from (17.10) that condition (17.8) is satisfied by any n.

In the case of spherical and cylindrical waves with $\mu = 1$ Equations (17.3) become, after linearizing with respect to the undisturbed state of rest with pressure p_1 and density ρ_1,

$$
\left.\begin{array}{l}
a_1\dfrac{\partial v}{\partial \xi} - \dfrac{1}{\rho_1}\dfrac{\partial p}{\partial \xi} = \dfrac{1}{\rho_1}\dfrac{\partial p}{\partial r}, \\[2em]
a_1\dfrac{\partial \rho}{\partial \xi} - \rho_1\dfrac{\partial v}{\partial \xi} = \dfrac{\rho_1}{r^{\nu-1}}\dfrac{\partial r^{\nu-1}v}{\partial r}, \\[2em]
\dfrac{p}{p_1} = \left(\dfrac{\rho}{\rho_1}\right)^\gamma \quad \text{or} \quad \dfrac{p-p_1}{p_1} = \gamma\dfrac{\rho-\rho_1}{\rho_1}.
\end{array}\right\} \quad (17.11)
$$

In contrast to the plane wave case with $\nu = 1$, Equations (17.11) have no solution which depend only on the single variable ξ.

Let us consider the solution of Equations (17.11) for travelling waves when the velocity v of the perturbed motion can be represented by the series

$$
v = a_1\left[\dfrac{\Phi_1(\xi)}{r^m} + \dfrac{\Phi_2(\xi)}{r^{2m}} + \ldots\right]. \qquad (17.12)
$$

where $\Phi_1(\xi)$ is an arbitrary function and $m > 0$ is a constant to be determined.

† Here and in what follows, $-a_1t_0$ will denote an arbitrary integration constant.

Let us put

$$\rho = \rho_1 \left[1 + \frac{\Psi_1(\xi)}{r^m} + \frac{\Psi_2(\xi)}{r^{2m}} + \ldots \right]. \tag{17.13}$$

From (17.11), the latter yields

$$p = p_1 \left[1 + \frac{\gamma \Psi_1(\xi)}{r^m} + \frac{\gamma \Psi_2(\xi)}{r^{2m}} + \ldots \right]. \tag{17.14}$$

Substituting these series in the first two equations of (17.11) leads to the relations

$$\left.\begin{array}{l} \dfrac{\Phi_1' - \Psi_1'}{r^m} + \dfrac{\Phi_2' - \Psi_2'}{r^{2m}} + \ldots + \dfrac{m\Psi_1}{r^{m+1}} + \dfrac{2m\Psi_2}{r^{2m+1}} + \ldots = 0, \\[4mm] \dfrac{\Phi_1' - \Psi_1'}{r^m} + \dfrac{\Phi_2' - \Psi_2'}{r^{2m}} + \ldots - \dfrac{(m+1-\nu)\Phi_1}{r^{m+1}} - \dfrac{(2m+1-\nu)\Phi_2}{r^{2m+1}} - \ldots = 0. \end{array}\right\} \tag{17.15}$$

Hence, it follows that

$$\Phi_1 = \Psi_1 + C, \tag{17.16}$$

where C is a constant of integration.

Since the function Φ_1 is arbitrary, it is necessary that the terms of order $m+1$, $2m+1$, ... in (17.15) should be reducible to one of the terms of order $2m$, $3m$, ... Therefore, we must have

$$m + 1 = km$$

where k is a certain integer.

Equating the coefficients of r^{km} to zero in (17.15) yields

$$\Phi_k' - \Psi_k' + m\Psi_1' = 0,$$

$$\Phi_k' - \Psi_k' - (m+1-\nu)\Phi_1 = 0.$$

Hence, it follows from (17.16) that

$$2m + 1 - \nu = 0 \quad \text{and} \quad C = 0.$$

Hence, we find in the cylindrical wave case with $\nu = 2$, when $m = 0 \cdot 5$ and $k = 3$

$$\Psi_n = \frac{2-n}{n}\Phi_n, \quad \Phi_{n+2}' = \frac{n^2 - 4}{4(n+1)}\Phi_n, \quad \Phi_2' = 0, \tag{17.17}$$

in which the function $\Phi_1(\xi)$ remains arbitrary and the functions $\Phi_{2n+1}(\xi)$ with the odd subscripts ($n = 1, 2, 3, \ldots$) are expressed in terms of the function $\Phi_1(\xi)$ by using the n-tuple indefinite integral

$$\int^{\xi}\int^{\xi_1}\int^{\xi_2} \cdots \int^{\xi_{n-1}} \Phi_1\, d\xi_{n-1} \cdots d\xi_1\, d\xi,$$

the functions $\Phi_{2n}(\xi)$ are polynomials in ξ of degree $n-2$; each succeeding polynomial is obtained by integrating the preceding one by means of the recursion formula (17.17).

In the spherical wave case with $\nu = 3$, we have

$$
\left.
\begin{aligned}
& m = 1 \qquad\qquad\qquad k = 2 \\[2mm]
& \Psi_n = \frac{2-n}{n}\,\Phi_n \qquad (n = 1, 2, 3, \ldots) \\[2mm]
& \Phi_n' = \frac{n(n-3)}{2(n-1)}\,\Phi_{n-1} \quad (n = 2, 3, 4, \ldots)
\end{aligned}
\right\} \quad (17.18)
$$

The function $\Phi_1(\xi)$ remains arbitrary. It follows from (17.18) that

$$\Phi_2(\xi) = -\int \Phi_1(\xi)\, d\xi$$

and that succeeding functions $\Phi_n(\xi)$ are polynomials of degree $n-3$ in ξ when $n > 2$; each succeeding polynomial is obtained from the preceding one by integration in the recursion formula (17.18).

Hence, the series (17.12), (17.13) and (17.14) represent a class of solutions of the linearized equations: these contain the arbitrary function $\Phi_1(\xi)$ and arbitrary constants which appear in the determination of coefficients of higher powers of $1/r$.

It follows from formulas (17.12), (17.13) and (17.14) that $a+v$ may be expanded in the form

$$a+v = \sqrt{\left(\frac{\gamma p}{\rho}\right)} + v = a_1\left[1 + \frac{\gamma+1}{2}\,\frac{\Phi_1(\xi)}{r^{(\nu-1)/2}} + \ldots\right]. \qquad (17.19)$$

Equation (17.11) can be analysed independently of (17.4) in the acoustic approximation if it is assumed that $\mu = 1$ and $\xi = r - a_1 t$, which is equivalent to replacing the quantity $a+v$ in (17.4) and (17.2) by a_1. More precise expressions for $t(\xi, r)$ and $\mu(\xi, r)$, corresponding to the first terms in the expansions of the functions v, ρ, p in (17.12), (17.13) and (17.14), can be found from Equations (17.4) and (17.2), with the use of (17.19).

The expansions (17.12), (17.13) and (17.14) change when the nonlinear equations (17.3) are solved. The first terms retain their form but the relation between r, t and ξ and the function $\mu(\xi, r)$ must be made more precise. These relations in the cylindrical wave case with $\nu = 2$ are

$$\left.\begin{aligned} \mu &= 1 + \sqrt{(r)}\left[(\gamma+1)\,\Phi_1'(\xi) + 0\!\left(\frac{\ln r/r^*}{\sqrt{r}}\right)\right], \\ a_1 t + \text{const} &= r - \sqrt{r}\left[(\gamma+1)\,\Phi_1(\xi) - \frac{(\gamma+1)^2}{4}\Phi_1^2(\xi)\frac{\ln r/r^*}{\sqrt{r}} + \ldots\right]. \end{aligned}\right\} \quad (17.20)$$

Here and subsequently, $0(f(r))$ will denote quantities which approach zero as r increases; the order of the approach to zero is $f(r)$, and r^* denotes a certain constant.

It follows from (17.2) and (17.20) that for all $\Phi_1'(\xi) \neq 0$

$$\left.\frac{\partial\xi}{\partial r}\right|_{t=\text{const}} = \frac{1}{\mu} = \frac{1}{(\gamma+1)\,\Phi_1'(\xi)\sqrt{r}}\left[1 + 0\!\left(\frac{\ln r/r^*}{\sqrt{r}}\right)\right]. \quad (17.21)$$

Similar formulas in the spherical wave case, with $\nu = 3$, are

$$\left.\begin{aligned} \mu &= 1 + \frac{\gamma+1}{2}\Phi_1'(\xi)\ln r/r^* + 0\!\left(\frac{\ln r/r^*}{r}\right), \\ a_1 t + \text{const} &= r - \frac{\gamma+1}{2}\Phi_1(\xi)\ln r/r^* + 0\!\left(\frac{\ln r/r^*}{r}\right), \end{aligned}\right\} \quad (17.22)$$

and for $\Phi'(\xi) \neq 0$

$$\left.\frac{\partial\xi}{\partial r}\right|_{t=\text{const}} = \frac{1}{\mu} = \frac{2}{(\gamma+1)\,\Phi_1'(\xi)\ln r/r^*}\left[1 + 0\!\left(\frac{1}{\ln r/r^*}\right)\right]. \quad (17.23)$$

It follows from formula (17.12)†

$$\left.\begin{aligned} \left.\frac{\partial v}{\partial r}\right|_{t=\text{const}} &= a_1\frac{\Phi_1'(\xi)}{\sqrt{r}}\left(\frac{\partial\xi}{\partial r}\right)_{t=\text{const}}\left[1 + 0\!\left(\frac{1}{\sqrt{r}}\right)\right] \quad \text{for } \nu = 2, \\ \left.\frac{\partial v}{\partial r}\right|_{t=\text{const}} &= a_1\frac{\Phi_1'(\xi)}{r}\left(\frac{\partial\xi}{\partial r}\right)_{t=\text{const}}\left[1 + 0\!\left(\frac{\ln r/r^*}{r}\right)\right] \quad \text{for } \nu = 3. \end{aligned}\right\} \quad (17.24)$$

From (16.46), (17.21) and (17.24) the asymptotic laws in the cylindrical wave case ($\nu = 2$) are defined by

$$\frac{a_1}{\gamma+1} = -\frac{2a_1}{\gamma+1} + \frac{4a_1}{\gamma+1}\frac{r_2}{1-q}\frac{dq}{dr_2}.$$

† Formulas (17.24) show the orders of the vanishing terms when additional terms to the first in the series (17.12) are retained. The series defined waves in the solution of the nonlinear equations (17.3).

Hence, we have after integration

$$1-q = \left(\frac{r_0}{r_2}\right)^{3/4} + \dots \quad \text{or} \quad a_1^2\left(\frac{dt}{dr_2}\right)^2 = 1 - \left(\frac{r_0}{r_2}\right)^{3/4} + \dots \quad (17.25)$$

where r_0 is a certain constant.

In the cylindrical wave case, from (17.25) and the shock conditions (12.6), we derive the following asymptotic laws of shock motion and the values of the dependent variables of the gas motion behind the shock

$$\left. \begin{aligned} a_1(t-t_0) &= r_2 - 2r_0^{3/4}r_2^{1/4} + \dots, \\[1.2em] v_2 &= \frac{2a_1}{\gamma+1}\left(\frac{r_0}{r_2}\right)^{3/4} + \dots, \\[1.2em] \rho_2 &= \rho_1 + \frac{2\rho_1}{\gamma+1}\left(\frac{r_0}{r_2}\right)^{3/4} + \dots, \\[1.2em] p_2 &= p_1 + \frac{2\gamma p_1}{\gamma+1}\left(\frac{r_0}{r_2}\right)^{3/4} + \dots. \end{aligned} \right\} \quad (17.26)$$

Formulas (17.26) agree with (17.12), (17.13) and (17.14) which represent the gas motion behind the shock front.

We obtain the following asymptotic equation in the spherical wave case ($\nu = 3$) from (16.46), (17.23) and (17.24)

$$\frac{2a_1}{\gamma+1}\frac{1}{\ln r_2/r^*} = -\frac{4a_1}{\gamma+1} + \frac{4a_1}{\gamma+1}\frac{r_2}{1-q}\frac{dq}{dr_2}.$$

Hence, we find after integration

$$1-q = \frac{k}{r_2\sqrt{(\ln r_2/r^*)}} + \dots; \; a_1^2\left(\frac{dt}{dr_2}\right)^2 = 1 - \frac{k}{r_2\sqrt{(\ln r_2/r^*)}} \dots, \quad (17.27)$$

where k is a certain constant with the dimensions of a length. By using (17.27) and the conditions on the shock (12.6), the appropriate asymptotic laws in the spherical symmetry case become

$$\left. \begin{aligned} a_1(t-t_0) &= r_2 - k\sqrt{(\ln r_2/r^*)} + \dots, \\[1.2em] v_2 &= \frac{2a_1}{\gamma+1}\frac{k}{r_2\sqrt{(\ln r_2/r^*)}} + \dots, \\[1.2em] \rho_2 &= \rho_1 + \frac{2\rho_1}{\gamma+1}\frac{k}{r_2\sqrt{(\ln r_2/r^*)}} + \dots, \\[1.2em] p_2 &= p_1 + \frac{2\gamma}{\gamma+1}p_1\frac{k}{r_2\sqrt{(\ln r_2/r^*)}} + \dots. \end{aligned} \right\} \quad (17.28)$$

Formulas (17.28) are in agreement with formulas (17.12), (17.13) and (17.14) which determine the gas motion behind the shock.

The velocity v_2 decreases more rapidly in both the cases considered ($\nu = 2$ and $\nu = 3$) than in the acoustic case; consequently, we have $\xi \to \xi_0$ for $r_2 \to \infty$, and we must have $\Phi_1(\xi_0) = 0$. It follows from (17.20) and (17.22) that the appropriate limiting characteristics approach straight lines in the r, t plane. The shock wave and the characteristic lines which differ from the limiting, diverge from any fixed line by an infinite distance.

The method just explained can be used to obtain asymptotic laws or to improve the accuracy of the above laws when the gas motion behind the shock front is governed by different and more precise laws than those used above.

CHAPTER V

Application to Astrophysical Problems

§1. CERTAIN OBSERVATIONAL RESULTS

To explain and analyse the internal motion in stars, the evolution of stars and of nebulae, modern astrophysics has to extend beyond the dynamics of systems of particles and of the hydrostatics of a liquid mass. These are the theories which were used up to the present as the basis for models and conceptions in classical astronomy. At the present time, a study of the motion of celestial objects as gaseous bodies must give the key to the solution of the main problems of cosmology and the explanation and interpretation of a number of observable effects can be found only by this means. It has now become evident that the formulation and solution of a number of dynamic problems of gas motion, which can be considered as theoretical models encompassing the essential peculiarities of the motion and evolution of stars and nebulae, must underlie the conceptions used in the investigation of celestial phenomena. In order to construct and to investigate such models, methods, apparatus and conceptions of modern theoretical gas dynamics must be used, namely aerodynamics, and the appropriate mechanical problems applicable to astrophysics must be formulated and solved.

In connection with this requirement, which had already been clearly perceived, an international conference was called in Paris in 1949, at which astrophysicists met the foremost representatives of aerodynamics, who had been previously occupied mainly with the problems of aviation and the theory of explosions.

As a result of this conference, a collection of notes and discussions were published (Burgers and van de Hulst, 1949) in which it was stated that underlying the investigations of the most important celestial phenomena must be problems of motion of gaseous masses with high relative velocities and with shock waves in a gravitational field, of turbulent gas motion taking electromagnetic fields into account, and so on.

A preliminary analysis showed that insufficient use has been made of the known phenomena of gas motion in the explanation of celestial phenomena; new models had to be constructed and new problems of gas dynamics had to be posed and solved.

305

We studied certain gas motions, on the basis of similarity methods developed in the preceding chapter, in application to the theory of one-dimensional gas motion with spherical symmetry, which apparently can reflect the essential features detected in astronomical observations.

Below, we shall consider certain applications to the theory of the luminosity and internal structure of stars, to the theory of the change in the luminosity of Cepheids and to the theory of the eruption of novae and supernovae stars. But first let us explain certain of the fundamental data on these phenomena which are known from astrophysics.

A basic characteristic of stars is their luminosity or brightness which is related directly to the perception of the eye during observations. The luminosity can be measured by using special instruments called photometers. In recent years, the intensity of stellar luminosity has been the basis of the "stellar magnitude" concept.

At first, the classification of stars was made simply by eye so that stars of the first magnitude seemed to be several times brighter than stars of the second magnitude, which, in turn, seemed to be very much brighter than stars of the third magnitude, and so on. Hence, all the stars could be arranged in a series according to their visual luminosity. Even in ancient times, stars were subdivided into six classes when observed by the naked eye. The very brightest stars were in the first class and were called stars of the first magnitude, then stars of the second magnitude followed, etc. Evidently, many stars fell into intervals between classes and, consequently, visual magnitudes were attributed to them which were expressed by fractional numbers. It is obvious from the above definition that the visual luminosity decreases as the stellar magnitude increases. At the present time, weak stars of the twenty-second or even the twenty-third magnitude can be studied on long-exposure photographs obtained by using modern powerful telescopes.

The arrangement of the stars described above depends on the methods of determining the brightness. Different conclusions can be arrived at depending on whether the visual magnitude is determined by eye or by using a preliminary photograph. The human eye is more sensitive to red and yellow light rays, while the photographic plate possesses greater sensitivity to blue and violet light. Consequently, a distinction is made between the visual magnitude of stars, measured by eye, by photographic, photoelectric or other means.

Stellar magnitudes which are determined by using apparatus with the same sensitivity to all wave lengths, with a correction for absorption in the terrestrial atmosphere and in the optics of the instrument introduced, are called bolometric. The bolometric magnitudes characterize the total stellar radiation arriving at the upper layers of the terrestrial atmosphere. Later, we shall use the concept of the bolometric stellar magnitude.

It is evident that the stellar magnitude m is determined by the quantity of radiant energy E_m incident on the earth from the star under consideration per unit time. This energy is proportional to the total energy radiated by the star in unit time, i.e. the luminosity of the star $\mathfrak{L}^*$, and is inversely proportional to the distance l from the earth to the star.

On the basis of the properties of the human eye, defined by the Weber-Fechner law, it would appear that the appropriate intensities of the exciter E_m vary in a geometric progression as the stellar magnitude m, defined by the intensity of perception, varies according to an arithmetical progression.

Accordingly, the following relation has been established

$$m = -2\cdot5\log E_m + \text{const,} \tag{1.1}$$

which can be written as

$$\frac{E_{m_2}}{E_{m_1}} = 10^{(m_1-m_2)/2\cdot5} = (2\cdot512)^{m_1-m_2}. \tag{1.2}$$

The visible magnitude of the sun is $m_2 = -26\cdot7$. The visible magnitude of Sirius, the brightest of the visible stars, is $m_1 = -1\cdot6$; hence, it follows that $m_1 - m_2 = 25\cdot1$ and $E_{\text{Sun}}/E_{\text{Sirius}} \doteq 10^{10}$. In other words, the visible luminosity of the sun is ten times greater than the visible luminosity of Sirius.

The visible magnitude of stars depends essentially on the distance from the star to the earth, consequently, the concept of the absolute visual magnitude M, which equals the visible stellar magnitude defined above if the star being considered is 10 parsecs from the earth,† is introduced in order to obtain the comparable characteristics of stars.

On the basis of the formulas

$$m = -2\cdot5\log\frac{\mathfrak{L}^*}{l^2} + \text{const} \quad \text{and} \quad M = -2\cdot5\log\frac{\mathfrak{L}^*}{10^2} + \text{const}$$

we obtain

$$M = m + 5 - 5\log l; \tag{1.3}$$

where the distance of the star to the earth l in (1.3) is expressed in parsecs.

The absolute magnitude of the sun is $4\cdot9$; this means that the sun would be represented as a weak stellar spot of the fifth magnitude to us

† One parsec equals $3\cdot26$ light years $= 3\cdot08 \times 10^{13}$ km. and corresponds to the distance to the earth from such a star for which the parallax is $1''$.

at a distance of 10 parsec, hardly observable with the naked eye. The absolute magnitude of Sirius is $1 \cdot 3$. Hence, it follows that:

$$\frac{\mathfrak{L}^*_{\text{Sirius}}}{\mathfrak{L}^*_{\text{Sun}}} = 10^{(4 \cdot 9 - 1 \cdot 3)/2 \cdot 5} = 28.$$

Therefore, the luminosity of Sirius is twenty-eight times larger than the

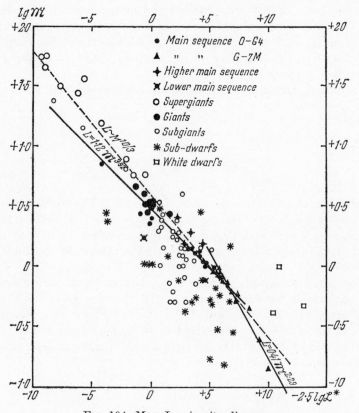

FIG. 104. Mass-Luminosity diagram.

luminosity of the Sun. The magnitude of the visible stellar brightness m can be measured directly; if we were able to find the absolute value of M from any other additional data, then the distance l to the star can be calculated easily by using formula (1.3).

In addition to the stellar luminosity $\mathfrak{L}^*$, the most important characteristics are the mass $\mathfrak{M}$ and the radius $\mathfrak{R}$ of a star. Using various kinds of observations and suitable analysis, astronomers have arranged results on the magnitudes of $\mathfrak{L}^*$, $\mathfrak{M}$ and $\mathfrak{R}$ for a considerable number of stars. An analysis of these results shows that the following relations exist for various groups of stars (Parenago and Masevich, 1951)

$$\mathfrak{L}^* = f_1(\mathfrak{M}) \quad \text{and} \quad \mathfrak{R} = f_2(\mathfrak{M}). \tag{1.4}$$

Figures 104 and 105 are diagrams taken from the work of P. P. Parenago and A. G. Masevich, from which such relations can be

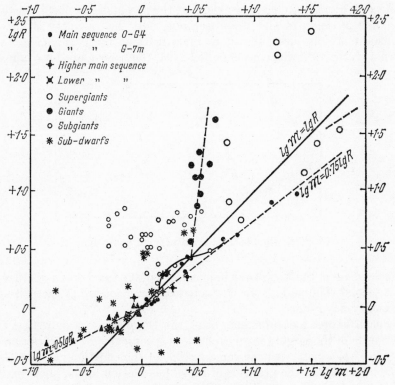

FIG. 105. Mass-Radius diagram.

deduced. We shall give certain theoretical support for the relations (1.4) in §3.

Now let us consider certain results on so-called variable stars. The astronomer Goodricke (1786) first published a report on observations in which he disclosed that a star denoted by the letter δ, which is near the head of Cepheid, changes its brightness periodically. At present, such variable stars, called cepheids, are well known; they are found in various regions of our galaxy and in other galaxies.

Empirical relations established between the periods and the absolute visual magnitudes of the cepheids permitted astronomers to find the distance to other galaxies† by using (1.3) and to obtain very valuable material to investigate the problems of the evolution of stars and to solve many other questions.

Figure 106 shows the results of photometric measurements of the brightness m of the δ star of Cepheid with time. The appropriate curves for the other cepheids retain the same general character. It has been detected, by modern observations, that the period of δ Cepheid is 5 days 8 hours 53 minutes and 27·46 seconds, and this quantity is almost constant. It has been shown in investigations by astronomers that although the variations in the period exist, they are small; for example,

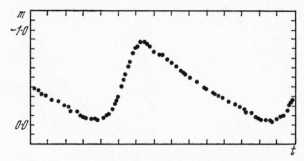

FIG. 106. Photo-visual brightness curve of Cepheid.

the decrease in the period of δ Cepheid in twenty years is of the order of one second. The periods for characteristic cepheids are in the half-day to eighty-day range.

Spectroscopic measurements using the Doppler effect show that the variation of the brightness of cepheids is accompanied by a variation of the ray velocities, with the same period, of the emitting gas particles, which testifies to the presence of radial gas motion in the photosphere of the cepheid. The amplitude of the oscillations of the ray speeds has the order of several tens of kilometres per second; this amplitude is 39 km./sec. for δ Cepheid. The brightness and the ray velocity curves are very similar.

Hence, the observational results undoubtedly indicate the presence of considerable gas motion in the photosphere of the cepheid. Moreover, it has been established by the observations that the brightness oscillations are accompanied by a change in their spectrum and effective

† At the present time, this is the only method of determining such large distances which are of the order of millions of light years.

temperatures. The change in the temperature and radius of the photosphere specifies a change in the brightness. The ratio of the luminosity (total luminous energy emitted by the star) of the cepheid varies several times during the oscillation; this ratio is 2 for δ Cepheid.

Table 1 gives certain magnitudes for typical cepheids.

TABLE 1

Data for Characteristic Cepheids†

Star		Spectral Class	Period (days)	M, Stellar Magnitude	Amplitude M, Stellar Magnitude	10^6 km. $\delta\Re$	10^6 km. $\Re$	$\mathfrak{M}/\mathfrak{M}_0$	Asymmetry of Brightness Curve
l	Car	F8–K0	35·52	−5·52	1·2	8·59	145	50	1·7
Y	Oph	F8–G7	17·12	−4·00	0·61	1·79	59	23	1·4
X	Cyg	F8–K0	16·39	−3·92	0·69	6·12	41	19	1·7
ζ	Gem	cG 1v	10·15	−3·16	0·42	1·82	43	15	1·0
S	Sge	F8–G7	8·38	−2·87	0·50	1·47	33	13	2·0
W	Sgr	F2–G5	7·59	−2·72	0·85	1·93	24	11	2·5
η	Agl	F2–G9	7·18	−2·62	0·51	1·77	25	11	2·0
X	Sgr	F5–G9	7·01	−2·60	0·67	1·33	28	11	2·0
Y	Sgr	F5–G8	5·77	−2·30	0·74	1·35	27	10	1·4
δ	Cep	F4–G6	5·37	−2·19	0·61	1·27	23	9	2·5
T	Vul	F3–G5	4·44	−1·95	0·71	0·97	17	8	1·7
α	UMi	cFV	3·97	−1·81	0·08	0·17	20	8	1·7
SU	Cyg	F0–G1	3·85	−1·78	0·74	0·71	12	7	2·5
RT	Aur	F1–G5	3·73	−1·74	0·80	0·86	14	7	3·3
SZ	Tau	F4–G2	3·15	−1·58	—	0·46	14	7	1·0
SU	Cas	F2–F9	1·95	−1·15	0·33	0·30	9·2	5	1·0
RR	Lyr	B9–F2	0·567	−0·35	0·85	0·17	4·3	4	2·0

† All the numerical data and this table are taken from Rosseland (1952).

In this table, $\mathfrak{M}$ is the mass of the star; $\mathfrak{M}_0$ is the mass of the sun; $\Re$ is the radius of the star (the radius of the sun is $\Re_0 = 6 \cdot 26 \times 10^5$ km.); $\delta\Re$ is one-half the difference between the greatest and least values of the radius of the star; M is the average absolute visual magnitude. The ratio of the persistence of the drop in brightness from the maximum to the minimum to the persistence of its growth is given in the column defining the brightness asymmetry curve.

The "spectral class" column is related to the temperature characteristic of the stars. As with all variable stars, the Cepheids are considerably larger than our sun in mass and dimensions.

The magnitudes of the stellar radius are calculated in this table under the assumption that the photosphere (expanding or contracting with the star) consists of identical gas particles. Hence, it is seen that changes in the radius δ are of the order of millions of kilometres. If the boundaries of the expanding photosphere should coincide with the front of a shock wave, then the change in the photosphere radius will be larger since the shock velocity is larger than the measurable velocities of the gas particles outside the shock. Evidently, the variation in the luminosity of a star due to the change in the photosphere area can be very significant for Cepheids. This effect is aggravated when there is a layer of cold gas in front of the shock-wave photosphere boundaries, which absorbs a very great deal of the emitted energy at the moment when the photosphere radius is a minimum.

Novae are stars which abruptly increase in brightness by 10–12 stellar magnitudes and even by 15 magnitudes in certain cases; this means that

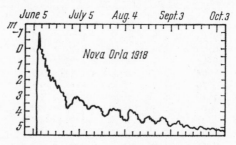

Fig. 107. Curve of Nova Orla brightness variation.

the visual brightness of such stars increases 10,000–100,000 times. A characteristic peculiarity of a flare-up of novae is their extreme suddenness; the brightness build-up time is of the order of one or two days and the energy liberation caused by the flare apparently lasts for several hours. After reaching a maximum, the brightness drops slowly to its original magnitude.

Figure 107 shows the change in the brightness of the Nova Orla star which flared in 1918.

The change in brightness is accompanied by sharp variations in the spectra. The shift of all the lines toward the violet, caused by the formation of high radial velocities of the emitting gas, is characteristic. These velocities are of the order of several hundred to three or four thousand kilometres per second. Clear forbidden lines, which are characteristic for the emission of a very rarefied gas and for the spectra of gas nebulae, appear in the spectra of novae for a certain time after the flare. The spectra of novae prior to and after the flare-up have belonged to the 0 class of the hottest stars for many years. Using the method of Zanstra

for the He-II line on Nova Orla, it has been found that the stellar
temperature was 65,000° (the temperature of the sun is 6000°) for three
months after the flare-up.

The diameter of a nova at its maximum is comparable to the diameter
of the Earth's orbit. According to various computations, astronomers
estimate the energy liberated in the flare of novae to be 10^{45}–10^{46} erg:
this energy equals the energy emitted by the sun in 10,000–100,000
years. About 100 novae per year flare up in our galaxy.

Between the novae and the Cepheids an intermediate type of variable
star exists—"secondary nova", which flares several times, sometimes
for several years. An example of such a variable star is Mira Kita
(Wonderful) for which the time between flare-ups is 320–370 days; the
brightness changes by a factor of 1500 during the flare-up.

The so-called dwarf novae of the U twins type which repeat their weak

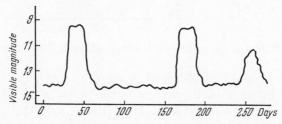

FIG. 108. Curve of the brightness variation of novae of the U-twins type.

flare-ups—stellar sneezing—for one–two weeks should be mentioned.
Figure 108 shows a curve of the brightness of such stars.

Finally, let us mention the R Silver Crown star for which the brightness
variation is shown in Fig. 109. The sudden drops in the brightness are
specified by the light absorption by carbon masses ejected from the star.

The flares of novae are wonderful, colossal cosmic explosions; however,
still more gigantic are the flare-ups of the so-called supernovae which
are monstrous catastrophes difficult to conceive. The luminosity during
flare-ups of supernovae is truly enormous and is comparable to the total
luminosity of all the stars in the galaxy.

A very brilliant supernova flared in NGC 5253 (it reached an absolute
magnitude of 18·4). This star gives light 15×10^{10} times brighter than
our sun at maximum brightness.

Figure 110 shows a typical curve of the brightness variation of a
supernova. Curves of the drop in brightness of supernovae are monotonic; the brightness is observed to oscillate as it drops for novae. The
flare-ups of supernovae are rare phenomena. Apparently, only two

supernovae have been observed in our galaxy in the last 900 years: the first, according to Chinese annals, flared in 1054 in the constellation Teltsa. Now the Crab nebula is observed there; at present this nebula

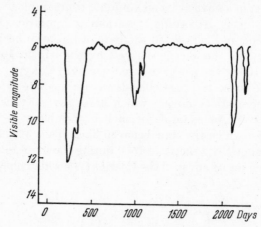

FIG. 109. Curve of the brightness variation of the R Silver Crown star.

continues to expand at high speed. A weak, but very hot star with a temperature higher than 100,000°, is in the centre of the Crab nebula. The second supernova, observed by Tycho Brahe, is the star which flared

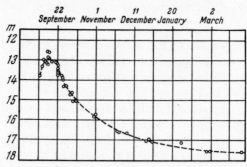

FIG. 110. Curve of the brilliance of the supernova in the spiral nebula NGC 1003.

up in 1572 in Cassiopeia and blazed in the sky for about a year (it was even seen during the day in total sunlight).

Astronomers have observed about forty supernovae in other galaxies. Because of their great distance supernovae in other galaxies can only be seen at the time of their maximum brightness with modern instruments.

According to the estimates of astronomers, the diameters of supernovae at their maximum can exceed the corresponding diameter of the usual novae by 250 times and the diameter of our solar system, the diameter of Pluto's orbit, by five to six times. Energy, estimated at about 10^{50} ergs, is liberated during the flare-ups of supernova within the brief interval of the order of several days; such energy is emitted by the sun in one milliard years. It has been disclosed, by using spectroscopic results, that the gas particles in the emitting photosphere move with an enormous velocity of the order of 6000 km./sec. during supernovae flare-ups.

The observations described above of the flare-ups of supernovae, novae and the luminosity oscillation of Cepheids, show that the existence of these phenomena is very directly related to the motion of the colossal gas mass from which the appropriate variable stars are formed.

Below, we analyse the formulation of the problem and the solution, in certain cases, of the unsteady motion of a gas which can be considered as an approximate model of the Cepheid oscillations and the flare-ups of novae and supernovae.

§2. On the Equations of Equilibrium and Motion of a Gaseous Mass Simulating a Star

To give a quantitative description of the behaviour of gaseous masses forming stars, equations of equilibrium and motion must be established. Below, we shall give the equilibrium equation and the Newtonian equation of motion of a gravitational mass of gas. We consider only radial gas motion with spherical symmetry.

The continuity equation for motion with spherical symmetry is expressed by the law of conservation of mass as

$$\frac{\partial \rho}{\partial t} + \frac{\partial \rho v}{\partial r} + \frac{2\rho v}{r} = 0 \qquad (2.1)$$

where ρ is the density, v is the radial velocity of a gas particle, r is the distance to the centre of symmetry and t is the time. This equation is satisfied identically in the equilibrium state.

Neglecting the gas viscosity is quite legitimate in view of the huge linear and velocity scales of the motions of the mass of gas; consequently, the momentum equation taking gravitational forces into account can be written under the assumption that the gas is inviscid

$$\frac{\partial v}{\partial t} + v\frac{\partial v}{\partial r} + \frac{1}{\rho}\frac{\partial p}{\partial r} + \frac{f\mathcal{M}}{r^2} = 0, \qquad (2.2)$$

where $f = 6\cdot670 \times 10^{-8}$ cm³/gm. sec² is the gravitational constant, p is total pressure equal to the sum of the molecular and light pressures, $\mathcal{M}(r, t)$ is the mass of gas within the sphere of radius r under consideration. As is known, the effect of all the masses on a particle at a distance r

from the centre of symmetry, when spherical symmetry exists, equals the force of attraction of a particle placed at the centre of symmetry and with a mass equal to $\mathcal{M}(r,t)$. In order to determine $\mathcal{M}(r,t)$, we have the equation

$$\frac{\partial \mathcal{M}}{\partial r} = 4\pi r^2 \rho \quad \text{or} \quad \mathcal{M} = 4\pi \int_0^r r^2 \rho \, dr. \tag{2.3}$$

The temperature of the material within the star is very high, of the order of millions of degrees, and near the outer surface of the star it is of the order of several thousand degrees; the stellar material at such temperatures can be considered as a perfect gas even when the pressure and the density are extremely high. Consequently, assuming that local thermodynamic equilibrium occurs, we take the equation of state as

$$P = P_{\text{mol}} + P_{\text{rad}} = \frac{\mathbf{R}\rho T}{\mu} + \frac{a}{3} T^4, \tag{2.4}$$

where T is absolute temperature; $\mathbf{R} = 8\cdot3144 \times 10^7$ erg/deg. mol is the absolute gas constant; μ is the molecular weight (mass of one mole of gas, $[\mu] = $ gm./mol) determined by the chemical composition of the gas,† and $a = 7\cdot569 \times 10^{-15}$ erg/cm³ deg⁴ is the constant emission density (Stefan's constant).

The light pressure is comparable to the molecular only for very large temperatures or for extremely low densities. Consequently, the light pressure can usually be neglected in studying the structure and internal gas motion in stars in comparison with the molecular pressure, which is equivalent to replacing (2.4) by Clapeyron's equation

$$p = \frac{\mathbf{R}\rho T}{\mu}. \tag{2.4'}$$

We use now this equation and the theorem of kinetic energy instead of the conservation of energy equation—the heat flow equation, which is, in general form,

$$\frac{dE}{dt} + \frac{dA^{(i)}}{dt} = \frac{dQ^{(e)}}{dt}, \tag{2.5}$$

where E is the internal particle energy excluding the internal potential energy due to gravitation $dA^{(i)}/dt$ is the work per unit time of the internal stresses (the work of the internal gravitational forces does not enter

† Total ionization of the volatile elements occurs because of the high temperature within the star. For total ionization μ is close to two for all elements except for hydrogen $\mu = 0\cdot5$ and for helium $\mu = 1\cdot33$.

here†), $dQ^{(e)}/dt$ is the external heat flow per unit time. Because of the assumption made above, the heat flow equation can be written as:

$$\frac{d}{dt}(c_v T) + p \cdot \frac{d}{dt}\frac{1}{\rho} = \epsilon, \qquad (2.5')$$

where c_v is the specific heat of a unit mass of gas for constant volume depending on the chemical composition of the material; ϵ is the total energy liberated per unit mass of gas per unit time. The quantity ϵ can differ from zero because of emission, absorption, because of nuclear or chemical reactions and because of heat conduction. If the star-gas is in equilibrium, then the general energy balance reduces to the equation

$$\epsilon = 0.$$

Since the star emits energy to the ambient space, then energy sources must be inside the star for equilibrium. The nature of these energy sources and their distribution within the stars are still not completely clear at present. However, an investigation of the equilibrium of the star for various distributions of the energy sources shows that the pressure and density distributions within the star and, in particular, their values at the centre of the star depend only slightly on the distribution of the energy sources. Computations show that the physical variables are obtained very closely if the source distribution is assumed to be uniform over the whole mass of the star or if it is assumed that the same quantity of energy is liberated at one point, the centre of the star. We may add that the quantity of liberated energy due to physical chemical processes depends very sensitively on the temperature. The temperature is very high at the centre of the star, consequently, the major part of the energy is liberated near the centre of the star. As computations show, this statement is justified in practice (Ambartsumian, Mustel', Severnyi and Sobolev, 1958, Chandrasekhar, 1939).

Consequently, we shall later sometimes use schemes in which it is assumed that the energy is liberated only in the centre of the star.

At the present time available scientific evidence points to the existence of pulsations of stars of long duration, deriving energy either from gravitational energy acquired in compressing the star (see §3, p. 173, Equation (3.13)) or from a reacting core inside the star.

The possible types of reacting core depend to a marked extent on the temperature and the nature of the combustible material.

Physical considerations and data concerning the chemical structure of the star show that the reaction of the core is essentially a transformation from a hydrogen core to a helium core. Calculations show

† The work of the internal gravitational forces and the increment of the internal gravitational energy are contained in the left side of (2.5).

that for a star of solar type the balance between energy available and energy used at any time can be explained by a reacting core which burns hydrogen. The effect of the reaction on the mass of the star is very small, causing a change of about $0 \cdot 007$ of the original mass, although each second the sun emits energy, which from the formula $E = mc^2$ (c = speed of light) means a loss in mass of 4 million tons and this process has already continued for 5–6 milliard years.

The energy can be transmitted by radiation from the centre to the periphery of the star; the energy distribution over the frequency spectrum can vary in this process because of absorption and intrinsic emission, but the energy emitted, absorbed and transmitted by heat conduction† gives a common balance at equilibrium which equals zero. Later, we shall assume as an approximate condition that such a statement still applies in nonstationary processes, in other words, we shall consider adiabatic gas motions ($\epsilon = 0$).

It is not difficult to show that the total energy emitted by the cepheids during the period of change in their brilliance is small in comparison with the total reserve of gravitational and internal thermal energy of the whole star: this also explains the slight influence of the distribution of the stellar energy sources on the density and pressure distributions in the stellar depths for the usual stars and for cepheids. Consequently, we can assume that the energy liberated in the centre and radiated into outer space during a period of oscillation does not play an essential part in the unsteady motion of stars. As a last assumption in considering the unsteady motion, we assume that the molecular weight μ and the heat conduction coefficient c_v are constant in the whole stellar mass.

It follows from the assumptions made above that we can use the following system of equations to describe the unsteady motion of gas models of stars:

$$
\left.
\begin{aligned}
&\frac{\partial \rho}{\partial t} + \frac{\partial \rho v}{\partial r} + \frac{2\rho v}{\partial r} = 0, \\[2mm]
&\frac{\partial v}{\partial t} + v\,\frac{\partial v}{\partial r} + \frac{1}{\rho}\frac{\partial p}{\partial r} + \frac{\partial f \mathcal{M}}{r^2} = 0, \\[2mm]
&\frac{\partial \mathcal{M}}{\partial r} - 4\pi r^2 \rho = 0, \\[2mm]
&\frac{\partial \dfrac{p}{\rho^\gamma}}{\partial t} + v\,\frac{\partial \dfrac{p}{\rho^\gamma}}{\partial r} = 0,
\end{aligned}
\right\} \quad \text{(I)}
$$

† The energy transmitted by heat conduction is small in comparison with the energy transmitted by radiation.

where $\gamma = c_p/c_v$ is the specific heat ratio. The last of Equations (I) follows directly from Equations (2.4′) and (2.5′) when $\epsilon = 0$. The system (I) consists of four equations with four unknown functions ρ, v, p, $\mathscr{M}$.

If $\gamma \neq c_p/c_v$ and $\gamma = $ constant, then Equations (I) apply to gas motions with polytropic processes.

In the equilibrium case, Equations (I) are simplified and reduce to one equation

$$\frac{1}{\rho}\frac{dp}{dr} + \frac{4\pi f}{r^2}\int_0^r r^2\rho\,dr = 0,$$

containing the two unknown functions ρ and p. To find the temperature distribution, it is necessary to rely on the data on the distribution of the energy sources within the star and on the equations of the theory of radiant energy transport.

For spherical symmetry and reversible thermodynamic processes the system of equations determining the physical variables within the star at equilibrium can be taken as (Ambartsumian, Mustel', Severnyi and Sobolev, 1958),

$$\left. \begin{array}{l} \dfrac{1}{\rho}\dfrac{dp}{dr} + \dfrac{f\mathscr{M}}{r^2} = 0, \\[2.5ex] \mathscr{M} = 4\pi\displaystyle\int_0^r r^2\rho\,dr, \\[2.5ex] p = \dfrac{R\rho T}{\mu} + \dfrac{a}{3}\,T^4, \\[2.5ex] \dfrac{d}{dr}\dfrac{R^4\,T^4}{3} = -\dfrac{\kappa_1\,\mathfrak{L}\rho}{4\pi r^2}, \quad \kappa_1 = \dfrac{\kappa R^4}{ac}, \\[2.5ex] \dfrac{d\mathfrak{L}}{dr} = 4\pi r^2\rho\epsilon^*, \end{array} \right\} \quad \text{(II)}$$

where κ is the absorption coefficient, c is the speed of light, ϵ^* is the intensity of the sources of energy liberated by a unit mass of the stellar material per unit time, $\mathfrak{L}(r)$ is the energy flow through the sphere of radius r.

In order to obtain the solution of the equilibrium equations (II), the mass $\mathscr{M}$ can be taken as independent variable and the following boundary conditions can be used:

$$\mathfrak{L} = \mathfrak{L}_0 \quad \text{at the centre of the star for } \mathscr{M} = 0. \qquad (2.6)$$

$$\rho = 0,\ T = 0 \quad \text{at the surface of the star for } \mathscr{M} = \mathfrak{M}. \qquad (2.6')$$

If the energy sources are distributed continuously within the star, then the quantity $\mathfrak{L}_0$ must be put equal to zero. We have $\mathfrak{L}_0 \neq 0$ if a concentrated energy source is present at the centre of the star.

Since the temperature within the star is somewhat larger than on its surface, many astrophysicists usually assume that $T = 0$ on the surface of the star. The solution of the system (II) for given μ, κ, ϵ^* depends on three arbitrary constants determined by the single boundary condition (2.6) at the centre and the two (2.6') on the surface of the star.

If the molecular weight μ, the absorption coefficient κ_1 and the intensity of the energy sources ϵ^* are given as functions of ρ and T:

$$\mu = \mu(\rho, T), \quad \kappa_1 = \kappa_1(\rho, T) \quad \text{and} \quad \epsilon^* = \epsilon^*(\rho, T), \qquad (2.7)$$

then the system (II) and the boundary conditions (2.6) and (2.6') completely determine the solution which is

$$\left.\begin{aligned}
\rho &= \rho(\mathscr{M}, \mathfrak{M}, \mathfrak{L}_0, a_1, a_2, \ldots), \\
T &= T(\mathscr{M}, \mathfrak{M}, \mathfrak{L}_0, a_1, a_2, \ldots), \\
\mathfrak{L} &= \mathfrak{L}(\mathscr{M}, \mathfrak{M}, \mathfrak{L}_0, a_1, a_2, \ldots), \\
r &= r(\mathscr{M}, \mathfrak{M}, \mathfrak{L}_0, a_1, a_2, \ldots),
\end{aligned}\right\} \qquad (2.8)$$

where $a_1, a_2, \ldots$ are physical constants entering in the system (II) and the functional relation (2.7).

If $\mathfrak{L}_0 = 0$, then the following formulas for the luminosity and the radius of the star follow from (2.8)

$$\left.\begin{aligned}
\mathfrak{L}^* &= \mathfrak{L}(\mathfrak{M}, \mathfrak{M}, a_1, a_2, \ldots), \\
\mathfrak{R} &= r(\mathfrak{M}, \mathfrak{M}, a_1, a_2, \ldots).
\end{aligned}\right\} \qquad (2.9)$$

Hence, the relations between the luminosity of the star and its mass, and between the radius of the star and its mass can be determined by solving the problem formulated and, therefore, the way is open to give a theoretical meaning to the empirical results described in §1 of this chapter.

Relations (2.9) depend parametrically on the constants $a_1, \ldots$ which can have different values for different stars. Relations (2.9) determine the unique dependence of the luminosity on the mass of the star for those groups of stars taking identical values of such constants.

If $\mathfrak{L}_0 \neq 0$, and is given independently of $\mathfrak{M}$, then (2.9) depend on $\mathfrak{L}_0$ as an additional parameter. If all the energy sources are concentrated at the centre, a model with a point energy source, then we obtain the following instead of (2.9)

$$\left.\begin{aligned}
\mathfrak{L}^* &= \mathfrak{L}_0, \\
\mathfrak{R} &= f(\mathfrak{M}, \mathfrak{L}_0, a_1, a_2, \ldots).
\end{aligned}\right\} (2.10)$$

Thus, it is impossible to establish two relations of the type $\mathfrak{L}^*(\mathfrak{M})$ and $\mathfrak{R}(\mathfrak{M})$ for a point energy source as a result of solving (II). These relations can be established in this case if the additional relation $\mathfrak{L}_0(\rho_0,\, T_0)$ is given, where ρ_0 and T_0 are the density and temperature at the centre of the star.

§3. Theoretical Formulas Relating Luminosity with Mass and Radius with Mass

In order to state the problem formulated in the preceding paragraph, an actual examination of the validity of (2.9) is of great interest. Strömgren (1936, 1937) considered this problem under the following assumptions†:

(i). The term $aT^4/3$, corresponding to the radiation pressure, is discarded in the equation of state.

(ii). The molecular weight μ is constant everywhere within the star.

(iii). The following formulas are true for the absorption coefficient and for the energy sources

$$\kappa_1 = B\rho(RT)^{-3-s} \quad \text{and} \quad \epsilon^* = \epsilon_0 \rho^\alpha (RT)^\beta,$$

where B, s, ϵ_0, α, β are constants.

Relying on these assumptions, Strömgren established the formula

$$\mathfrak{L}^* = \mathfrak{D}\frac{1}{B}\frac{\mathfrak{M}^{5+s}}{\mathfrak{R}^s}\mu^{7+s}, \tag{3.1}$$

where the constant $\mathfrak{D}$ depends only on f, s, α, β.

However, it appears (Sedov, 1954) that by using dimensional analysis, retaining assumption (i) and with more general assumptions than (ii) and (iii) such as

$$\mu = \mu_0 \rho^{1-\xi}(RT)^{1-\eta}, \quad \kappa_1 = B\rho^w(RT)^v, \quad \epsilon^* = \epsilon_0 \rho^\alpha (RT)^\beta, \tag{3.2}$$

where ξ, η, w, v, α, β are constants, two simple relations such as (2.9) can replace the single relation (3.1).

Let us introduce the new variables

$$p_1 = \frac{p}{f}, \quad \tau = \frac{RT}{(f\mu_0)^{1/\eta}}, \quad x = \frac{\mathscr{M}}{\mathfrak{M}}$$

and the notation

$$\frac{a(f\mu_0)^{4/\eta}}{3fR^4} = \Omega \quad \text{and} \quad B\epsilon_0(f\mu_0)^{(v+\beta-4)/\eta} = \omega.$$

It is not difficult to verify that the system of equations (II) and the

† An explanation of many results derived from the theory of stellar gas models can be found in Chandrasekhar (1939).

boundary condition (2.6) for $\mathfrak{L}_0 = 0$ are equivalent to the following system

$$
\left.
\begin{array}{l}
\dfrac{dp_1}{dx} = -\dfrac{x\mathfrak{M}^2}{4\pi r^4}, \quad r^3 = \dfrac{3}{4\pi}\mathfrak{M}\displaystyle\int_0^x \dfrac{dx}{\rho}, \\[4mm]
p_1 = \rho^\xi \tau^\eta + \Omega\tau^4, \quad \tau^3\dfrac{d\tau}{dx} = -\dfrac{3\omega\mathfrak{M}^2}{16\pi^2 r^4}\rho^w \tau^v \displaystyle\int_0^x \rho^\alpha \tau^\beta\, dx,
\end{array}
\right\}
\tag{3.3}
$$

and the formula to calculate the luminosity

$$
\mathfrak{L} = \epsilon_0 \mathfrak{M}(\mu_0 f)^{\beta/\eta}\int_0^x \rho^\alpha \tau^\beta\, dx. \tag{3.4}
$$

Hence, the problem is reduced to solving the system (3.3) for the following boundary conditions† (see conditions (2.6′)): for $x = 1$ on the surface of the star, we have

$$
\rho = 0 \quad \text{and} \quad \tau = 0. \tag{3.5}
$$

Hence it follows that if the solution of the formulated mathematical problem exists and has physical meaning then the desired functions

$$
\tau, \rho, r, p_1\left([p_1] = \dfrac{M^2}{L^4} \quad \text{and} \quad [\tau] = M^{(2-\xi)/\eta}\, L^{(3\xi-4)/\eta}\right)
$$

are determined by the quantities

$$
x, \mathfrak{M}, \omega, \Omega,
$$

related by the following dimensional formulas:

$$
[\mathfrak{M}] = M, \quad [\omega] = M^{k_1} L^{k_2}, \quad [\Omega] = M^{2-4[(2-\xi)/\eta]} L^{-4-4[(3\xi-4)/\eta]}, \tag{3.6}
$$

where

$$
\left.
\begin{array}{l}
k_1 = -2 - w - \alpha + \dfrac{(4 - v - \beta)(2 - \xi)}{\eta}, \\[4mm]
k_2 = 4 + 3w + 3\alpha + \dfrac{(4 - v - \beta)(3\xi - 4)}{\eta}.
\end{array}
\right\}
\tag{3.7}
$$

Furthermore, we assume that $k_2 \neq 0$. If it is assumed that $\Omega = 0$ (assumption (i)) or if $\xi = \eta = 1$ (assumption (ii)) then no quantity depending on the linear dimensions is among the characteristic quantities in the $k_2 = 0$ case while the desired quantities τ, ρ, r, p_1 are related to the linear dimensions. In this case, the formulation of the problem mentioned above requires improvement.

---

† Boundary conditions (3.5) can be changed. If new dimensional physical constants do not appear under the altered conditions, then all the subsequent conclusions remain valid.

Only two independent combinations can be formed from the characteristic parameters:

$$x = \frac{\mathcal{M}}{\mathfrak{M}} \quad \text{and} \quad \delta = \Omega \frac{\omega^{[4\eta+4(3\xi-4)]/[(4+3w+3\alpha)\eta+(4-v-\beta)(3\xi-4)]}}{\mathfrak{M}^{k_3}}$$

where

$$k_3 = \frac{1}{\eta}\left\{2\eta + 4(\xi-2) - \frac{[4\eta+4(3\xi-4)][(2+w+\alpha)\eta - (4-v-\beta)(2-\xi)]}{(4+3w+3\alpha)+(4-v-\beta)(3\xi-4)}\right\}$$

Evidently, for $\mu = $ const, i.e. $\xi = \eta = 1$, we have

$$k_3 = -2 \quad \text{and} \quad \delta = \Omega\mathfrak{M}^2.$$

If the conditions $\xi = 1$ and $\eta = 1$ are not satisfied simultaneously, then the following conditions can be satisfied for certain values of the exponents $\xi, \eta, \alpha, \beta, w, v$

$$k_3 = 0.$$

We find that the abstract parameter δ is independent of the mass $\mathfrak{M}$ for $k_3 = 0$.

From the usual reasoning of dimensional analysis, it follows that the desired solution will be

$$\left.\begin{aligned}
r &= \left(\frac{\omega}{\mathfrak{M}^{k_1}}\right)^{1/k_2} r_1(x, \delta), \\
\rho &= \mathfrak{M}\left(\frac{\omega}{\mathfrak{M}^{k_1}}\right)^{-3/k_2} \rho_1(x, \delta), \\
\tau &= \mathfrak{M}^{(2-\xi)/\eta}\left(\frac{\omega}{\mathfrak{M}^{k_1}}\right)^{(3\xi-4)/k_2\,\eta} \tau_1(x, \delta), \\
p_1 &= \mathfrak{M}^2\left(\frac{\omega}{\mathfrak{M}^{k_1}}\right)^{-4/k_2} (\rho_1^\xi \tau_1^\eta + \delta\tau_1^4).
\end{aligned}\right\} \quad (3.8)$$

Substitution of (3.8) into the system (3.3) and the boundary condition (3.5) leads to the equations

$$\left.\begin{aligned}
\frac{d}{dx}(\rho_1^\xi \tau_1^\eta + \delta\tau_1^4) &= -\frac{x}{4\pi r_1^4}, \\
r_1^3 &= \frac{3}{4\pi}\int_0^x \frac{dx}{\rho_1}, \\
\tau_1^3 \frac{d\tau_1}{dx} &= -\frac{3}{16\pi^2 r_1^4}\rho_1^w \tau_1^v \int_0^x \rho_1^\alpha \tau^\beta dx,
\end{aligned}\right\} \quad (3.9)$$

and the boundary conditions

$$\tau_1 = 0 \quad \text{and} \quad \rho_1 = 0 \quad \text{for } x = 1. \tag{3.10}$$

Equations (3.9) and conditions (3.10) do not contain the dimensional constants $\mathfrak{M}$, ω and they define the nondimensional functions $\tau_1(x, \delta)$, $\rho_1(x, \delta)$, and $r_1(x, \delta)$. If assumption (i) is used, the parameter δ does not enter. In the latter case, the functions $r_1(x)$, $\rho_1(x)$, and $\tau_1(x)$ are universal functions with numerical values dependent only on the exponents ξ, η, w, v, α, β. From Equations (3.8) and (3.4) and combining the relation $\omega = B\epsilon_0(f\mu_0)^{(v+\beta-4)/\eta}$ with (3.7) we obtain, for the radius $\mathfrak{R}$ and the luminosity $\mathfrak{L}^*$ of the star

$$\left.\begin{aligned}
\mathfrak{R} &= [(B\epsilon_0)^\eta (f\mu_0)^{v+\beta-4} \\
&\quad \times \mathfrak{M}^{(2+w+\alpha)-(4-v-\beta)(2-\xi)}]^{\frac{1}{(4+3w+3\alpha)\eta+(4-v-\beta)(3\xi-4)}} r_1(1, \delta), \\[8pt]
\mathfrak{L}^* &= \epsilon_0(B\epsilon_0)^{\frac{(3\xi-4)\beta-3\alpha\eta}{(4+3w+3\alpha)\eta+(4-v-\beta)(3\xi-4)}} \\
&\quad \times (f\mu_0)^{\frac{\beta}{\eta}+\frac{v+\beta-4}{\eta}\left[\frac{(3\xi-4)\eta-3\alpha\eta}{(4+3w+3\alpha)\eta+(4-v-\beta)(3\xi-4)}\right]} \\
&\quad \times \mathfrak{M}^{1+\alpha+\frac{2-\xi}{\eta}+\frac{[(2+w+\alpha)\eta-(4-v-\beta)(2-\xi)][(3\xi-4)\beta-3\alpha\eta]}{[(4+3w+3\alpha)\eta+(4-v-\beta)(3\xi-4)]\eta}} \times \int_0^1 \rho_1^\alpha \tau_1^\beta dx.
\end{aligned}\right\} \tag{3.11}$$

If the light pressure is neglected in the equation of state (assumption (i)) then it is necessary to put $\delta = 0$ after which (3.11) determine completely the dependence of $\mathfrak{R}$ and $\mathfrak{L}^*$ on ϵ_0, $f\mu_0$, B and on the mass of the star $\mathfrak{M}$.

If $\delta \neq 0$ but $k_3 = 0$, then δ is independent of the mass of the star and, consequently, (3.11) also determine completely the relation between $\mathfrak{R}$ and $\mathfrak{L}^*$ and the mass of the star $\mathfrak{M}$ in this case.

If ϵ_0 is eliminated from (3.11), then we obtain the relation

$$B\mathfrak{L}^* = \mathfrak{D} \cdot (f\mu_0)^{(4-v)/\eta} \cdot \mathfrak{R}^{4+3w+(4-v)[(3\xi-4)/\eta]} \cdot \mathfrak{M}^{-1-w+(4-v)[(2-\xi)/\eta]} \tag{3.12}$$

where $\mathfrak{D}$ is an abstract constant which can depend only on ξ, η, α, β, w, v for $\delta = 0$.

It is curious to note that only the constant $\mathfrak{D}$ can depend on the energy liberation in (3.12).

Formula (3.1) is obtained as a particular case of (3.12).

It is easy to see directly that (3.12) remains valid for a star model with point energy sources at the centre of the star when the intensity of the point energy sources $\mathfrak{L}^*$ is given arbitrarily.

Formulas (3.12) and (3.11) can be used to process observations. When these formulas are compared with empirical data, we are able to test the soundness of the laws of (3.2) and also the validity of the formulation of the problem.

§4. Certain Simple Solutions of the System of Equations of Stellar Equilibrium

Data on the distribution of the variables of state within the gas at equilibrium are needed as initial conditions when investigating the unsteady motions of a gaseous stellar mass in the theory of novae flare-ups. To use the solution of the system (II) with the boundary conditions (2.6) and (2.7) is inconvenient for this purpose because of their complexity and it eliminates the possibility of obtaining effective solutions of the unsteady motions.

Moreover, it is useful to consider the solutions of (II) independently of the boundary conditions on the surface of the star for a deeper under-standing of the role of the various physical factors.

We shall analyse the exact solution of Equations (II) (in which we neglect the light pressure); hence, let us rely on certain additional hypo-theses instead of on the boundary conditions on the surface of the star. Starting from dimensional considerations, let us analyse the very simple hypothesis that the distribution of the variables of state, related to the value of gravitational constant f in addition to the gravitational force, depends essentially on some physical law whose influence can be realized by means of just one characteristic physical constant which we will denote by a.

More specifically, let us consider the equilibrium of a gas for which the density and pressure distribution depend only on the following three dimensional parameters

$$[r] = L, \quad [f] = \frac{L^3}{MT^2}, \quad [a] = ML^k T^s. \tag{4.1}$$

The dimensions of the parameter a (the values of the constants k and s) are fixed by the above hypothesis, which we shall presently state more precisely. We shall consider the case when the dimensions of the ad-ditional physical constant a contains the mass symbol. Evidently, without loss of generality, it can be assumed that the mass symbol enters linearly in the formula giving the dimensions of the constant a.†

---

† If the additional assigned constant a^* is a kinematic quantity, then a constant equal to a^*/f can be taken as a.

It is easy to verify that the following formulas result from the hypothesis (4.1)

$$
\left.
\begin{aligned}
\rho &= \alpha_1 \frac{a^{2/(2-s)} f^{s/(2-s)}}{r^{(2k+6)/(2-s)}}, \\[2mm]
\mathcal{M} &= \alpha_4 \, a^{2/(2-s)} \cdot f^{s/(2-s)} \cdot r^{-(2k+3s)/(2-s)}, \\[2mm]
p &= \alpha_2 \frac{a^{4/(2-s)} f^{(2+s)/(2-s)}}{r^{(8+2s+4k)/(2-s)}}, \\[2mm]
RT &= \alpha_3 \frac{a^{2/(2-s)} f^{2/(2-s)}}{r^{2[(k+s+1)/(2-s)]}}.
\end{aligned}
\right\}
\qquad (4.2)
$$

where α_1, α_2, α_3, α_4 are abstract constants and $s \neq 2$. If $s = 2$ and $k \neq -3$, then it is impossible to establish a dependence on r since the abstract combination af/r^{k+3} arises in this case; if $s = 2$ and $k = -3$, then the dimensions of f and a are independent and, consequently, the system (4.1) is incomplete in this case and cannot define ρ and p.

The physical meaning of $\mathcal{M}(r)$ is a positive monotonic quantity which does not decrease as r increases; consequently, we must have

$$
-\frac{2k+3s}{2-s} \geqslant 0. \qquad (4.3)
$$

Formulas (4.2) show that the centre of symmetry is a singular point at which the density, pressure and temperature are generally infinite. On the one hand, this fact can reflect the real state of affairs in that the density, pressure and temperature have maximums at the centre of the star and attain very large values. On the other hand, it follows from physical reasoning that the pressure, density and temperature are essentially finite at the centre of the star. Hence, it follows that the accepted hypothesis on the existence of only two physical constants f and a is inapplicable directly near the centre of the star in those cases when the variables of state become infinite. This also serves as an indication that additional physical factors to f and a become important near the centre of the star. However, if it is admitted that such refinements are necessary only directly near the centre of the star, then (4.2) can be used to simulate the actual equilibrium outside the neighbourhood of the centre of the star.† If $w = (2k+6)/(2-s) < 3$ then as $r \to 0$ we have $\mathcal{M} \to 0$.

† In passing, let us note that the gravitational constant near the centre of the star is apparently not essential in explicit form since the equivalent force of Newtonian gravitation is almost zero at the centre of the star. However, we will not dwell here upon the much deeper phenomena near the centre of the star. In §6 pp. 336–337 we find that, for the solution of (4.2), the total energy $H(r)$ is finite for $w < 2 \cdot 5$ and $H(r) \to 0$ as $r \to 0$.

Formulas (4.2) also show that the pressure and the density can become zero only at infinity and that the mass approaches infinity together with r if $-(2k+3s)/(2-s) > 0$. This shows that the hypothesis under consideration also requires correction at remote distances from the centre of the star.

If $(2k+3s)/(s-2) = 0$ then $k = -(3/2)s$; in this case the combination $a^{2/(2-s)} f^{s/(2-s)} = \mathcal{M}_0$ has the dimensions of a mass. This constant can be taken instead of a and, consequently, (4.2) become †

$$\rho = \alpha_1 \frac{\mathcal{M}_0}{r^3}, \quad \mathcal{M} = \mathcal{M}_0, \quad p = \alpha_2 \frac{\mathcal{M}_0^2 f}{r^4}, \quad RT = \alpha_3 \frac{\mathcal{M}_0}{r} f. \quad (4.4)$$

Evidently, in this case the equation

$$\mathcal{M} = 4\pi \int_0^r r^2 \rho \, dr$$

is not satisfied, consequently, this case cannot lead to an exact solution of the system (II). It is not difficult to see that we satisfy the first three equations of the system (II)

$$\frac{1}{\rho} \frac{dp}{dr} + \frac{f\mathcal{M}}{r^2} = 0, \quad \mathcal{M} = 4\pi \int_0^r r^2 \rho \, dr, \quad p = \frac{\rho RT}{\mu}, \quad (4.5)$$

where μ is a constant, if we define the constants α_2, α_3 and α_4 in terms of the constant α_1 by means of the formulas

$$\alpha_4 = -4\pi \frac{2-s}{2k+3s} \alpha_1, \quad (4.6)$$

$$\alpha_2 = -2\pi \frac{(2-s)^2 \alpha_1^2}{(2k+3s)(4+s+2k)}, \quad (4.7)$$

$$\alpha_3 = \frac{\mu \alpha_2}{\alpha_1} = -2\pi\mu \frac{(2-s)^2 \alpha_1}{(2k+3s)(4+s+2k)}. \quad (4.8)$$

It is not difficult to see that the exponent in the formula for the pressure $2(4+s+2k)/(2-s)$ must be non-zero since, otherwise, the pressure would be found to be constant along the radius, which eliminates the possibility of satisfying the equilibrium equations.

Now, let us turn to the question of constructing a solution of the desired type which will satisfy the last two equations of the system (II)— the radiation equations

$$\frac{d}{dr} R^4 T^4 = -\frac{3\kappa_1 \mathcal{L}\rho}{4\pi r^2} \quad \text{and} \quad \frac{d\mathcal{L}}{dr} = 4\pi r^2 \rho \epsilon^*. \quad (4.9)$$

† We put $\mathcal{M} = \mathcal{M}_0$ and, therefore $\alpha_4 = 1$.

These equations can be satisfied by using a solution of the (4.2) type if the energy sources are concentrated at the centre of the star, $\epsilon^* = 0$ for $r > 0$, and if we take as the formula for the absorption coefficient

$$\kappa_1 = B\rho^w (RT)^v \qquad (4.10)$$

where v, w and B are constants. It is easy to verify that the dimensions of the constant B are given by

$$[B] = M^{-(2+w)} L^{10+3w-2v} T^{7 2v-5}.$$

We obtain from (4.9)

$$\mathfrak{L} = \mathfrak{L}_0 = \text{const} \quad \text{and} \quad \kappa_1 = -\frac{4\pi}{3\mathfrak{L}_0} \frac{r^2}{\rho} \frac{d}{dr}(RT)^4. \qquad (4.11)$$

A comparison of (4.10) and (4.11) when using (4.2) shows that the radiation equations are satisfied if we have

$$k(2w+2v-6)+s(2v-9)+6w+2v = 0. \qquad (4.12)$$

and

$$\frac{32\pi}{3} \cdot \frac{\alpha_3^{4-v}}{\alpha_1^{1+w}} \cdot \frac{s+k+1}{2-s} = \frac{B\mathfrak{L}_0}{a^{(6-2w-2v)/(2-s)} f^{(8-s-sw-2v)/(2-s)}}. \qquad (4.13)$$

Equation (4.12) gives the relation between k, s, w and v. The coefficients α_1, α_2, α_3 and α_4 are expressed in terms of B, $\mathfrak{L}_0$, μ, a, f, w, v by means of (4.6), (4.7), (4.8), (4.12) and (4.13). After substituting these expressions in (4.2), we find the exact solution of the complete system of equations of gas equilibrium (II) as

$$\rho = \left\{ \frac{(3-w-v)}{16\pi(2w+3)} \left[\frac{(3w+v)(v-w-6)}{(3-w-v)^2 \, 2\pi\mu f} \right]^{4-v} \frac{3B\mathfrak{L}_0}{r^{9-2v}} \right\}^{1/(3-w-v)},$$

$$\mathscr{M} = 4\pi \frac{w+v-3}{3w+v} \left\{ \frac{3-w-v}{16\pi(2w+3)} \right.$$
$$\left. \times \left[\frac{(3w+v)(v-w-6)}{(3-w-v)^2 \, 2\pi\mu f} \right]^{4-v} \frac{3B\mathfrak{L}_0}{r^{3w+v}} \right\}^{1/(3-w-v)},$$

$$p = \left\{ \frac{(3-w-v)\mu^{v-4}}{16\pi(2w+3)} \left[\frac{(3w+v)(v-w-6)}{(3-w-v)^2 \, 2\pi f} \right]^{(5+w-v)/2} \right.$$
$$\left. \times \frac{3B\mathfrak{L}_0}{r^{6+w-v}} \right\}^{2/(3-w-v)},$$

$$RT = \left\{ \frac{(3-w-v)}{16\pi(2w+3)} \left[\frac{(3w+v)(v-w-6)}{(3-w-v)^2 \, 2\pi\mu f} \right]^{w+1} \frac{3B\mathfrak{L}_0}{r^{3+2w}} \right\}^{1/(3-w-v)}.$$

$$\left. \right\} \quad (4.14)$$

The laws of variation of p, ρ and RT along the radius in the exact solution, constructed in this manner, of the problem of equilibrium of a Newtonian gravitational mass of gas are determined completely by the exponents w and v in the formula for the absorption coefficient and by the product of the constant B, in the law for the absorption coefficient, with the intensity of the source of radiation $\mathfrak{L}_0$ at the centre of symmetry.

The constant a does not figure in the solution obtained (4.14); this is explained by the fact that the constants α_1 and a enter only in the combination

$$\alpha_1 \, a^{2/(2-s)}$$

which is expressed in terms of the product $B\mathfrak{L}_0$ because of the equilibrium equations in the problem. The degree of the dependence on $B\mathfrak{L}_0$ is obtained from the exponents w and v.

The solution (4.14) permits the influence of the laws for the absorption coefficient κ on the equilibrium of gases with a radiation source to be estimated.

If the dimensions of the constant a (the exponents k and s) are given, then (4.2) determines all the laws for the distribution of the variables of state along the radius while the exponents w and v in (4.10) can be varied to satisfy the condition (4.12). If the exponents w and v are given, then the laws of variation along the radius are also determined but the constants k and s can be varied to satisfy the same condition (4.12).

According to the solution (4.14), the appropriate gas model of a star occupies the whole of space and has an infinite mass. Evidently, the mass is finite within any sphere S of finite radius on which the surface pressure p_S is very small. The presence of an infinite mass outside the sphere S does not exert any influence on the equilibrium of the mass within the sphere S for a fixed p_S on the surface of S. Hence, the equilibrium of a finite mass within the sphere S is not related essentially to the distribution of the equilibrium variables beyond the sphere S. In order to obtain approximate solutions with a finite mass, a solution of the type (4.14) can be used within a certain sphere S and the solution can be continued outside this sphere with a continuous pressure variation and with a certain law for the density variation which would guarantee a finite and given magnitude of the mass.[†]

In particular, if the constant a has the dimensions of energy, i.e., $[a] = ML^2 T^{-2}$ (from which $k = 2$ and $s = -2$), then condition (4.12) yields

$$v = -3 - 5w. \tag{4.15}$$

[†] A discontinuity in the density distribution can be admitted in the approximate construction of the equilibrium of a finite mass of gas.

11*

In this case formulas (4.14) become

$$\left. \begin{aligned}
\rho &= \frac{1}{2^{2\cdot5}}\left[\left(\frac{3}{2\pi\mu f}\right)^{7+5w}\frac{3B\mathfrak{L}_0}{4\pi}\right]^{1/(6+4w)}\cdot\frac{1}{r^{2\cdot5}}, \\[2mm]
\mathscr{M} &= 2\pi\left[\left(\frac{3}{2\pi\mu f}\right)^{7+5w}\frac{3B\mathfrak{L}_0}{2^{5+2w}\,\pi}\right]^{1/(6+4w)}\cdot r^{0\cdot5}, \\[2mm]
p &= \frac{1}{8}\left[\left(\frac{3}{2\pi\mu f}\right)^{4+3w}\frac{3B\mathfrak{L}_0}{\mu^{3+2w}\,4\pi}\right]^{1/(3+2w)}\cdot\frac{1}{r^3}, \\[2mm]
RT &= \frac{1}{\sqrt{2}}\left[\left(\frac{3}{2\pi\mu f}\right)^{1+3w}\frac{3B\mathfrak{L}_0}{4\pi}\right]^{1/(6+4w)}\frac{1}{r^{0\cdot5}}.
\end{aligned} \right\} \tag{4.16}$$

In order that the solution shall be completely defined, the value of the exponent w must be selected, or in accordance with (4.15), the value of v.

Using (4.14), it is easy to write the solution corresponding to the various particular formulas for the absorption coefficient required in astrophysics.

For example, if Kramers' formula is taken for the absorption coefficient

$$\kappa_1 = B\rho T^{-7/2} \quad (w = 1, v = -3\cdot5) \tag{4.17}$$

then the solution (4.14) becomes

$$\left. \begin{aligned}
\rho &= 0\cdot00457\frac{(B\mathfrak{L}_0)^{2/11}}{(\mu f)^{15/11}}\frac{1}{r^{32/11}}, \qquad
\mathscr{M} = 0\cdot6319\frac{(B\mathfrak{L}_0)^{2/11}}{(\mu f)^{15/11}}r^{1/11}, \\[2mm]
p &= 0\cdot000756\frac{(B\mathfrak{L}_0)^{4/11}}{\mu(\mu f)^{19/11}}\frac{1}{r^{42/11}}, \quad
RT = 0\cdot1654\frac{(B\mathfrak{L}_0)^{2/11}}{(\mu f)^{4/11}}\frac{1}{r^{10/11}}.
\end{aligned} \right\} \tag{4.18}$$

Let us consider the case when $[a] = [\mathfrak{L}] = ML^2T^{-3}$ $(k = 2, s = -3)$, which corresponds to the assumption that the state along the radius is determined by the total luminosity. This case is singular and it must be analysed separately since we obtain from (4.2), (4.6), (4.7) and (4.8)

$$\left. \begin{aligned}
\rho &= \alpha_1\frac{a^{2/5}f^{-3/5}}{r^2}, \qquad
\mathscr{M} = 4\pi\alpha_1 a^{2/5}f^{-3/5}\cdot r, \\[2mm]
p &= 2\pi\alpha_1^2\frac{a^{4/5}f^{-1/5}}{r^2}, \quad
RT = 2\pi\alpha_1\mu(af)^{2/5} = \text{const.}
\end{aligned} \right\} \tag{4.19}$$

Hence, it follows that the temperature is constant along the radius. In this case, the equation

$$\frac{d(RT)^4}{dr} = -\frac{\kappa_1 \mathfrak{L}\rho}{4\pi r^2},$$

can only be satisfied under the assumption that $\kappa_1 = 0$ if $\mathfrak{L} \neq 0$. Afterwards the second equation of radiation theory

$$\frac{d\mathfrak{L}}{dr} = 4\pi r^2 \rho \epsilon^*$$

can always be satisfied and used to determine $\mathfrak{L}$ if ϵ^* is known as a function of r, ρ, and T.†

In particular, if $\epsilon^* = 0$, then $\mathfrak{L} = \mathfrak{L}_0 = \text{const}$. Only one essential arbitrary parameter $\alpha_1 a^{2/5}$ is contained in (4.19).

§5. On the Relation Between the Period of Variation of the Brightness and the Average Density for Cepheids

The observed results on the period of variation of the brightness and on the velocities of the gas of radiating particles provide convincing evidence that the oscillations of the observed luminosity depend on the radial pulsations of the gas mass forming the star, with large amplitude and velocity.

Using the reasoning explained in §2, we can assume that these pulsations are an adiabatic process described by the system of Equations (I) of §2.

The motion under consideration must be an essentially nonlinear process. Shock waves propagating to the centre of the star and from the centre to the periphery are possible within the gas. It is possible that the boundary of the luminous photosphere coincides with the shock wave in a certain time interval during its expansion. Since the shock wave velocity is larger than the velocity of the gas particles beyond its front, its maximum diameter during the expansion of the photosphere bounded by the shock wave will be considerably larger than the transverse linear dimension calculated by integration of the ray velocities measured in observations.

As the shock waves propagate towards the centre of the star, the focusing motion discussed in §4 of Chapter IV will develop at the instant

---

† It is evident that these considerations can also refer to the more general case when
$$s+k+1 = 0$$
which leads to a constant temperature according to (4.2).

of reflection from the centre. The change in the brightness of the Cepheid
can be explained by the occurrence of large changes in the photosphere
temperature during pulsations and by the changes in the magnitude of
the absorption in the gaseous atmosphere surrounding the Cepheid. The
length of the path of the light ray in the ambient atmosphere will vary
for different photosphere diameters.

In order to describe the pulsations of the gas sphere of the Cepheid
model, we solve the system of Equations (I) with the following boundary
conditions and the shock conditions, which are of the same form as those
in the absence of gravitational forces (see §2, Chapter IV)

$$\left.\begin{array}{l} \mathcal{M} = 0 \quad \text{and} \quad v = 0 \text{ at the centre of the star for } r = 0, \\[2em] \mathcal{M} = \mathfrak{M} \quad \text{and} \quad p = 0, \rho = 0 \text{ on the surface of the star for} \\ \qquad\qquad r = \mathfrak{R} + \varDelta\mathfrak{R}. \end{array}\right\} \quad (5.1)$$

Let us assume that the steady pulsation region is determined completely by the equations of motion, the conditions on the shock, the
boundary and other conditions which do not contain new dimensional
constants. Under such assumptions, it is not difficult to write down the
system of determining parameters. This system is given by the set

$$r, t, \gamma, f, \mathfrak{R}, \mathfrak{M}. \qquad (5.2)$$

The average radius of the star $\mathfrak{R}$ and the mass $\mathfrak{M}$ in this case are
considered as independent parameters; as was already shown in §2 for
equilibrium, the radius $\mathfrak{R}$ is a function of the mass $\mathfrak{M}$. However, such a
relation between $\mathfrak{R}$ and $\mathfrak{M}$ cannot be the reason for the absence of
equilibrium, the reason for the pulsations of the Cepheids.[†]

The mathematical nonlinear problem of finding u, ρ and p in this
formulation is very difficult. There are only a few isolated results from
investigations of this problem, which were obtained by using important
additional assumptions and were based on the linearization of the
equations of motion in the majority of cases.[‡]

However, fundamental results relating the period of the Cepheid
pulsation to its mass and radius can be established in the very general
formulation of the problem without giving a solution but by using the
reasoning of dimensional analysis.

---

 † In the case of the existence of such a relation in the form of a power law, a dimensional
constant besides the mass $\mathfrak{M}$ would appear. Instead of $\mathfrak{M}$ and this constant (which can
differ for different series of Cepheids) $\mathfrak{M}$ and $\mathfrak{R}$ can be used directly in the system of
determining parameters.

 ‡ The explanation of these results is contained in Rosseland (1952).

In fact, let us denote the period of Cepheid pulsation by τ; since this is a characteristic of motion in the large, we conclude from the set of determining parameters established in (5.2), that

$$\tau = f(\gamma, f, \mathfrak{R}, \mathfrak{M}). \tag{5.3}$$

Since f, $\mathfrak{R}$ and $\mathfrak{M}$ have independent dimensions, it follows that

$$\frac{1}{\tau} = f_1(\gamma) \Big/ \sqrt{\left(\frac{f\mathfrak{M}}{4\pi\mathfrak{R}^3}\right)} = f_1(\gamma) \sqrt{(f\rho_s)} \tag{5.4}$$

where ρ_s is the density of the gas sphere.

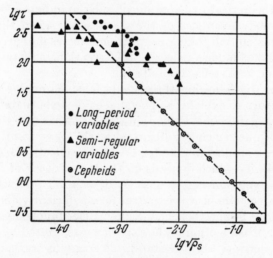

Fig. 111. Empirical data on the period-density relation for various variable stars.

The simple considerations leading to relation (5.4) based on sufficiently general assumptions, but without using a detailed mathematical solution, were first established by using empirical observations.

A diagram of the results of processing observations is shown in Fig. 111 in the logarithmic $\log \tau$ and $\log \sqrt{(\rho_s)}$ scales; the separate points are superposed as the average values for a group of stars, the dashed line is drawn according to the theoretical formula (5.4). The location of the points obtained from the observations of Cepheids shows a remarkable agreement with the slope of the theoretical dashed line.

Evidently, relation (5.4) reflects the fundamental regularity in the pulsations of the gaseous mass forming the variable state.

The processing of observations of semi-regular and long-period variables shows that the corresponding points on analogous diagrams are located along certain lines parallel to the corresponding line for the cepheids.

§6. On the Theory of Flare-ups in Novae and Supernovae

1. EQUATIONS OF MOTION AND BOUNDARY CONDITIONS

According to observational data given in §1, flare-ups in novae and supernovae result from the unsteady motion of large masses of gas accompanied by a sudden release in energy.

To find a theoretical explanation of such grandiose, cosmic catastrophes, we study examples of the exact solutions of the equations of unsteady gas motion taking gravitational forces into account. These can be regarded as models reflecting the essential characteristics of the actual phenomena of stellar flare-ups.

Solutions of the following three kinds must be analysed and each considered in turn to explain the phenomena of stellar flare-ups.

(i). The propagation of a detonation wave from the core to the surface of a star, with nuclear energy liberated at the wave front. Motions of such a kind were studied in Chapter IV, §8. The effects of increasing the detonation velocity which depends on the law of decay of density, and the complete explosion of the gaseous detonation products with the formation of a vacuum near the centre† for a sufficiently abrupt drop in the density, have been established (Sedov, 1956; Iavorskaia, 1956).

(ii). The disturbed motion in the explosive gas, caused by the sudden liberation of energy within the stars; this energy is transported to the periphery along with the shock wave.

(iii). Motion of explosive type without liberation of energy (Sedov, 1957), a result of instability in equilibrium of the masses of gas forming the star.

Solutions corresponding to types (ii) and (iii) are given in this section taking gravitation into account. We shall show below that disturbances due to the complete explosion of masses of gas, initially at rest,‡ are also possible in certain cases.

---

† The appropriate self-similar solutions without gravity have been developed; gravitational forces and the absence of self-similarity introduce certain quantitative corrections.

Analogous solutions can be constructed for the spherical problem of the propagation of a rarefaction shock with energy liberation, a flame front kind of shock, into a gas at rest with a variable initial density.

‡ In order to apply the conclusions obtained to the treatment of observations, it is necessary to investigate unsteady effects in stellar photospheres. It is also necessary to explain the significance of the electromagnetic field in stellar flare-ups.

Consider the solution of the system of Equations (I), which govern unsteady gas motion with spherical symmetry as described in §2. These are

$$
\left.
\begin{aligned}
&\frac{\partial \rho}{\partial t}+\frac{\partial \rho v}{\partial r}+\frac{2\rho v}{r} = 0, \quad \frac{\partial \mathcal{M}}{\partial r} = 4\pi r^2 \rho, \\[2mm]
&\frac{\partial v}{\partial t}+v\frac{\partial v}{\partial r}+\frac{1}{\rho}\frac{\partial p}{\partial r}+\frac{f\mathcal{M}}{r^2} = 0, \\[2mm]
&\frac{\partial \dfrac{p}{\rho^\gamma}}{\partial t}+v\frac{\partial \dfrac{p}{\rho^\gamma}}{\partial r} = 0.
\end{aligned}
\right\}
\quad (6.1)
$$

The unsteady motion starts at time $t = 0$ in the undisturbed gas as a result of liberation of energy at the centre of symmetry under adiabatic conditions. Additional energy is liberated in the disturbed gas under polytropic conditions.

The initial behaviour of the gas determined by the equilibrium conditions is given by formulas (4.2), (4.6), (4.7) and (4.8) which, putting

$$
\frac{2k+6}{2-s} = \omega
$$

and

$$
\alpha_1 a^{2/(2-s)} f^{s/(2-s)} = A > 0
$$

can be written in the following simple form:

$$
v = 0; \quad \rho = A/r^\omega;
$$

$$
\mathcal{M} = \frac{4\pi A}{3-\omega} r^{3-\omega};
$$

$$
\left.
\begin{aligned}
&p = \frac{2\pi A^2 f}{(\omega-1)(3-\omega)}\frac{1}{r^{2\omega-2}}; \\[3mm]
&RT = \frac{2\pi A f}{(\omega-1)(3-\omega)}\frac{1}{r^{\omega-2}}.
\end{aligned}
\right\}
\quad (6.2)
$$

The characteristic constant a in (4.2) which has the dimensions $[a] = ML^k T^s$ is replaced in (6.2) by the constant A with the dimensions

$$
[A] = ML^{\omega-3}.
$$

The family of solutions (6.2) of the equations of equilibrium depends on a dimensional constant A and on a characteristic parameter ω, determined by the dimensions of the constant A, in addition to the gravitational constant f.

It follows from the condition that the mass near the centre of symmetry be finite, that $\omega < 3$. For the pressure and the temperature to be positive, we must have $\omega > 1$. For the temperature to increase as the centre is approached, we must have $\omega > 2$, the gas temperature is independent of r for $\omega = 2$ and isothermal conditions prevail. If the constant a has the dimensions of energy $k = 2$, $s = -2$, then $\omega = 2 \cdot 5$.

According to (3.12) of Chapter IV, the total energy of gas between spheres of radii r' and r'' is determined by

$$\mathscr{E} = \int_{r'}^{r''} \left[\frac{\rho v^2}{2} + \frac{p}{\gamma^* - 1} - \frac{\rho f(\mathscr{M} - \mathscr{M}')}{r} \right] 4\pi r^2 \, dr,$$

where $\gamma^* = c_p/c_v$. Using this formula and (6.2), it is not difficult to calculate the total initial energy H enclosed within a sphere of radius r. We have for $1 < \omega < 2 \cdot 5$:

$$H = \frac{8\pi^2}{\gamma^* - 1} \frac{1 - 2(\omega - 1)(\gamma^* - 1)}{(\omega - 1)(3 - \omega)(5 - 2\omega)} f A^2 r^{5 - 2\omega}. \tag{6.3}$$

If $\omega > 2 \cdot 5$, then the initial energy for any γ^* and r is equal to $\pm \infty$.
    If

$$1 < \omega < \frac{2\gamma^* - 1}{2(\gamma^* - 1)} \quad \text{and} \quad \omega < 2 \cdot 5$$

then the initial energy is finite and positive: $H > 0$.

The initial energy is finite and negative† $H < 0$ (Fig. 112) for

$$\frac{2\gamma^* - 1}{2(\gamma^* - 1)} < \omega < 2 \cdot 5.$$

After evaluating the integral, we obtain for $\omega = 2 \cdot 5$:

$$\mathscr{E}_{r'}^{r''} = 32\pi^2 f A^2 \left\{ \frac{4 - 3\gamma^*}{3(\gamma^* - 1)} \ln \frac{r''}{r'} + 2 - 2 \sqrt{\left(\frac{r'}{r''} \right)} \right\}, \tag{6.4}$$

from which it is clear that for $\omega = 2 \cdot 5$ and $\gamma^* = 4/3$:

$$H = 64\pi^2 f A^2. \tag{6.5}$$

In this case the initial energy is finite, positive and independent of the radius r. If $\omega = 2 \cdot 5$ and $\gamma^* > 4/3$, then $H = -\infty$.

---

† If $H > 0$, then the internal thermal energy is larger than the absolute value of the gravitational energy; it is less than the gravitational energy for $H < 0$.

Unsteady gas motions with the initial conditions (6.2) will be self-similar if all the dimensional constants in the additional conditions have dimensions dependent on the dimensions of the gravitation constant f and of the constant A.

If the self-similar disturbed motion is caused by the liberation of energy at time $t = 0$, then the law of energy liberation is determined by just three dimensional quantities with the independent dimensions t, A and f; consequently

$$E = \alpha f^{(5/\omega)-1} A^{5/\omega} t^{[2(5-2\omega)]/\omega}, \tag{6.6}$$

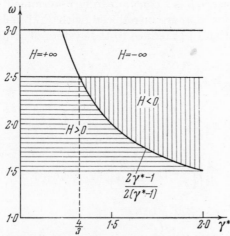

FIG. 112. Sign of the initial energy H as a function of ω and γ^*. H is finite in the cross-hatched regions; H is infinite for $\omega > 2 \cdot 5$.

where α is an arbitrary coefficient independent of time. The quantity α can be infinite in particular cases. If $\omega = 2 \cdot 5$, then (6.6) becomes

$$E = \alpha f A^2. \tag{6.7}$$

The case when α is finite corresponds to the instantaneous liberation of finite energy at the centre of the star. If the energy liberated instantaneously is infinite, then the coefficient α is also infinite.

According to (6.6), continuous energy liberation must occur for $\omega < 2 \cdot 5$ and $\alpha > 0$ so that the magnitude of E increases with time.

The field of disturbed motion is determined by the system of dimensional parameters

$$A, f, r, t; \tag{6.8}$$

The motion still depends on the abstract constants γ, which appears in the equation of motion, and α, which enters in (6.6).

From (6.8), it follows that the gas motion is self-similar, and the laws of motion can be expressed by means of the formulas:

$$\lambda = \frac{r}{(\beta A f)^{1/\omega}\, t^{2/\omega}}; \quad v = \frac{r}{t} V(\lambda); \quad \rho = \frac{1}{ft^2} R(\lambda);$$

$$\mathscr{M} = \frac{r^3}{ft^2} M(\lambda); \quad p = \frac{r^2}{ft^4} P(\lambda); \quad z(\lambda) = \frac{\gamma P}{R}. \tag{6.9}$$

where V, R, M, z are nondimensional functions of λ, containing certain arbitrary parameters, β is a constant to be determined.

When Equations (6.9) are introduced, with $P(\lambda)$ replaced by $z(\lambda)$, the equations of motion (6.1) reduce to the system of ordinary differential equations:

$$\lambda\left\{\left(\frac{2}{\omega}-V\right)V' - \frac{z}{\gamma}\left(\frac{z'}{z}+\frac{R'}{R}\right)\right\} = M + V^2 - V + \frac{2z}{\gamma},$$

$$\lambda\left[V' - \left(\frac{2}{\omega}-V\right)\frac{R'}{R}\right] = 2 - 3V,$$

$$\lambda\left(V-\frac{2}{\omega}\right)\left[\frac{z'}{z}-(\gamma-1)\frac{R'}{R}\right] = -2(V+\gamma-2),$$

$$\lambda M' = -3M + 4\pi R. \tag{6.10}$$

The required functions $V(\lambda)$, $R(\lambda)$ and $z(\lambda)$ are similar to the corresponding functions determined from Equations (2.1), (2.2) and (2.3) of Chapter IV, in which $\delta = 2/\omega$; however, Newtonian gravitational force has been taken into account in (6.10) and the single function $M(\lambda)$ enters in addition.

According to the general theory explained in §3 of Chapter IV, the order of the system (6.10) can be lowered by two by using the integrals (3.7) and (3.9) of Chapter IV, which hold for any motion. In this case, it is necessary to put $\nu = 3$, $\delta = 2/\omega$ and $k = -3$, $s = 2$ since

$$[1/f] = ML^{-3}T^2.$$

These integrals are represented by the relations

$$\lambda^3\left[\left(1-\frac{3}{\omega}\right)M - 2\pi R\left(V-\frac{2}{\omega}\right)\right] = C_1, \tag{6.11}$$

$$z = R^{\gamma-1} M^{[\gamma+(2/\omega)-2]/[(3/\omega)-1]} \lambda^{[3\gamma-4]/[(3/\omega)-1]} C_2. \tag{6.12}$$

Moreover, the integral (3.15) (Chapter IV; the energy integral) still holds for $\omega = 2 \cdot 5$

$$\lambda^5 \left[\frac{VzR}{\gamma} + \left(V - \frac{4}{5} \right) \left(\frac{RV^2}{2} + \frac{zR}{\gamma(\gamma-1)} - RM \right) \right] = C_3. \qquad (6.13)$$

As we mentioned in §3 of Chapter IV, these integrals are also correct for polytropic processes in which the entropy per particle is variable and there is an external heat flow.

A shock wave, travelling from the centre to the periphery of the star, is formed when energy is liberated suddenly at the centre of the star.

The following conditions must be satisfied on the surface of the shock wave being propagated into the gas at rest:

$$\left. \begin{aligned} v_2 &= \frac{2}{\gamma+1} c \left(1 - \frac{a_1^2}{c^2} \right), \\[2mm] \rho_2 &= \frac{\gamma+1}{\gamma-1} \rho_1 \left(1 + \frac{2a_1^2}{(\gamma-1)c^2} \right)^{-1}, \\[2mm] \mathcal{M}_2 &= \mathcal{M}_1, \\[2mm] p_2 &= \frac{2\gamma}{\gamma+1} p_1 \frac{c^2}{a_1^2} \left(1 - \frac{\gamma-1}{2\gamma} \frac{a_1^2}{c^2} \right). \end{aligned} \right\} \qquad (6.14)$$

Here c is the velocity of shock propagation and $a_1^2 = \gamma p_1/\rho_1 = RT_1$ is the square of the speed of sound in the undisturbed state. The shock wave radius r_2 is determined by just three dimensional quantities with the independent dimensions f, A, t; consequently

$$r_2 = C (\beta A f)^{1/\omega} t^{2/\omega}, \quad \lambda_2 = C = \text{const.} \qquad (6.15)$$

The constant value of λ, equal to C, on the shock wave can be taken equal to unity and the constant factor β can be determined from this condition. We have in this case:

$$\lambda_2 = 1, \quad \lambda = \frac{r}{r_2}. \qquad (6.16)$$

It also follows from (6.15) that

$$c = \frac{dr_2}{dt} = \frac{2}{\omega} \frac{r_2}{t} = \frac{2}{\omega} (\beta A f)^{1/\omega} t^{(2/\omega)-1} = \frac{2}{\omega} (\beta A f)^{1/2} r_2^{(2-\omega)/2}. \qquad (6.17)$$

The shock wave is retarded for $\omega > 2$; acceleration of the shock wave occurs for $\omega < 2$.† In particular, we obtain for $\omega = 2 \cdot 5$

$$r_2 = (\beta A f)^{2/5} t^{4/5}; \quad c = \frac{4}{5} (\beta A f)^{1/4} \frac{1}{r_2^{1/4}}. \tag{6.18}$$

The abstract quantity

$$q = \frac{a_1^2}{c^2} = \frac{\gamma p_1}{\rho_1 c^2}$$

appears in the shock conditions. From general dimensional considerations, it follows that the abstract quantity q, dependent only on three dimensional quantities f, A, r_2, must be a constant. It follows from (6.2) and (6.17) that

$$q = \frac{\pi \gamma \omega^2}{2(\omega - 1)(3 - \omega)\beta}. \tag{6.19}$$

The value of q for fixed ω is determined by the constant β and conversely.

After transformation to nondimensional form by using (6.9), the conditions on the shock wave (6.14) yield:

$$\left.\begin{aligned}
V_2 &= \frac{4}{\gamma + 1} \frac{1 - q}{\omega}, \\[2mm]
z_2 &= \frac{4}{\omega^2} \frac{[2\gamma - (\gamma - 1)q][\gamma - 1 + 2q]}{(\gamma + 1)^2}, \\[2mm]
R_2 &= \frac{2(\gamma + 1)(3 - \omega)(\omega - 1)}{\pi \gamma \omega^2} \frac{q}{\gamma - 1 + 2q}, \\[2mm]
M_2 &= \frac{8(\omega - 1)}{\gamma \omega^2} q,
\end{aligned}\right\} \tag{6.20}$$

in which

$$\lambda_2 = 1.$$

† The effects of retardation or acceleration of shock wave propagation in media of variable density are not determined solely by the laws of decay in density. The pressure gradient and external disturbances within the waves can influence the wave motion substantially.

It has been shown in §14 of Chapter IV that the formula $c = \mathrm{const}/[r_2^{(3-\omega)/2}]$ is correct for self-similar motions, determined by the constants $[A] = ML^{\omega-3}$ and $[E] = ML^2 T^{-2}$ and, in particular, for a violent explosion in a medium with a variable initial density. Therefore, retardation is obtained for all $\omega < 3$.

In the present problem, with the initial variable density $\rho_1 = A/r^\omega$ in which the self-similar motion is determined by the constants $[A] = ML^{\omega-3}$ and $[f] = M^{-1}L^3 T^{-2}$, the shock wave is accelerated for $\omega < 2$ and retarded for $\omega > 2$.

Conditions (6.20) for a given q are the Cauchy data for the system of Equations (6.10) and, therefore, completely determine the whole gas motion within the shock wave. The quantity β is determined from (6.19) by fixing the formula for the variable λ. It is also evident that the constant α is determined by assigning q and hence, the law of energy liberation (6.6) is determined completely in advance. Hence, the value of q fixes the law of energy liberation at the centre of symmetry.

Using boundary conditions (6.20), the constants in the integrals (6.11), (6.12) and (6.13) are easily determined. After simple calculations, we obtain:

$$C_1 = 0;$$

$$C_2 = \frac{\pi^{\gamma-1} 2^{[\omega(3-2\gamma)-3(\gamma-1)]/(3-\omega)}}{\omega^2 (\gamma+1)^{\gamma+1} (3-\omega)^{\gamma-1}}$$

$$\times \left[\frac{\gamma\omega^2}{q(\omega-1)} \right]^{(3\gamma-1-\omega)/(3-\omega)} [(2\gamma - (\gamma-1)q][\gamma-1+2q]^{\gamma}, \quad (6.21)$$

and for $\omega = 2\cdot 5$

$$C_3 = \frac{384(3\gamma-4)}{3125\pi\gamma^2(\gamma-1)} q^2. \qquad (6.22)$$

If $\gamma = 4/3$, then $C_3 = 0$ and the variable λ is eliminated from the energy and adiabatic integrals, which simplifies the problem considerably. If $\omega = 2\cdot 5$ and $\gamma = 4/3$, then the above Cauchy problem is solved by means of a single quadrature which can be evaluated in finite form.

The results of numerical computations are given below for $\gamma = 4/3$, $\omega = 2\cdot 5$, and for $\gamma = 5/3$ and $\omega = 2\cdot 5$.†

Analysis of the solution for $\gamma = 4/3$ and $\omega = 2\cdot 5$ gives the disturbance due to a point explosion with liberation of a finite amount of energy at time $t = 0$ at the centre of symmetry. The magnitude of the energy liberated increases as q decreases, i.e. as the shock wave intensity increases. A spherical vacuum of radius r^* forms at the centre of symmetry for $0 < q < 1/36$ in which

$$r^* = \lambda^* r_2,$$

where the constant λ^* depends only on q and equals the value of the parameter λ at the interior boundary where the pressure and density vanish. If $q > 1/36$, the disturbance created by the explosion occupies the whole interior of the spherical shock wave. The density and pressure become infinite at the centre for $q > 1/36$. The state of the gas motion for

† The problem considered was studied for $\gamma = 5/3$ and $\omega = 2\cdot 5$ in another approach by integrating the system (6.10) numerically in the paper by Carrus, Fox, Gaas, and Kopal (1951).

$q = 0$ agrees with that studied in §15 of Chapter IV, for a point explosion in a weightless mass with variable density and $p_1 = 0$.

We note that the transition from motion which extends to the centre of symmetry to motion with a spherical vacuum of increasing radius forming at the centre is the result of various factors. We showed in §15 of Chapter IV that this effect is due to the more rapid rate of decay in density, i.e. the growth of the exponent $\omega(\rho_1 = A/r^\omega)$ or due to the increase in the coefficient γ when ω is fixed.

In this case, such an effect occurs for constant $\omega = 2 \cdot 5$ and $\gamma = 4/3$ because of the change in the parameter q, which determines the quantity of energy to be liberated at the centre of symmetry. The variation of the

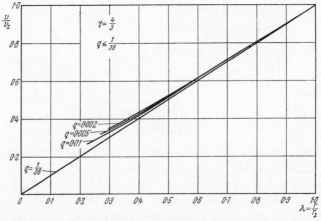

FIG. 113. Velocity distributions in the disturbed gas. An empty sphere with expanding radius $r^* = \lambda^*(q)r$ is formed about the centre.

physical variables defining the motion for $\gamma = 4/3$ and $\omega = 2 \cdot 5$ is shown in Figs. 113–120 for various q.

It is easy to verify that the system of ordinary differential equations (6.10) has the following solution for $\gamma = 4/3$:

$$V = \frac{2}{3}, \quad R = R_0 = \text{const}, \quad M = \frac{4\pi}{3}R_0, \quad z = \frac{4}{27} - \frac{8\pi}{9}R_0; \quad (6.23)$$

where R_0 is an arbitrary positive constant. According to (6.9), the appropriate exact solution of the partial differential equations (6.1) is

$$\left.\begin{array}{l} v = \frac{2}{3}\frac{r}{t}; \quad \rho = \frac{R_0}{ft^2}; \quad \mathscr{M} = \frac{r^3}{ft^2}\frac{4\pi}{3}R_0; \\[2mm] p = \frac{r^2}{ft^4}\left[\frac{R_0}{9} - \frac{2\pi}{3}R_0^2\right]. \end{array}\right\} \quad (6.24)$$

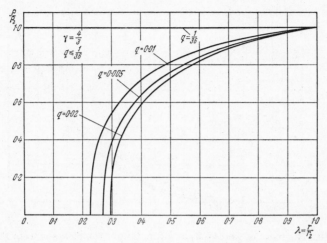

FIG. 114. Density distribution in the disturbed gas. The density is zero at the interior boundary.

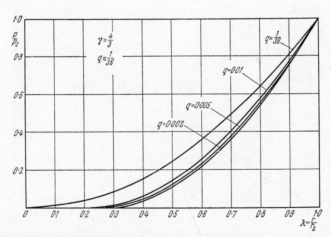

FIG. 115. Pressure distribution in the disturbed gas. The pressure is zero at the interior boundary.

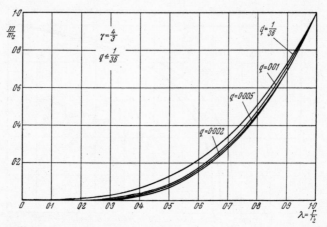

FIG. 116. The mass of gas $\mathcal{M}/\mathcal{M}_2$ as a function of the distance from the centre of symmetry. We have $\mathcal{M} = 0$ for $r = r^*$.

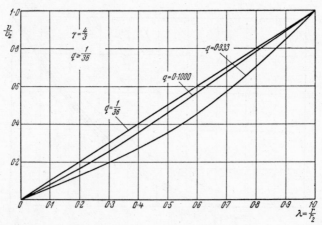

FIG. 117. Velocity distribution in the disturbed gas. The motion is continued to the centre of symmetry.

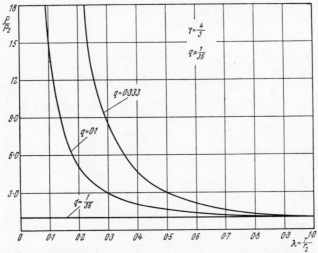

FIG. 118. Density distribution in the disturbed gas. The density is infinite at the centre for $q > 1/36$.

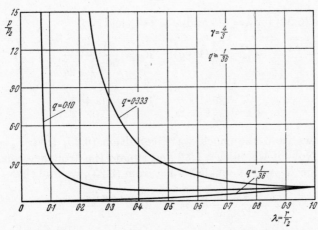

FIG. 119. Pressure distribution in the disturbed gas. Pressure is infinite at the centre for $q > 1/36$.

The density in the corresponding motion depends only on the time and is independent of the radius. The solution (6.24) is a particular case of the solution (15.24) of Chapter IV for $\gamma = 4/3$ and $\beta = 0$; the quantities χ and R_0 are related by means of

$$\chi = \frac{1}{9R_0} - \frac{2\pi}{3}.$$

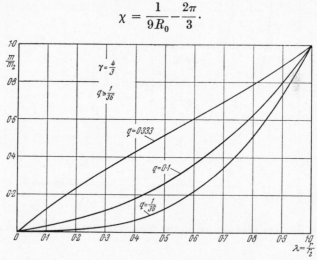

FIG. 120. Mass of gas $\mathscr{M}/\mathscr{M}_2$ as a function of the distance from the centre of symmetry.

Formulas (6.20) for $\lambda = 4/3$ and $\omega = 2\cdot5$ yield:

$$V_2 = \frac{24}{35}(1-q); \quad R_2 = \frac{21}{50\pi}\frac{q}{\frac{1}{3}+2q};$$

$$M_2 = \frac{36}{25}q; \quad z_2 = \frac{16(8-q)(1+6q)}{25\times49}. \quad \Bigg\} \quad (6.25)$$

It follows from the shock wave condition (6.25) that $V_2 = 2/3$ for $q = 1/36$. Now, if we put $q = 1/36$ and equate the arbitrary constant R_0 to the value R_2, which equals $3/100\pi$ from (6.25), we find that

$$z = \frac{4}{27} - \frac{8\pi}{9}R_0 = z_2 \quad \text{and} \quad \frac{4\pi R_0}{3} = M_2.$$

Therefore, (6.24) give the exact solution of the problem of flare-up in simple form for $\gamma = 4/3$, $\omega = 2\cdot5$ and $q = 1/36$. In this case, the solution (6.24) can be written in the following simple form

$$\frac{v}{v_2} = \lambda; \quad \frac{\rho}{\rho_2} = 1; \quad \frac{p}{p_2} = \lambda^2; \quad \frac{\mathscr{M}}{\mathscr{M}_2} = \lambda^3; \quad \lambda = \frac{r}{r_2}. \quad (6.26)$$

The solution obtained separates the solution with a vacuum forming for $q < 1/36$ and that with motion extending to the centre of symmetry for $q > 1/36$. Curves corresponding to the solution (6.26) are also plotted in Figs. 112–119.

Other exact solutions of explosion problems can be found in simple form by using the exact solution (15.24) of Chapter IV.

In fact we find from the solution (15.24) of Chapter IV and the boundary conditions on the shock wave (6.14) for u_2, ρ_2 and $\mathcal{M}_2$ that these conditions are satisfied by (6.2) if the constants q, R_0, γ_1 and ω satisfy the relations,†

$$q = \frac{\gamma_1 \omega^2 \pi R_0}{6(\omega - 1)}, \quad q = 1 - \frac{(\gamma_1 + 1)\omega}{6}, \tag{6.27}$$

in which the motion of the shock wave is determined by the formula

$$r_2^\omega = \frac{3fA}{(3 - \omega) R_0} t^2. \tag{6.28}$$

Using (15.24) of Chapter IV, (6.2), (6.14) and (6.28), the condition for p_2 at the shock wave gives

$$\kappa\left(\frac{3fAt_0^2}{3 - \omega}\right)^{\gamma - (4/3)} \frac{1}{r_2^{\omega[(2/3) + \gamma]}} + \beta r_0^2 \left(\frac{3fAt_0^2}{3 - \omega}\right)^{\gamma - 2} \frac{1}{r_2^{\omega\gamma + 2}}$$

$$= \frac{2\pi}{9} \frac{(3 - \omega)[6 + (\gamma_1 - 1)\omega]}{(\omega - 1)[6 - (\gamma_1 + 1)\omega]} \frac{1}{r_2^{2\omega}}; \quad \left(R_0 = \frac{1}{3\left(2\pi + \dfrac{\kappa}{\gamma - 1}\right)}\right). \tag{6.29}$$

If $\omega < 3$, then (6.29) can be satisfied in two cases:

(i) $\qquad \beta = 0; \quad \gamma = \dfrac{4}{3}; \quad \omega = \dfrac{12(2 + \gamma_1)}{4 + 9\gamma_1}; \quad q = \dfrac{2\gamma_1(2 - \gamma_1)}{4 + 9\gamma_1};$

$\qquad \kappa = \dfrac{2\pi}{9} \dfrac{(3 - \omega)[6 + (\gamma_1 - 1)\omega]}{(\omega - 1)[6 - (\gamma_1 + 1)\omega]}$ $\left.\begin{array}{c} \\ \\ \\ \\ \end{array}\right\}$ (6.30)

a family of solutions dependent on γ_1 is obtained; if $\gamma_1 = \gamma = 4/3$, the solution corresponding to (6.30) agrees with (6.26):

(ii) $\qquad \kappa = 0; \quad \gamma = 2\left(1 - \dfrac{1}{\omega}\right); \quad \gamma_1 = \dfrac{6(\omega - 1)(6 - \omega)}{\omega(7\omega - 6)};$

$\qquad q = \dfrac{\omega(6 - \omega)}{6(7\omega - 6)}; \quad \beta r_0^2 = \left(\dfrac{3fAt_0^2}{3 - \omega}\right)^{2/\omega} \dfrac{2\pi}{9} \dfrac{(3 - \omega)[6 + (\gamma_1 - 1)\omega]}{(\omega - 1)[6 - (\gamma_1 + 1)\omega]}.$ $\left.\begin{array}{c} \\ \\ \\ \\ \end{array}\right\}$ (6.31)

† The constant γ in the shock conditions (6.14) is replaced by γ_1 in deriving (6.27) and (6.28).

The quantity γ in (6.1) can sometimes be considered as the polytropic index, but in the shock conditions we should put $\gamma_1 = c_p/c_v \neq \gamma$.

If $\gamma_1 = \gamma$, then

$$\gamma = \frac{7}{6}; \quad \omega = 2 \cdot 4; \quad q = \frac{2}{15}; \quad \beta r_0^2 = (5fAt_0^2)^{5/6} \frac{16\pi}{21}. \qquad (6.32)$$

A family of particular solutions has been obtained in case (ii) for various γ, however, unequal values of γ and γ_1 are obtained for each ω and only for $\omega = 2 \cdot 4$ do we have $\gamma = \gamma_1 = 7/6$.

The variables determining the gas motion within the shock wave for all γ_1 in case (i) are given by (6.26). The following formulas are obtained in case (ii):

$$\frac{v}{v_2} = \lambda; \quad \frac{\rho}{\rho_2} = 1; \quad \frac{p}{p_2} = 1; \quad \frac{\mathscr{M}}{\mathscr{M}_2} = \lambda^3; \quad \lambda = \frac{r}{r_2}. \qquad (6.33)$$

Therefore, both density and pressure behind the wave front are constant in case (ii).

The motion extends right down to the centre of symmetry in both cases, where $p = 0$ at the centre in the first case and $p = p_2(t)$ in the second. If $\gamma = \gamma_1 = 7/6$ is the adiabatic index, then solution (6.33) corresponds to the breakdown of unstable equilibrium without energy liberation. The energy of the disturbed gas within an explosive wave equals the initial energy at equilibrium.

We now consider the numerical solution for $\gamma = 5/3$ corresponding to a monatomic or completely ionized gas.

The results of the computations for $\gamma = 5/3$ and different numerical values of q are shown in Figs. 121–124.

The lower limits of $\lambda = r/r_2$ on the curves correspond to $\mathscr{M} = 0$; consequently, these points can be considered as interior boundaries arising at the centre of symmetry at the initial instant. The disturbed gas motion can be considered to result from the expulsion of gas by a special spherical piston, a sphere whose radius r^* increases from zero according to the relation

$$r^* = \lambda^* r_2, \quad \text{where} \quad \lambda^*(\gamma, q, \omega) = \text{const.} \qquad (6.34)$$

The function $\lambda^*(\gamma, q)$ is shown in Fig. 125 for $\gamma = 4/3$ and $5/3$ and $\omega = 2 \cdot 5$.

The solution approaches that without gravity found in §14 of Chapter IV as $q \to 0$.

We have $\lambda^* \neq 0$ for $q = 0$ and λ^* depends only on γ; the corresponding limiting values $r^*/r_2 = \lambda^*(0, \gamma)$ for $\gamma = 5/3$ and $4/3$ for $\omega = 2 \cdot 5$ are also given in Fig. 125.

Let us find the asymptotic behaviour of the solution near the interior boundary. At this boundary, $\mathscr{M} = 0$ and since $r = r^*$ is a finite quantity, it follows from (6.9) that $\mathscr{M} \to 0$ as $r \to r^*$. Consequently, the equations

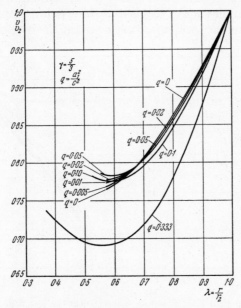

FIG. 121. Velocity distribution for $\gamma = 5/3$.

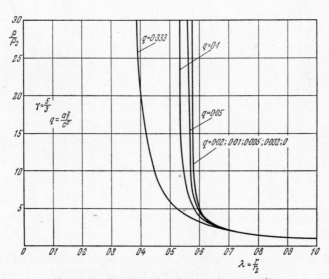

FIG. 122. Density distribution for $\gamma = 5/3$.

of motion (6.10) approach (5.10) and (5.11) of Chapter IV near $r = r^*$. The physical implication of this is that the effect of gravity is small near the interior boundary and approaches zero at the boundary.

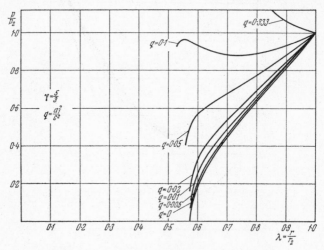

FIG. 123. Pressure distribution for $\gamma = 5/3$.

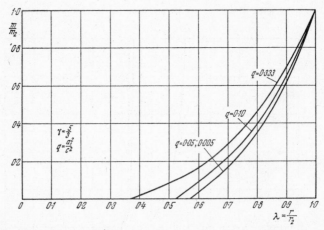

FIG. 124. Mass of gas M/M_2 as a function of distance from the centre of symmetry.

In order to obtain asymptotic formulas, it is sufficient to study the solutions of the ordinary equations (5.10) and (5.11) of Chapter IV near the singular point $V = \delta = 2/(5 - \omega)$, $z = 0$. In the case being considered, $\omega = 2 \cdot 5$, consequently, $\delta = 4/5$. The asymptotic formulas for $q = 0$ near the interior boundary $(r \to r^*)$ are given in §14 of Chapter IV (formulas (14.17)).

Only one integral curve through the singular point $V = 4/5$, $z = 0$, with the asymptotic form

$$z = \frac{2}{5}\gamma\left(\frac{4}{5} - V\right) \tag{6.35}$$

exists. This curve corresponds to the solution for a point explosion for $\omega = 2\cdot5$ and $q = 0$. However, this curve corresponds to $C_3 = 0$. In the present case $q \neq 0$, $C_3 \neq 0$; consequently, it is necessary to use other solutions. The asymptotic equations of the other integral curves in the

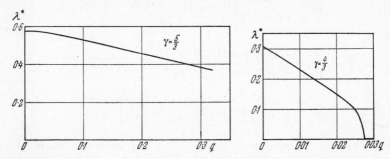

FIG. 125. Variation of the radius of the internal boundary with q for $\gamma = 5/3$ and $\gamma = 4/3$.

z, V plane which pass through the singular point $V = 4/5$, $z = 0$, are, in the neighbourhood of this point,

$$z = A\left(\frac{4}{5} - V\right)^{(5\gamma-6)/[6(\gamma-1)]}, \tag{6.36}$$

where A is an arbitrary constant. It is evident that $(5\gamma - 6)/[6(\gamma - 1)] < 1$. The validity of (6.36) is easily verified directly.

Using (6.36), we find from the integrals (6.11), (6.13):

$$\left.\begin{array}{ll} R = B\left(\dfrac{4}{5} - V\right)^{-(5\gamma-6)/[6(\gamma-1)]}, & M = 10\pi B\left(\dfrac{4}{5} - V\right)^{\gamma/[6(\gamma-1)]}, \\[3mm] P = \dfrac{zR}{\gamma} = AB\dfrac{1}{\gamma}, & B = \dfrac{5\gamma C_3}{4A}\lambda^{*-5}. \end{array}\right\} \tag{6.37}$$

Using the integral (6.12) and formulas (6.21) and (6.22), the constants A, C_2 and C_3 are expressed in terms of γ, q and λ^*. The formulas (6.37) are not valid for $\gamma = 4/3$ since $C_3 = 0$ in this case. Furthermore, using the second equations of (6.10) and (6.37) we easily find the asymptotic formula:

$$\frac{r - r^*}{r^*} = \frac{\lambda - \lambda^*}{\lambda^*} = \frac{5\lambda}{12(\gamma - 1)}\left(\frac{4}{5} - V\right). \tag{6.38}$$

The appropriate asymptotic formulas for the dimensional quantities where $\gamma \neq 4/3$ and $r \to r^*$ are, from (6.9), (6.20), (6.37) and (6.38):

$$\left.\begin{aligned}
\frac{v}{v_2} &= \frac{\gamma + 1}{2(1 - q)}\frac{r^*}{r_2}, \\[2mm]
\frac{\rho}{\rho_2} &= \frac{25\pi\gamma(\gamma - 1 + 2q)\,B}{6(\gamma + 1)q\left[\dfrac{12(\gamma + 1)}{5\gamma}\dfrac{r - r^*}{r^*}\right]^{(5\gamma - 6)/[6(\gamma - 1)]}}, \\[2mm]
\frac{p}{p_2} &= \frac{625\pi\gamma(\gamma + 1)\,A\,B}{96q[2\gamma - (\gamma - 1)q]}\left(\frac{r^*}{r_2}\right)^2, \\[2mm]
\frac{\mathscr{M}}{\mathscr{M}_2} &= \frac{\pi\gamma B}{0\cdot192q}\left(\frac{r^*}{r_2}\right)^3\left[\frac{12(\gamma - 1)}{5\gamma}\frac{r - r^*}{r^*}\right]^{\gamma/[6(\gamma - 1)]}.
\end{aligned}\right\} \tag{6.39}$$

The asymptotic behaviour of the solution has been studied in §14 of Chapter IV for $q = 0$; it has been shown there that the pressure at the interior boundary is zero in the case under consideration for $\omega = 2\cdot5 < 3$ and for the density we have

$$\frac{\rho}{\rho_2} = A\left(\frac{r - r^*}{r_2}\right)^{-(2\cdot5\gamma - 3\cdot5)/(3\gamma - 3\cdot5)}.$$

It follows from this formula that the ratio ρ/ρ_2 approaches zero as $r \to r^*$ when $\gamma = 4/3$. The ratio ρ/ρ_2 is finite for $\gamma = 1\cdot4$ and approaches infinity as $A[r_2/(r - r^*)]^{4/9}$ for $\gamma = 5/3$.

It follows from (6.39) that the mass is zero at the interior boundary for $\omega = 2\cdot5$, $q > 0$ and $\gamma \neq 4/3$, but the pressure is finite and not zero. This results from the fact that the work done by the pressure at the interior boundary (the work performed by a spherical piston) is not zero.

From the solution obtained, it follows that the total energy of the disturbed gas within the shock wave is finite. Since the initial energy is infinitely negative for $\gamma > 4/3$ and $\omega = 2\cdot5$, the idealized formulation of the problem implies that these motions are caused by liberation of an infinitely large energy at the centre of symmetry at the initial instant. This energy is consumed in compensating the initial negative infinite energy and in creating the disturbed gas motion. It is evident from dimensional considerations that the energy of disturbed motion is constant. The additional negative energy which the gas inside the shock wave possesses is compensated by the energy flow due to work done by the pressure at the interior spherical boundary.

If $\gamma = 4/3$, the initial energy and the disturbance energy are finite and independent of the shock wave radius; in this case, the pressure at the interior boundary is zero. The energy is only liberated instantaneously at the initial instant at the centre.

If $\omega < 2 \cdot 5$, then the initial energy, according to (6.3), is finite within any sphere of radius r_2. Calculations show that motions, produced by a spherical piston with non-zero pressure are obtained for $\gamma = 5/3$ and $\omega = 1 \cdot 4$ or $\omega = 2$, where the work of the piston W is comparable to the initial gas energy $H: H > 0$ for $\omega = 1 \cdot 4$ and $H < 0$ for $\omega = 2$.

The ratio $W/|H|$ increases as q decreases, i.e. as the shock wave intensity increases.

The ratio r^*/r_2, where r^* is the piston radius, also increases as q decreases. It is significant that the pressure on the piston is larger than the pressure on the shock wave p_2, where the ratio p^*/p_2 increases as q decreases.

Models of star flare-ups for small γ, for $\gamma = 7/6$ or $4/3$ say, can be constructed within the scope of the self-similar motions studied. The small values of γ can sometimes be considered as polytropic indices with continuous heat liberation from sources distributed over the volume of the gas.

We considered self-similar motions above. Various non-self-similar motions, which are almost self-similar (Lidov, 1955), can be considered in a linearized form. Hence, relations in finite form, similar to the above mass, adiabatic and energy relations, can be obtained for the unknowns defining the departure of the solution from the self-similar solution.

12

References

REFERENCES TO FOREWORD TO THIRD EDITION

Birkhoff, G. (1950). "Hydrodynamics, a Study in Logic, Fact and Similitude". Princeton University Press.

Drobot, S., and Warmus, M. (1954). Dimensional analysis in sampling inspection of merchandise. *Rozprawy Matematyczne, V, Warsaw.*

REFERENCES TO CHAPTER II

Bridgman, P. (1937). "Dimensional Analysis". Yale University Press.

Bucky, P. B. (1931). The American Institute of Mining and Metallurgical Engineers. *Tech. Publ. No. 413.*

Chaplygin, Y. S. (1940). *Trud. tsent. aero-gidrodin. Inst., Mosk.* **508**.

Davidenkov, N. N. (1933). *J. tech. Phys., Moscow* 1.

Doyère, Ch. (1917). "Theorie du navire". Paris.

Epshtein, L. A. (1940). *Trud. tsent. aero-gidrodin. Inst., Mosk.* **508**.

Hilsenrath, J., and Touloucian, Y. S. (1954). *Trans. Amer. Soc. mech. Engnrs* **76**, No. 6.

Jahnke, E., and Emde, F. (1938). "Funktionentafeln mit Formeln und Kurven". Berlin.

Jeans, J. (1925). "The Dynamical Theory of Gases", p. 284. Cambridge University Press.

Jonquières, E. de (1883). *C. R. Acad. Sci., Paris*, **23**, 1278.

Katanskii, V. V. (1936). Planning balloon-rigging constructions, etc., *O.N.T.I.*

Kochin, N. E. (1935). On the theory of Cauchy-Poisson waves. Communicated by V. A. Steklov, *Trud. Math. Inst.* 9.

Kochin, N. E. (1938). *Trud. tsent. aero-gidrodin. Inst., Mosk.* **356**.

Kreps, R. L. (1939). *Trud. tsent. aero-gidrodin. Inst., Mosk.* **438**.

Kreps, R. L. (1940). *Trud. tsent. aero-gidrodin. Inst., Mosk.*, **513**.

Pokrovskii, G. I. (1934). *J. tech. Phys., Moscow* 4.

Pokrovskii, G. I. (1940). *Vestn. Voyen-inzh. akademii R.K.K.A.* Communicated by V. V. Kuibyshev, No. 30.

Rayleigh, L. (1915a). *Nature, Lond.* **95**, 66.

Rayleigh, L. (1915b). *Nature, Lond.* **95**, 644.

Reynolds, O. (1883). *Phil. Trans.* **174**, 935.

Reynolds, O. (1885). *Phil. Trans.* **177**, 157.

Riabouchinsky, D. (1915). *Nature, Lond.* **95**, 591.

Sedov, L. I. (1936). *Trud. tsent. aero-gidrodin Inst., Mosk.* **252**.

Sedov, L. I. (1937). Proceedings of the Conference on the Theory of Wave Resistance, pp. 7–30. *Trud. tsent. aero-gidrodin Inst., Mosk.*

Sedov, L. I. (1940). *Air Force Engng* 4.

Sedov, L. I. (1942). *C. R. Acad. Sci. U.R.S.S.* **37**, 9.

Sedov, L. I. (1948). On the theory of waves on the surface of an incompressible fluid. *Vestnik. Moskov. Univ.* 11.

Sedov, L. I., and Vladimirov, A. N. (1941a). *C. R. Acad. Sci. U.R.S.S.* **33**, 2, 116.

Sedov, L. I., and Vladimirov, A. N. (1941b). *C. R. Acad. Sci. U.R.S.S.* **33**, 3, 194.
Sedov, L. I., and Vladimirov, A. N. (1943). *Bull. Acad. Sci. U.R.S.S., Otdel. Tekh. Nauk* **1**.
Taylor, D. W. (1933). "Speed and Power of Ships". Washington.
Vorob'ev, A. G. (1928). Hydrostatic testing of balloon models. *Sborn. Leningr. Inst. Engng Means of Communications* **18**.
Wagner, H. (1932). *Z. angew. Math. Mech.* **12**, 4, 193.

REFERENCES TO CHAPTER III

Blasius, H. (1908). *Z. Math. Phys.* **56**, 37.
Darcy, H. P. G. (1858). *Mém. prés. Acad. Sci., Paris*, **15**, 265.
Fage, A. (1934). Aeronautical Research Committee. Reports and Memoranda, No. 1580.
Friedman, A., and Keller, L. (1924). *Proc. Int. Congr. appl. Mech., Delft*, 395.
Fritsch, W. (1928a). *Z. angew. Math. Mech.* **8**, 3, 199.
Fritsch, W. (1928b). *Abh. aerodyn. Inst. Aachen* **8**, 45.
Hansen, M. (1928). *Z. angew. Math. Mech.*, **8**, 3, 185.
Iatseev, V. I. (1950). *J. exp. theor. Phys.* **11**, 1031.
Jahnke, E., and Emde, F. (1938). "Funktionentafeln mit Formeln und Kurven". Berlin.
Kármán, Th. von (1931). *N.A.C.A. T.M.* 611.
Kármán, Th. von, and Howarth, L. (1938). *Proc. roy. Soc.* A **164**, 192.
Keller, L. (1925). Aufstellung eines System von Charakteristiken der atmosphärischen Turbulenz. *Izv. glav. fiz. Obs.*
Kochin, N. E., Kibel', I. A., and Roze, N. V. (1948). "Theoretical Hydromechanics", Pt. II, Gostekhizdat. A detailed description and solution of this problem using dimensional reasoning is contained in this book.
Kolmogorov, A. N. (1941). *C. R. Acad. Sci. U.R.S.S.* **31**, 6, 538.
Landau, L., and Lifshitz, E. M. (1959). "Fluid Mechanics". Translated by Sykes, J. B., and Reid, W. H. Pergamon Press, London.
Loitsianskii, L. G. (1939). *Trud. tsent. aero-gidrodin. Inst., Mosk.* **440**.
Loitsianskii, L. G. (1941). "Aerodynamics of the Boundary Layer", p. 76. Gostekhizdat, Moscow.
Millionshchikov, M. D. (1939). *C. R. Acad. Sci. U.R.S.S.* **22**, 5, 231.
Millionshchikov, M. D. (1941). *C. R. Acad. Sci. U.R.S.S.* **32**, 9, 615.
Nikuradse, J. (1930). *Proc. Int. Congr. appl. Mech., Stockholm* **1**, 239.
Prandtl, L. (1904). *Int. Congr. Math., Heidelberg*, 484.
Prandtl, L. (1936). Mechanics of viscous fluids in "Aerodynamic Theory", vol. 3. Edited by W. F. Durand. Berlin.
Sedov, L. I. (1944). *C. R. Acad. Sci. U.R.S.S.* **42**, 3, 116.
Sedov, L. I. (1950). "Plane Problems of Hydro and Aerodynamics", p. 432. Gostekhizdat, Moscow.
Slezkin, N. A. (1934). "Lecture Notes", vol. 2. Moscow University.
Stanton, T. E. (1911). *Proc. roy. Soc.* A **85**, 366.
Taylor, G. I. (1935). *Proc. roy. Soc.* A **151**, 421, 444, 455, 465.
Taylor, G. I. (1936). *Proc. roy. Soc.* A **156**, 307.
Taylor, G. I. (1937). *J. aero. Sci.* **4**, No. 8, 311.
Töpfer, C. (1912). *Z. Math. Phys.* **60**, 397.

REFERENCES TO CHAPTER IV

Bam-Zelikovich, G. M. (1949a). Collapse of an arbitrary discontinuity in a combustible mixture. Collection of papers No. 4, "Theoretical Hydrodynamics". Edited by L. I. Sedov. Oborongiz, Moscow.

Bam-Zelikovich, G. M. (1949b). Propagation of intense explosive waves. Collection of papers No. 4, "Theoretical Hydrodynamics". Edited by L. I. Sedov. Oborongiz, Moscow.

Barenblatt, G. I. (1952). *Appl. Math. Mech. Leningr.* **16**, 6, 679.

Barenblatt, G. I. (1954a). *Appl. Math. Mech. Leningr.* **18**, 4, 351.

Barenblatt, G. I. (1954b). *Appl. Math. Mech. Leningr.* **18**, 4, 409.

Bechert, K. (1941). Differentialgleichungen der Wellenausbreitung in gasen. *Ann. Phys., Lpz.* **39**, 357.

Brode, H. J. (1955). *J. appl. Phys.* **26**, 6, 766.

Brushlinskii, D. N., and Solomakhova, T. S. (1956). Vol. 19, No. 7, in collection "Theoretical Hydromechanics". Edited by L. I. Sedov. Oborongiz, Moscow.

Crussard, M. L. (1913). *C. R.* **156**, 446, 611.

Iavorskaia, I. M. (1956). *C. R. Acad. Sci. U.R.S.S.* **111**, 4, 783.

Korobeinikov, V. P. (1955). *C. R. Acad. Sci. U.R.S.S.* **104**, 4, 509.

Korobeinikov, V. P. (1956a). *C. R. Acad. Sci. U.R.S.S.* **109**, 2, 271.

Korobeinikov, V. P. (1956b). *C. R. Acad. Sci. U.R.S.S.* **111**, 3, 557.

Korobeinikov, V. P. (1957). *C. R. Acad. Sci. U.R.S.S.* **113**, 5, 1006.

Krasheninikova, N. L. (1955). *Bull. Acad. Sci. U.R.S.S.*, OTN, **8**.

Landau, L. D. (1945). *Appl. Math. Mech., Leningr.* **9**, 4, 286.

Landau, L. D., and Lifshitz, E. M. (1954). "Fluid Mechanics". Translated by Sykes, J. B., and Reid, W. H. Pergamon Press, London.

Lidov, M. L. (1954). *C. R. Acad. Sci. U.R.S.S.* **97**, 3, 409.

Lidov, M. L. (1955a). *C. R. Acad. Sci. U.R.S.S.* **103**, 1, 35.

Lidov, M. L. (1955b). *C. R. Acad. Sci. U.R.S.S.* **102**, 6, 1089.

Lipschitz, R. (1887). *Z. reine angew. Math.* **100**, 89.

Liubimov, G. A. (1956). On possible kinds of one-dimensional unsteady viscous gas motions. Paper No. 19 in collection "Theoretical Hydromechanics", vol. 7. Oborongiz, Moscow.

Mel'nikova, N. S. (1954). *Zh. Mekhanika* **3**, 2535.

Neumann, J., and Goldstine, H. (1955). *Commun. pure appl. Math.* **8**, 2, 327.

Okhotsimskii, D. E., Kondrasheva, I. L., Vlasova, Z. P., and Kazakova, R. K. (1957). *Trud. Steklov Math. Inst. A.N. U.S.S.R.* **50**.

Riemann, B. (1953). Collected works. Dover Publications Inc., New York.

Rosseland, S. (1949). "The Pulsation Theory of Variable Stars". Oxford University Press.

Sakurai, A. (1953). *J. phys. Soc. Japan,* **8**, 5, 662.

Sakurai, A. (1954). *J. phys. Soc. Japan,* **9**, 2, 256.

Sedov, L. I. (1945a). On certain unsteady compressible fluid motions, *Appl. Math. Mech. Leningr.* **9**, 4, 294.

Sedov, L. I. (1945b). *C. R. Acad. Sci. U.R.S.S.* **47**, 2, 91.

Sedov, L. I. (1945c). *Appl. Math. Mech. Leningr.* **9**, 4, 293.

Sedov, L. I. (1946a). *Appl. Math. Mech. Leningr.* **10**, 2, 241.

Sedov, L. I. (1946b). *C. R. Acad. Sci. U.R.S.S.* **52**, 1, 17.

Sedov, L. I. (1950). "Plane Problems of Hydrodynamics and Aerodynamics". Gostekhizdat, Moscow.

Sedov, L. I. (1952a). *C. R. Acad. Sci. U.R.S.S.* **85**, 4, 723.

Sedov, L. I. (1952b). *C. R. Acad. Sci. U.R.S.S.* **87**, 1, 4.

Sedov, L. I. (1953). *C. R. Acad. Sci. U.R.S.S.* **90**, 5, 753.

Sedov, L. I. (1956). *C. R. Acad. Sci. U.R.S.S.* **111**, 4, 780.

Sidorkina, S. I. (1957). *C. R. Acad. Sci. U.R.S.S.*, **112**, 3, 398.

Staniukovich, K. P. (1945). On self-similar solutions of the hydronamics equations with central symmetry. *C. R. Acad. Sci. U.R.S.S.* **48**, 5, 331.

Staniukovich, K. P. (1949). *C. R. Acad. Sci. U.R.S.S.* **64**, 4, 467.

Staniukovich, K. P. (1955). "Unsteady Motion of Continuous Media", Moscow. English translation edited by M. Holt, Pergamon Press, 1959.

Taylor, G. I. (1950). *Proc. roy. Soc.* A **201**, 155, 175.

Zel'dovich, Y. B. (1942). *J. exp. theor. Phys.* **12**, 9, 389.

Zel'dovich, Y. B. (1948). "Introduction to the Theory of Shock Waves and Gas Dynamics". Gostekhizdat, Moscow.

Zel'dovich, Y. B., and Kompaneets, A. S. (1955). "Theory of Detonation". Gostekhizdat, Moscow.

REFERENCES TO CHAPTER V

Ambartsumian, V. A., Mustel', E. R., Severnyi, A. B., and Sobolev, V. V. (1952). "Theoretical Astrophysics". Moscow, Gostekhizdat; English translation by J. B. Sykes, New York, Pergamon Press (1958).

Burgers, J. M., and van de Hulst, H. C. (Editors). (1949). Proceedings of the Symposium on the Motion of Gaseous Masses of Cosmical Dimensions. U.S.A.F. Central Air Doc. Off., Dayton, Ohio.

Carrus, P., Fox, P., Gaas, F., and Kopal, Z. (1951). *Astrophys. J.* **113**, 496.

Chandrasekhar, S. (1951). "An Introduction to the Study of Stellar Structure". Dover Publications Inc., New York.

Goodricke, J. (1786). *Phil. Trans.* **76**, 48.

Iavorskaia, I. M. (1956). *C. R. Acad. Sci. U.R.S.S.* **111**, 4, 783.

Lidov, M. L. (1955). *Appl. Math. Mech., Leningr.* **19**, 5, 541.

Paranego, P. P., and Masevich, A. G. (1951). Investigation of the mass-radius-luminosity relations. *Trud. Gas Astron. Inst.* **20**. Communicated by P. K. Sternberg. Moscow University Press.

Rosseland, S. (1949). "The Pulsation Theory of Variable Stars". Oxford University Press.

Sedov, L. I. (1954). *C. R. Acad. Sci. U.R.S.S.* **94**, 4, 643.

Sedov, L. I. (1956). *C. R. Acad. Sci. U.R.S.S.* **111**, 4, 783.

Sedov, L. I. (1957). *C. R. Acad. Sci. U.R.S.S.* **112**, 2, 211.

Strömgren, B. (1936). *Handb. Astrophys.* **7**, 159.

Strömgren, B. (1937) *Ergebn. exact. Naturw.* **16**, 465.

The journal referred to as *C. R. Acad. Sci. U.R.S.S.* appeared in translated form, under this title, before the second world war. Since the war only the Russian edition has been issued. This is frequently referred to in libraries as *Doklady Akademii Nauk U.S.S.R.* Page numbers refer to the translated version in the few cases applicable and otherwise to the Russian edition. The journal *Appl. Math. Mech. Leningr.* is usually referred to in libraries as *Prikladnaia Matematika i Mekhanika*. It has been issued in translated form by the American Society of Mechanical Engineers since 1958.

Author Index

Subject Index

A

Aircraft, dimensions and long range flight, 68
  suitable dimensions of engines, 68
Astrophysics, 164, 305 et seq.
Asymptotic decay of shock waves, 295 et seq.
Atomic bomb, 213, 221, 233

B

Blasius formula, 139
Boltzmann constant, 6
Boundary layer on a flat plate, 106
Brightness of stars, 306

C

Centrifuge, 47
Cepheids, 281, 310 et seq.
  average density of, 331
Chapman–Jouguet condition, 162
Clapeyron's equation, 23
Combustion, 150, 160, 200
Completeness of system of characteristic parameters, 22
Conservation of energy, 14, 173
Correlation coefficient, 121
  moments, 121
Craters due to explosions, 256

D

Decay of arbitrary discontinuity in combustible mixture, 151, 206
Derivatives behind shock waves, 291
Detonation, 150, 160 et seq., 178
  spherical, 193
Diffusion of vorticity in a viscous fluid, 97
Dimensional formula, 8
  quantity, 1, 2
Dimensionless (abstract) quantity, 2, 3
  variables, 154

Dimensions, 4
Dispersion of gas from a point, 150, 191
Dynamic length, 246
Dynamic similarity, 43

E

Energy, of atomic bomb, 214
  gravitational, 336
  internal, 337
  kinetic, 14
  potential, 14
  total, for gas motion, 171
Entry of cone and wedge into liquid, 83
Equation of conservation of energy, 171
  of equilibrium and motion of mass of gas simulating star, 315
  of heat flow, 316
  of state for explosions, 234
Explosion with counterpressure, 238 et seq.
  in dusty atmosphere, 230
  in ideal incompressible fluid, 235
  point, in incompressible media, 235
  point, linearized problem, 235
  strong, 152, 177, 210 et seq.
  strong, with heat conduction, 232
  strong, in medium with variable density, 260

F

Fineness coefficient, 62
Flow of heavy liquid through spillway, 27
  past a body, 33 et seq., 106
Focussing of gas on a point, 150, 191
Force, 13
  inertial, 15
Friction coefficient, 63
Froude number, 64
  similarity law, 64

361